FOUNDATIONS OF PHYSICS

Mechanics, Waves and Heat

Volume - I

FOUNDATIONS OF PHYSICS

Mechanics, Waves and Heat

Volume - I

R.S. Gambhir
Associate Professor
Department of Physics, University of Jodhpur
Jodhpur, Rajasthan

D. Banerjee
Associate Professor
Department of Physics, University of Jodhpur
Jodhpur, Rajasthan

M.C. Durgapal
Professor of Physics
Kumaun University
Nainital, Uttarakhand

PUBLISHING FOR ONE WORLD

NEW AGE INTERNATIONAL (P) LIMITED, PUBLISHERS

New Delhi • Bangalore • Chennai • Cochin • Guwahati • Hyderabad
Jalandhar • Kolkata • Lucknow • Mumbai • Ranchi

Visit us at www.newagepublishers.com

Published by New Age International (P) Ltd., Publishers
First Edition: 1992
Reprint: 2008

Branches:

- 36, Malikarjuna Temple Street, Opp. ICWA, Basavanagudi, **Bangalore**. ✆ (080) 26677815
- 26, Damodaran Street, T. Nagar, **Chennai**. ✆ (044) 24353401
- Hemsen Complex, Mohd. Shah Road, Paltan Bazar, Near Starline Hotel, **Guwahati**. ✆ (0361) 2543669
- No. 105, 1st Floor, Madhiray Kaveri Tower, 3-2-19, Azam Jahi Road, Nimboliadda, **Hyderabad**. ✆ (040) 24652456
- RDB Chambers (Formerly Lotus Cinema) 106A, Ist Floor, S.N. Banerjee Road, **Kolkata**. ✆ (033) 22275247
- 18, Madan Mohan Malviya Marg, **Lucknow**. ✆ (0522) 2209578
- 142C, Victor House, Ground Floor, N.M. Joshi Marg, Lower Parel, **Mumbai**. ✆ (022) 24927869
- 22, Golden House, Daryaganj, **New Delhi**. ✆ (011) 23262370, 23262368

ISBN (13): 978-81-224-0416-6

Rs. 160.00

C-08-03-2359

2 3 4 5 6 7 8 9 10

Printed in India at Ramprintograph, Delhi.
Typeset at Laser Business Services, Delhi.

PUBLISHING FOR ONE WORLD
NEW AGE INTERNATIONAL (P) LIMITED, PUBLISHERS
4835/24, Ansari Road, Daryaganj, New Delhi-110002
Visit us at **www.newagepublishers.com**

Preface

Under the educational pattern emerging in India, a student will have to spend twelve years in schools. This book has been written to meet the requirements of physics syllabi of plus two classes of the Central schools, and schools of various States. Requirements of various entrance examinations after Class XII, e.g., IIT-JEE, have also been kept in mind.

The present belief is to convey to the students from the very beginning the basic concepts of physics so that they develop confidence to follow advanced courses in science and technology, and are enabled to make smooth transition from the school level courses to the graduate courses. The main emphasis in the book is therefore on the basic concepts. Various concepts have been developed from the very beginning. No previous knowledge on the part of the reader is assumed. But the applications of these concepts are quite varied so as to provide some challenge to intelligent students. As far as possible, calculus has been avoided to make the concepts equally comprehensible to the students having no background in it. Instead the emphasis is on graphical analysis. Moreover, non-calculus treatment gives better physical insight. However, the derivations involving calculus are given in foot-notes.

A large number of questions have been interspersed with the text to make the students pause and consolidate their learning. Solved examples have been given to train the students to apply the knowledge to actual situations, and details of instruments have been incorporated as a guide to the students working in the laboratory. The discussion and problems at the end of each chapter form an integral part of the course, as quite a few topics have been introduced through them. Some important topics could not fit in the run of the text and therefore are given in appendices. Though SI units have been followed, reference to other systems has also been made at appropriate places.

The book has been divided into two volumes. The present volume, which deals with mechanics, properties of matter, waves and heat, covers the entire course of class XI of Central schools. However, arrangement of various topics is not essentially the same. For example, thermal expansion of solids is not included in heat as is usually done, but in intermolecular forces to highlight the fact that it arises from the anharmonicity of the intermolecular potential function. The emphasis in the book is on conservation laws and frames of reference. All types of waves have been introduced in a general manner. Sound waves are taken up in detail. The concepts of thermodynamics and black-body radiation have been emphasized.

The school physics for many students going to medical, bio-medical and earth-science courses would be a terminal course. In the present book which has been written primarily for plus two students, future requirements of such students have also been kept in mind. Special theory of relativity, which does not make a part of plus two courses at the moment, has been included because of its importance in non-physical branches of science. Even liberal arts students may get fascinated by its results, as it has affected dramatically the perceptions of mankind. Moreover, there should be some thing more in a text book for bright students to delve into. Further, students should not get away with the impression that the syllabus contains all the physics there is to it. A wider coverage of topics allows flexibility in its use. The sections marked with asterisks in the list of contents may be omitted without loss of continuity.

We are thankful to our colleagues for enlightening discussions which helped us in straightening our concepts. We also wish to thank the publishers, Wiley Eastern Limited, for processing the manuscript meticulously.

R.S. Gambhir
D. Banerjee
M.C. Durgapal

Contents

Preface			*iii*
Chapter	**1**	**Introduction**	**1**
	1.1	Physics, Mankind and Universe *1*	
	1.2	Time *3*	
	1.3	Distance *4*	
	1.4	SI Units *6*	
	1.5	Units and Dimensions *7*	
	1.6	Accuracy and Errors in Measurement *10*	
	1.7	Vectors *12*	
	1.8	Sum of Vectors *13*	
	1.9	Components of a Vector *14*	
	1.10	Scalar Product *18*	
	1.11	Vector Product *19*	
Chapter	**2**	**Force**	**21**
	2.1	Motion along a Straight Line *21*	
	2.2	Motion in Space *23*	
	2.3	Angular Vectors *24*	
	2.4	Force *26*	
	2.5	Centripetal Force *27*	
	2.6	Weight of a Body *28*	
	2.7	Gravitational Force *29*	
	2.8	Kepler's Laws — Planetary Motion *32*	
	2.9	Newton's Third Law of Motion *33*	
	2.10	Frictional Force *34*	
	2.11	Equilibrium of Concurrent Forces *36*	
Chapter	**3**	**Frames of Reference**	**39**
	3.1	Introduction *39*	
	3.2	Frames of Reference Moving with Uniform Velocity with respect to Each Other *40*	
	3.3	Galilean Transformations *41*	
	3.4	Accelerated Frames *42*	
	3.5	Fictitious Force in Circular Motion and Weightlessness in a Satellite *44*	
	3.6*	Constancy of Speed of Light *45*	
	3.7*	Simultaneity *48*	
	3.8*	Time Dilation *49*	
	3.9*	Length Contraction *50*	
	3.10*	Lorentz Transformations *52*	

	3.11*	Relativistic Addition of Velocities *54*	
	3.12	General Theory of Relativity *55*	
Chapter	**4**	**Linear Momentum**	**60**
	4.1	Linear Momentum *60*	
	4.2	Conservation of Momentum *62*	
	4.3	A Rocket *63*	
	4.4	Centre of Mass *64*	
	4.5*	Relativistic Variation of Mass *67*	
Chapter	**5**	**Work, Power and Energy**	**71**
	5.1	Work *71*	
	5.2	Graphical Representation *73*	
	5.3	Work in Gravitational Field *74*	
	5.4	Kinetic Energy *75*	
	5.5	Potential Energy *76*	
	5.6	Collision of Two Particles *80*	
	5.7	Conservative and Non-conservative Forces *83*	
	5.8	Power *85*	
	5.9*	Mass and Energy *86*	
	5.10	Binding Energy *88*	
Chapter	**6**	**Rotational Dynamics**	**91**
	6.1	Torque *91*	
	6.2	Angular Momentum *92*	
	6.3	Conservation of Angular Momentum *94*	
	6.4	Equilibrium of Rigid Bodies *95*	
	6.5	Moment of Inertia *96*	
	6.6	Angular Momentum of a Rigid Body *98*	
	6.7	Comparison of Linear and Rotational Motions *98*	
	6.8	Theorems of Parallel and Perpendicular Axes *101*	
Chapter	**7**	**Simple Harmonic Motion**	**105**
	7.1	Introduction *105*	
	7.2	Characteristics of Simple Harmonic Motion *105*	
	7.3	Loaded Spring *106*	
	7.4	Analysis of SHM *107*	
	7.5	Simple Pendulum *111*	
	7.6	Angular SHM and Physical Pendulum *112*	
	7.7	Lissajous Figures *113*	
	7.8	Damped SHM *115*	
	7.9	Forced Vibrations and Resonance *116*	
Chapter	**8**	**Fluid Mechanics**	**119**
	8.1	Fluids and Pressure — Pascal's Law *119*	
	8.2	Streamline and Turbulent Motions *121*	
	8.3	Bernoulli's Equation *122*	
	8.4	Viscosity *125*	
	8.5	Flow of a Liquid through a Tube *125*	
	8.6	Motion of Bodies through Fluids *126*	

Chapter 9		**Molecular Properties of Matter**	**130**
	9.1	Intermolecular Forces *130*	
	9.2	Atomic and Molecular Bindings *131*	
	9.3	Three States of Matter *133*	
	9.4	Elasticity *136*	
	9.5	Thermal Expansion *138*	
	9.6	Surface Tension *139*	
	9.7	Angle of Contact *141*	
	9.8	Excess Pressure Inside a Curved Surface *142*	
	9.9	Rise of a Liquid in a Capillary Tube *143*	
	9.10*	Jaeger's Method *145*	
	9.11*	Detergent Action *145*	
Chapter 10		**Waves in Elastic Media**	**147**
	10.1	Types of Waves *147*	
	10.2	Progressive Waves *150*	
	10.3	Wave Parameters *151*	
	10.4	Intensity of a Wave *152*	
	10.5	Superposition Principle *152*	
	10.6	Beats *153*	
	10.7	Group Velocity and Wave Velocity *155*	
	10.8	Fourier Analysis *156*	
Chapter 11		**Sound Waves**	**159**
	11.1	Velocity of Longitudinal Waves *159*	
	11.2	Intensity of Sound *162*	
	11.3	Reflection, Refraction, Interference and Diffraction of Sound *165*	
	11.4	Doppler Effect *168*	
	11.5	Ultrasonics *171*	
	11.6*	Architectural Acoustics *173*	
Chapter 12		**Vibrations of Strings and Air Columns**	**176**
	12.1	Boundary Restrictions on SHM *176*	
	12.2	Stationary Waves *178*	
	12.3	Sonometer *180*	
	12.4	Melde's Experiment *181*	
	12.5	Open and Closed Pipes *182*	
	12.6*	Kundt's Tube *184*	
	12.7*	Musical Scales *184*	
Chapter 13		**Temperature**	**187**
	13.1	Introduction *187*	
	13.2	Concept of Temperature *188*	
	13.3	Measurement of Temperature *189*	
	13.4	Choosing Temperature Scale *189*	
	13.5	Triple Point of Water *190*	
	13.6	Constant Volume Gas Thermometer *190*	
	13.7	Ideal Gas Temperature Scale *191*	
	13.8	Platinum Resistance Thermometer *192*	
	13.9	An Ideal (Perfect) Gas *193*	

Chapter 14 Laws of Thermodynamics **196**
14.1 Introduction *196*
14.2 External Work Done by a System *196*
14.3 First Law of Thermodynamics *199*
14.4 Principle of Calorimetry *199*
14.5 Specific Heats of a Gas *200*
14.6 Carnot Cycle — Second Law of Thermodynamics *202*
14.7 Steam Engine *205*
14.8 Internal Combustion Engines *206*

Chapter 15 Kinetic Theory of Gases **210**
15.1 Introduction *210*
15.2 Model for a Gas *210*
15.3 Pressure Exerted by a Perfect Gas *211*
15.4 Deduction of Gas Laws *212*
15.5 Kinetic Theory Interpretation of Temperature *213*
15.6 Degrees of Freedom and Principle of Equipartition of Energy *215*
15.7 Specific Heats of Gases *216*
15.8 Brownian Motion *217*
15.9 Distribution of Molecular Velocities *218*
15.10 Kinetic Temperature *219*

Chapter 16 Imperfect Gases **222**
16.1 Introduction *222*
16.2 van der Waals Equation of State *222*
16.3 Joule-Thomson Expansion *224*
16.4 Refrigerators *225*
16.5 Liquefaction of Gases *226*
16.6 Change of Phase *227*
16.7 Superconductivity *228*
16.8 Superfluidity *229*

Chapter 17 Heat Transfer **231**
17.1 Conduction *231*
17.2 Convection *232*
17.3 Radiation *232*
17.4 Some Definitions *232*
17.5 Black-body *233*
17.6 Emission and Absorption of Radiation *233*
17.7 Black-body Radiation, or Temperature Radiation *234*
17.8 Stefan-Boltzmann Law *235*
17.9 Spectral Distribution of Black-body Radiation *237*
17.10 Pyrometers *239*

Appendices **243**
A.1 Vacuum Technique *243*
A.2 Trigonometric Functions *246*
A.3 Symbols and Notations *249*
A.4 Numerical Constants *251*

Answers to Selected Problems **253**

Index **255**

Chapter 1

Introduction

1.1 Physics, Mankind and Universe

The discipline of physics deals with the basic laws which govern the behaviour of nature. The development of this branch of science started with the exploration of the universe. The primitive man was primarily concerned about his survival. As he developed skills and became more secure on this account, he diverted his attention towards understanding and controlling nature. By the middle of the fourth millenium B.C. he had learned to read and write. By being able to record observations, these early people noticed that nature followed definite patterns from which they could predict future happenings, *e.g.*, eclipses.

The first branch of physics to develop was mechanics, through the study of planetary motions. Ptolemy (A.D. 150), who accepted Aristotle's idea that the earth was at the centre of the universe, developed in detail the motions of the moon, planets and the sun about the earth. Copernicus (1473-1543), however, suggested that the sun is at the centre of a system with the planets (including earth) revolving in circles about it. He was severely opposed by the theologians of the time for denying the earth its supreme central position in the universe. Even the contemporary scientists did not accept this system, so great were the forces of psychological inertia of 'common sense' ! Giordano Bruno (1548-1600) who was an active supporter and popularizer of this system, went further and gave the idea of the infinity of the universe and a multitude of worlds existing without spiritual motive forces. For this he paid with his life and was burned in the Square of Flowers in Rome on the 17th of February, 1600, after eight years of prison and torture to make him to renounce his convictions! In spite of this the pursuit of truth went on.

The next in line was Brahe (1546-1601) who made a great number of observations of the stars and planets. He rejected both the Copernican and Ptolemaic systems, and proposed that the planets revolved around the sun, and this system and the moon revolved around the earth. But, Kepler (1571-1630), who was Brahe's assistant, after careful analysis of the data compiled by Brahe, became convinced that the sun was really the centre of the planetary system. He also formulated three basic laws which described the motions of the planets.

Galileo (1564-1642) constructed a telescope which opened new vistas for the exploration of the universe. From his observations he was convinced that Copernicus and Kepler were right—the sun was the centre of the planetary system and earth moved around it. The theologians and many scientists declared the discoveries to result from 'defects of the telescope'! Under the pressure of these social forces, Galileo signed on June 22, 1633, his renouncement of this system. But he was never convinced. He also experimented with balls rolling down inclined planes, and established the laws of accelerated motion. He rejected Aristotle's theory of falling objects and proposed, and demonstrated that heavy and light objects fall at the same rate. Newton (1642-1727) carried ahead the work of Galileo. He formulated the three laws of motion. He also gave the inverse square law of gravitation, and on the basis of this law he could explain the motion of the moon (a satellite) around the earth, and the motion of planets (including earth) around the sun (a star).

The universe has been explored further with the help of very powerful telescopes, like the 200-in. telescope on Mount Palomar (USA) and a larger instrument at Crimea peninsula (Black Sea). We not only see billions of stars, but we also see them clustered together in groups. Such a group of stars is called a galaxy. Like the sun, we have so many other stars ($\sim 10^{11}$) in our galaxy—the Milky Way —and they may have their own planets. Our nearest star, Proxima, is 4.3 light-years (LY) away. The Milky Way is a flat disc with a huge central ball of stars and spiral arms. It is about 80,000 light-years in diameter and 10,000 light-years thick.*

Galaxies are often called nebulae because they appear as nebulous clouds. The great spiral Nebula in Andromeda is about 1.5 million light-years from us, and it is approximately 140,000 light-years in diameter. About 10^{10} galaxies have come within the range of modern telescopes. This may only be a small part of the number in existence. Like a cluster of stars in a galaxy, galaxies too occur in groups.

Through the study of Fraunhofer lines a great amount of information about the temperature, chemical composition, rotational motion, atmosphere and magnetic field of stars has been obtained. Next to hydrogen, helium is the most abundant element in our sun. Hydrogen is continuously being converted into helium through fusion, and energy is being released. When the light from distant galaxies is analysed, it is found that all these galaxies exhibit a red shift in their lights. This decrease in frequency indicates that they are receding from us (Doppler effect). Further, the magnitude of the red shift—and, hence, the magnitude of the recessional speed—is greater for the more distant galaxies. From this we conclude that our universe is expanding.

Man has also tried to formulate theories about the origin of the universe. The "big bang" theory was proposed by Gamow in 1948. According to this theory, about 10^{10} years ago all the matter and energy now in the universe were concentrated in a single fire ball. The density of this fire ball was $\geq 10^{28}$ kg/m^3 and the temperature was $\geq 10^{16}$ K. Due to the enormously high radiation pressure, the ball expanded with explosive rapidity causing the big bang. Though there is presently a lot of evidence in favour of this picture, more sophisticated measurement techniques designed to test this theory are still being worked on. The universe is presently expanding. Will this expansion continue for ever, or will it stop and then motion will reverse is yet an open question.

Apart from visible light, stars also emit other radiations, like radio waves, infra-red and ultraviolet radiation, X-rays, γ-rays, electrons, neutrinos and heavy particles (cosmic rays). These are also the probes used to explore the universe. Radio telescopes were built in 1936, and since then, many radio sources have been discovered. Quasars (quasi-stellar objects) were discovered in the early 1960s. The radiations emitted by them show variations with time that have periods of the order of a day. Because the radiation from an object cannot change significantly in the duration less than the time required for light to cross the object, these unusual objects can be no larger than about one light-day ($\sim 3 \times 10^{13}$m or about 200 astronomical units). These star-like objects have an unusual blue colour and emit a very large amount of energy—about 10^{12}-10^{13} times the output of the sun. To account for such a huge expenditure of energy, quasars must have masses that are $\sim 10^9$ times the sun's mass.

---

*As the distances involved in the outer space are extremely large, very big units are taken to denote them. One of these units is the astronomical unit. It is the mean distance of the earth from the sun, and

$$1 \text{ astronomical unit (AU)} = 1.496 \times 10^{11} \text{ m}$$

Another unit is the light-year. It is equal to the distance travelled by light in one year. As one year has 3.156×10^7 s and the speed of light is 2.998×10^8 m/s,

$$1 \text{ light-year (LY)} = 9.461 \times 10^{15} \text{ m}$$

One more unit which is frequently used is the parsec. A parsec is the distance at which one astronomical unit subtends an angle of one second of arc, and

$$1 \text{ parsec} = 3.084 \times 10^{16} \text{ m}$$

Pulsars were discovered in 1968. These are the radio sources which pulsate quite rapidly, with periods ranging from 0.033 s to just over 3 s. These objects are no larger than about 30 km. As per a recent model, a pulsar is a rapidly rotating neutron star. Electrons and protons get trapped in the very high magnetic field of the neutron star. As the star spins, the magnetic field and the trapped particles spin with it. Because the motion is circular, the particles are accelerated and, therefore, emit radiation.

Various radiations coming to earth from outer space – called cosmic rays – are now-a-days studied with the help of instruments placed on earth-satellites. While coming to the earth surface these radiations are absorbed to a large extent by the atmosphere. Earth's magnetic field also deflects away some of them – the charged ones*. But the atmosphere is almost negligible above the satellite orbits and, therefore, the study will be more exact. Magnetic field is also weak there. At moon there is no atmosphere, and the magnetic field is very weak.

Tremendous strides have also been made in understanding the ultimate building blocks of the matter. The atom has been analysed and a large number of elementary particles identified. Tremendously large distance, mass and time ranges are involved in the exploration of this universe. The distances range from 10^{-15}m (nuclear radius) to 10^{27} m (edge of the universe); the masses vary from 10^{-30} kg (mass of the electron) to 10^{50} kg (mass of the universe); time intervals range from 10^{-24} s (time taken by light to travel nuclear diameter) to 10^{18} s (age of the universe). Man, the most intelligent and the fiercest of all animals known so far, is sitting near the mid-position (height ~1 m, mass ~10^2 kg, life ~10^9 s), and is looking in all directions.

*Cosmic rays are harmful to our health. The atmosphere by absorbing them and the magnetic field by deflecting some of them — the charged ones — provides us with a safe umbrella.

During the last four centuries (taking Galileo's period as the beginning of the scientific approach) man has succeeded in collecting a large volume of information concerning various physical processes. A great deal of this information has been analysed and physical theories developed. The history of science shows that a physical theory holds good for a long time, and then suddenly some experimental facts are discovered which do not agree with that theory. The new theories are then formulated or the old theories improved upon. In any case, theories can only be disproved and not proved. More questions are generated than the answers provided. But to know more about what we know little is a sign of progress.

It can be well-appreciated from the above that physics has helped a lot in developing rational thinking in the society. Threats from orthodoxy have not deterred the physicists from the path of exploration of truth.

As physics deals with basic laws operating in nature, any innovations and discoveries in it are bound to influence other branches of science and technology in particular, and human society in general. Discoveries of Newton's law of gravitation, Einstein's mass-energy equivalence, X-rays, lasers, holography, superconductivity, are a few examples. Development of various sources of energy for welfare of the human race and of deadly weapons for its destruction have raised questions of social and political values, ecology, economics, etc. The decisions how to utilize scientific discoveries are not made by scientists but by governmental leaders who might not have any understanding of science. Leaving aside the question of interaction of science with society in a life and death way, to physicists understanding the world around us is a valuable goal in itself. Scientific understanding is an important part of our modern culture.

1.2 Time

The term 'time' is very familiar and obvious to us yet it is very difficult to define it. It is something the passage of which tells us about the sequence in which events occur. It does not mean that if life becomes eventless, time would not exist. Time exists even if nothing else happens. Time is like a river flowing endlessly through the universe.

Let us use the standard yardstick–if we can measure something, we understand it. Time can be measured with the help of devices which have regular periodicity, like a pendulum. Long time intervals can be measured in terms of days and years. Ordinary time intervals can be expressed in terms of the periods of sand clocks, rotations of the hands of watches, the period of a pendulum, etc.

We can pick up some periodic event and assign some arbitrary unit of time to it. For example, a day is divided into 24 hours. We can adjust the period of a simple pendulum by adjusting its length in such a way that it makes 3600 oscillations in one hour. Then the time interval of one oscillation of this pendulum will be called a second. Similarly a second can be further divided, and so on.

Mechanical pendulums cannot be very fast. For fast vibrations we can depend on electrical pendulums, called oscillators. Crystal (quartz) controlled oscillators have periods of the order of a microsecond. We can get still smaller periods with the help of microwave oscillators, masers, lasers, etc.

We can go down to the measurement of very short time intervals by constructing a series of oscillators, each with a period ten times shorter than the previous one. Periods of two oscillators can be compared with the help of an oscilloscope. For a given sweep rate, if for one oscillator there is one full vibration on the oscilloscope screen, then for the oscillator having a period ten times smaller than this there will be ten vibrations displayed on the screen.

Extremely small time intervals can also be measured in another way. Suppose two events take place on an object moving with a known speed. If we find out the distance between the points where the events occurred, the time interval can be found out. The measurement of the time period of a light wave from the measurement of its wavelength comes in this category. This procedure can also be used to find out the lifetimes of elementary particles from the knowledge of their track lengths.

Very long time intervals, like the age of our earth, can be measured with the help of radioactive clocks. A detailed account of this method has been given in Chapter 34.

1.3 Distance

Motion, the study of which is basic to all physics, requires for its specification both time and distance. Distance is a measure of separation between two locations in space around us. Ordinary distances can be measured with the help of a meter stick. Vernier caliper can go upto 0.1 mm or so. Distances up to a micron (10^{-6} m) can be measured with the help of a good micrometer and a microscope, *e.g.*, microscopic comparator. Very small objects can first be magnified with the help of a microscope with a large magnification, *e.g.*, electron microscope, along with the scale or an object of known dimensions, and then their sizes can be determined. A scale can be fitted inside a microscope with a suitable eyepiece.

Still smaller distances are measured by studying the scattering of electromagnetic radiations (visible light, X-rays, γ-rays, etc., depending upon the sizes involved) from the objects under study. First the wavelength of the radiation used is found out by studying the interference or diffraction from a pattern of marks of known separation (gratings, crystals, etc.). Nuclear sizes are of the order of a fermii (10^{-15} m). These can be measured by studying the scattering of beams of high energy particles when these are passed through thin foils of the material under consideration.

Inaccessible and large distances are generally measured by using the method of *triangulation.* In a triangle, if we know the length of one side and two angles, the lengths of the other two sides can be easily computed.

Suppose the width of a river is to be measured. Draw a base line *AB* of length *c* along one of the banks [Fig. 1.1 (a)]. Select a tree or some such

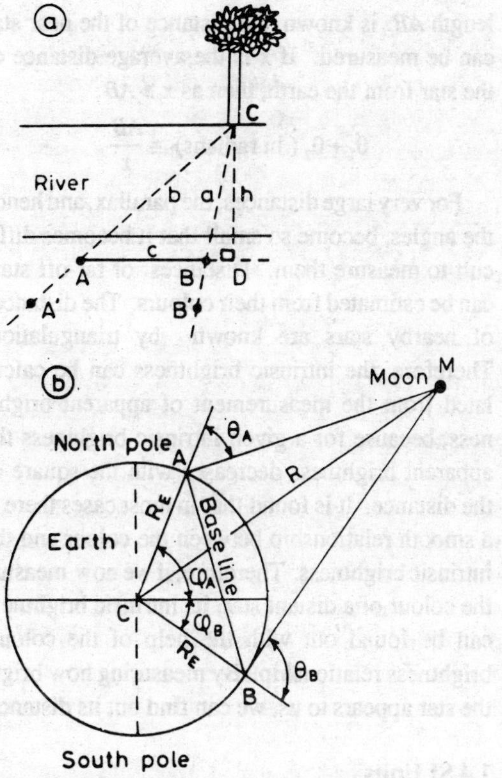

Fig. 1.1 ***Measurement of large distances* : (a) *Measurement of the width of a river by the method of triangulation* (b) *Measurement of distance of moon from earth with base line AB***

thing C on the other bank and measure the angles CAB and CBA. Directions $A'AC$ and $B'BC$ can be located using stakes. If $AC = b$ and $BC = a$, then from the law of sines:

$$\frac{a}{\sin CAB} = \frac{b}{\sin CBA} = \frac{c}{\sin ACB}$$

$$\therefore \quad a = c \times \frac{\sin CAB}{\sin ACB}, \; b = c \times \frac{\sin CBA}{\sin ACB}$$

$$\text{But } \angle ACB = 180° - (\angle CAB + \angle CBA)$$

Therefore, a and b can be readily evaluated. If CD is perpendicular to AB, then its length h (= CD) is given by,

$$\sin CAB = \frac{h}{b},$$

or

$$h = b \sin CAB$$

The above method of triangulation is extensively used by surveyors. The same technique can also be used to find the distance of moon from earth [Fig. 1.1 (b)]. For this purpose two stations on the earth's surface near the same geographic meridian (one in each hemisphere) are chosen. Angles θ_A and θ_B are measured simultaneously at stations A and B— θ_A is the angle between the vertical at A and the line of sight of the moon, and similarly θ_B is the angle between the vertical at B and the line of sight of the moon. The length of base line AB can be easily evaluated from the isosceles triangle CAB with knowledge of radius R_E of the earth and the geocentric latitudes ϕ_A and ϕ_B. Further, from the geometry of Fig. 1.1 (b), the angle at M can be calculated. Then, using the law of sines, AM and BM are found out. Finally, the distance CM can be calculated from one of the triangles ACM and BCM.

For measuring still larger distances we need to have a much longer base line for better accuracy. For distances within our solar system the equatorial diameter of our earth can be taken as base line. For measuring distances of nearer stars the diameter of the earth's orbit around the sun is used for this purpose (Fig. 1.2). The two angles are measured in summer and winter. Most distant stars are used to fix the reference direction. Being very far off they do not change their positions with respect to one another. If you watch through the window of a rapidly moving train, the trees and other objects near you apparently move rapidly past, but the distant objects do not seem to move at all. Distant stars which do not change their positions are called 'fixed' stars, and can be used to give us known directions from any point on the earth's orbit. The nearby stars appear to move relative to the distant ones as we see them from different positions on earth's orbit. This is like the shift you see when you hold out a finger before your eyes and look at it against some background (say, wall of the room) first with one eye open and then with the other. The finger appears to shift its position along the wall. This shift is more if the finger is nearer to the eye. This apparent shift of one object relative to the other is called parallax. But if both the objects are distant, or are at the same position, they do not show this parallax.

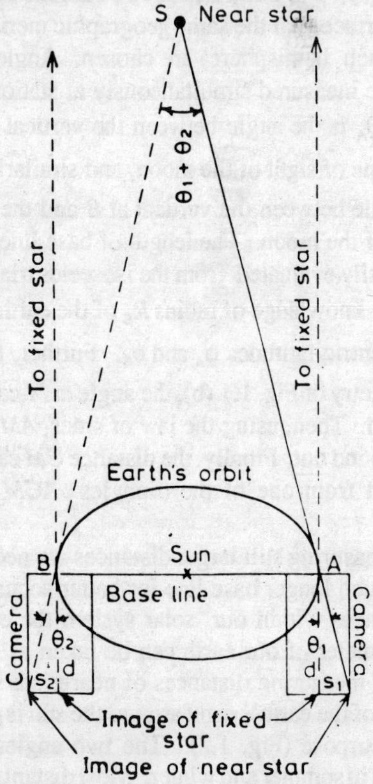

Fig. 1.2 ***Determination of the distance of a near star from the measurement of its parallax with respect to a 'fixed' star. Here the base line AB is the diameter of earth's orbit around sun***

The parallax between the near star and a fixed star is measured as shown in Fig. 1.2. The picture of the two stars is taken at *A* by pointing the camera towards the fixed star. Another picture is taken at *B* after six months by again pointing the camera towards the fixed star. Thus the direction of the camera in both the exposures remains the same. Angles θ_1 and θ_2 can be calculated from the distances between the images of the two stars and the focal length of the camera lens. The sum of θ_1 and θ_2 is equal to angle *ASB*, the angle subtended by the diameter of the earth's orbit at the star. As length *AB* is known, the distance of the near star can be measured. If x is the average distance of the star from the earth, then as $x \gg AB$,

$$\theta_1 + \theta_2 \text{ (in radians)} \cong \frac{AB}{x}$$

For very large distances, the parallax, and hence the angles, become so small that it becomes difficult to measure them. Distances of far-off stars can be estimated from their colours. The distances of nearby stars are known by triangulation. Therefore, the intrinsic brightness can be calculated from the measurement of apparent brightness, because for a given intrinsic brightness the apparent brightness decreases with the square of the distance. It is found that in most cases there is a smooth relationship between the colour and the intrinsic brightness. Therefore, if we now measure the colour of a distant star, its intrinsic brightness can be found out with the help of the colour-brightness relationship. By measuring how bright the star appears to us, we can find out its distance.

1.4 SI Units*

SI is the abbreviation of *Le Systeme International d' Unites* (International System of Units). It is the combination of MKS (metre, kilogram and second) system in mechanics and RMKSA (rationalized metre, kilogram, second and ampere) system in electricity and magnetism; expanded by the addition of kelvin from the field of heat, candela from the field of light, and the mole, unit of amount of substance used in chemistry. The main connecting link between mechanics and electricity is the joule, the 'common' unit of energy. This system has seven base and two supplementary units (Tables 1.1 and 1.2), and a large number of derived units, some of which are given in Table 1.3.

---

*For details refer : International Union of Pure and Applied Physics : Symbols, Units and Nomenclature in Physics. Physica A, 39A, Nos. 1+2, August 1978

Table 1.1 *SI Base Units*

S No	*Base quantity*	*Unit*	*Symbol for the unit*
1	Length	metre	m
2	Mass	kilogram	kg
3	Time	second	s
4	Electric current	ampere	A
5	Thermodynamic temperature	kelvin	K
6	Luminous intensity	candela	cd
7	Amount of substance	mole	mol

Table 1.2 *SI Supplementary Units*

S No	*Physical quantity*	*Unit*	*Symbol of the unit*
1	Plane angle	radian	rad
2	Solid angle	steradian	sr

The definitions of the base units are :

Metre : The metre is the length equal to 1,650,763.73 wavelengths in vacuum of the radiation corresponding to the unperturbed transition between the levels $2p_{10}$ and $5d_5$ of krypton-86.

Kilogram : It is equal to the mass of the international kilogram kept at Sevres, France.

Second : It is the duration of 9, 192, 631, 770 vibrations of the radiation corresponding to the unperturbed hyperfine transition 4,0-3,0 of the fundamental state $^2S_{1/2}$ of caesium-133 atom.

Ampere : It is the constant current which, when maintained in two, straight, infinitely long parallel conductors of negligible cross-section, and placed one metre apart in vacuum, would produce between these conductors a force equal to 2×10^{-7} newton per metre length.

Candela : It is the luminous intensity, in perpendicular direction, of a surface of 1/600,000 m^2 of a black body at the temperature of freezing platinum under a pressure of 101325 N m^{-2}.

Kelvin : The kelvin of thermodynamic temperature is the fraction 1/273.16 of the thermodynamic temperature of the triple point of water.

Mole : The mole is the amount of substance of a system which contains as many elementary entities as there are atoms in 0.012 kg of carbon-12. When the mole is used, the elementary entities must be specified and may be atoms, molecules, ions, electrons, other particles or specified groups of such particles.

Table 1.3 *Derived SI Units with Special Names*

S No	*Physical quantity*	*Unit*	*Symbol of the unit*
1	Energy, work	joule	J
2	Force	newton	N
3	Power, radiant flux	watt	W
4	Electric charge	coulomb	C
5	Electric potential, electric potential difference	volt	V
6	Electric resistance	ohm	Ω
7	Electric capacity	farad	F
8	Magnetic flux	weber	Wb
9	Magnetic field-*B*, magnetic flux density, or magnetic induction	tesla	T
10	Inductance	henry	H
11	Luminous flux	lumen	lm
12	Illuminance	lux	lx
13	Frequency	hertz	Hz
14	Customary temperature	celsius	°C
15	Pressure, stress	pascal	Pa
16	Electric conductance	siemens	S
17	Activity	becquerel	Bq
18	Absorbed dose	gray	Gy

1.5 Units and Dimensions

In mechanics, any physical quantity, like area, velocity, energy, etc., can be written in terms of mass, length and time. In addition to these three, we require current to express various quantities in electricity and magnetism; temperature to express various quantities in heat; luminous intensity to express various quantities in light; amount of substance in chemistry. These seven quantities —length, mass, time, current, temperature, luminous intensity, and amount of substance—are called *fundamental quantities* of the SI system and their units are known as *fundamental units*. All other quantities, which are expressed in terms of these quantities, are called derived quantities, and corresponding units are called derived units.

The dependence of a derived unit on various fundamental units can be well understood if we know how the latter appear in its expression. For example, area is length × length or $(\text{length})^2$. If we double the side of a square, or the radius of a circle, the area will increase four times. Further, if the unit of length is changed by certain amount, the number expressing area will change according to the square of it. Thus, if the unit of length is changed from inch to centimetre, the number expressing area will increase by $(2.54)^2$. Also, area does not involve other fundamental units. Therefore, it does not at all depend upon units of mass, time, etc. Similarly, velocity involves length and time only, and charge involves current and time only.

We can express the above dependences of derived units on various fundamental units with the help of *dimensions*. In other words, a relation of any physical quantity with the fundamental units is established through its dimensions. The dimensions of a derived unit are defined as the powers of various fundamental units to which these are raised in order to represent it. Thus, the dimensions of area are 2 in length only; dimensions of velocity are 1 in length and −1 in time; and dimensions of charge are 1 in current and 1 in time. $[L^2]$, $[LT^{-1}]$ and $[IT]$ are the dimensional formulae of area, velocity and charge, respectively. The relations of the type $[\text{area}] = [L^2]$, are called *dimensional equations*. For the purpose of dimensional analysis, the fundamental units —length, mass, time, current, temperature, luminous intensity, and amount of substance—may be represented by capital letters L, M, T, I, K, C and N, respectively. The square brackets simply indicate that we are concerned only with the basic dimensionality of the quantities enclosed, with no reference to numbers or magnitudes.

The dimensions of various quantities can be found from their defining relations. For example :

$$\begin{aligned}\text{Work} &= \text{force} \times \text{distance}\\ &= \text{mass} \times \text{acceleration} \times \text{distance}\\ &= \text{mass} \times \frac{\text{distance}}{\text{time}^2} \times \text{distance}\end{aligned}$$

Therefore

$[\text{Work}] = [ML^2T^{-2}]$

These dimensions are useful for : (1) checking a given relation, and (2) conversion of units from one system to another.

Checking of a given relation : It is based on the principle of homogeneity, which says that in an equation, each term represents the same kind of quantity. Thus :

2 metres + 3 metres = 5 metres, is valid;
while 2 metres + 3 kg = ? has no meaning.

Similarly the equation :

6 pencils × 50 p per pencil + 2 pens × Rs. 5.00 per pen + 5 erasers × 40 p per eraser = Rs. 15.00,

is a homogeneous equation because each term represents a quantity of money. Thus to check whether a particular equation is correct (or not), one has simply to find out the dimensions of various terms which add on both sides, and to see whether each of them has the same dimensions in each of the fundamental units. In any meaningful physical equation the dimensions of all the terms must be the same. But one will not be able to check a numerical multiplier which has no dimensions. As an illustration let us take the relation :

$$s = ut + \frac{1}{2}at^2$$

The dimensions of s are one in length only. Therefore, the dimension of each of the right-hand terms should also be one in length only. Now,

$$ut = \frac{\text{distance}}{\text{time}} \times \text{time} = \text{distance},$$

and $$at^2 = \frac{\text{distance}}{\text{time}^2} \times (\text{time})^2 = \text{distance}$$

Therefore, the equation is dimensionally correct.

Conversion of Units : Expression of every physical quantity requires two things : magnitude and unit in which it has been measured. If the distance between two points is 10 m, it means that it has been measured in terms of a fixed distance called metre and it contains 10 such units, *i.e.* the distance =10 × 1 metre. Let us now measure the same distance in feet. If it is x feet, then x feet = 10 metres, because each of these expresses the same distance. In general, any given quantity = $n_1a_1 = n_2a_2$ where n_1 is the numerical factor and a_1 is the unit in one system of units, and n_2 is the numerical factor, and a_2 is the unit in some other system of units. Let us convert 10 yards into centimetres.

Let :

$$x \text{ cm} = 10 \text{ yards} = 30 \text{ feet}$$

$$\therefore \quad x = 30 \text{ feet/1cm}$$

$$= \frac{30 \times 12 \times 2.54 \text{ cm}}{1 \text{ cm}} = 914.4,$$

or 10 yards = 914.4 cm

In general, let the dimensional formula of a quantity be $M^aL^bT^cI^dK^eC^fN^g$, where a, b, c, d, e, f and g are any numbers. Let the units in one system be $M_1^aL_1^bT_1^cI_1^dK_1^eC_1^fN_1^g$ and that in other system be $M_2^aL_2^bT_2^cI_2^dK_2^eC_2^fN_2^g$. Further, let n_1 and n_2 be the numerical magnitudes in the systems '1' and '2' respectively. Therefore :

$$n_1[M_1^aL_1^bT_1^cI_1^dK_1^eC_1^fN_1^g]$$

$$= n_2\,[M_2^aL_2^bT_2^cI_2^dK_2^eC_2^fN_2^g]$$

Thus, if n_1 is known n_2 can be calculated or vice-versa. Of course, the relationship between respective fundamental units is essential.

As an example let us find the number of dynes in one newton. We know that the physical quantity involved is force, the dimensional formula of which is $[MLT^{-2}]$. Let :

$$x \text{ dynes} = 1 \text{ newton, or}$$

$$x[\text{g} \times \text{cm} \times \text{s}^{-2}] = 1[\text{kg} \times \text{m} \times \text{s}^{-2}]$$

$$\therefore \quad x = 1\left[\frac{\text{kg}}{\text{g}}\right]\left[\frac{\text{m}}{\text{cm}}\right]\left[\frac{\text{s}^{-2}}{\text{s}^{-2}}\right]$$

$$= 1\left[\frac{1000 \text{ g}}{\text{g}}\right]\left[\frac{100 \text{ cm}}{\text{cm}}\right] = 10^5$$

Hence, 1 newton = 10^5 dynes

From the above we see that while checking equations, and while changing units from one system to another, the dimensional labels attached to various quantities may be treated like algebraic quantities. These may be combined, cancelled, etc., as if they were factors in an equation.

Another use of the dimensional analysis is to derive functional relationships between various physical quantities. Let us take as an illustration the time-period of a simple pendulum. If we somehow guess that the time-period depends on length of the pendulum, acceleration due to gravity, and angle of swing, then :

$$[T] = [L^a g^b \theta^c] \text{ say}$$

$$\therefore \quad [T] = [L^a L^b T^{-2b}]$$

because θ has no dimensions as it is the ratio of two lengths. Equating dimensions of each fundamental unit on both the sides :

$$1 = -2b, \text{ or } b = -\tfrac{1}{2},$$

and $$a + b = 0, \text{ or } a = -b = \tfrac{1}{2},$$

therefore $T = \text{constant} \sqrt{l/g}$

As already pointed out the value of the constant cannot be found out.

Exercise: Find an expression for the centripetal force on a particle of mass m moving with a speed v in a circle of radius r.

The dimensions of various physical quantities which appear in the text are given in Table 1.4.

1.6 Accuracy and Errors in Measurement

It is highly desirable that the measurements be carried out carefully, and the observations recorded should reflect the accuracy of their measurement. For example, if the length of a rod is measured using a meter scale then the record will be, say, 9.3 cm, because the least count of the scale is only 0.1 cm. The record of 9.30 cm is erroneous. When we read 9.3 cm, we know that the true value is nearer 9.3 cm than 9.2 cm or 9.4 cm, but we do not know that it is 9.30 cm. Since we do not know that the value of the hundredths digit is zero, it is not permissible to assume that it is. We can only say that the length of the rod lies between 9.25 cm and 9.35 cm. The record of 9.30 cm would mean that the measurement was carried out using an instrument which had the least count of 0.01 cm, *e.g.* a vernier caliper, but it so happened that the value of the hundredths digit came out to be zero. Any figure in a measurement, including zero, is a significant figure only if it contributes to the accuracy with which that measurement is carried out.

One has to be careful in computations as well. Suppose two observations 78.3 cm and 1.47 cm are to be added. Straightforward answer would be 79.77 cm. But as we are not sure about the hundredths digit in 78.3 cm, we cannot be sure about it in the sum also. Hence, we must round off the result to 79.8 cm. The second decimal place is insignificant.

Similar considerations also hold good in the calculation of the product. Suppose the length and the breadth of a rectangle are measured as 1001 cm and 8.3 cm, respectively. Straightforwardly, its area comes out to be 8308.3 cm^2. But this expression of the result is not correct. There could be an error of about 0.1 cm in the measurement of width, as it could be any thing between 8.25 cm and 8.35 cm. Accordingly, the area could be any thing between 8258.25 cm^2 and 8358.35 cm^2. Therefore, the appropriate expression of the result would be 8.3×10^3 cm^2. For better accuracy of the result the width should have been measured using a more accurate instrument, say a vernier caliper.

A thumb rule for the calculations is: 'no chain is stronger than its weakest link'. The 'weakest link' in multiplication is the number of significant figures; and the same is true for division. However, in both addition and multiplication, until the final result is obtained one uncertain digit should be carried in each value (*i.e.*, one more than the least number of significant figures).

It may also be emphasized here that the decimal point as such has nothing to do with accuracy of measurement. Two measurements 0.0083 cm and 8.3 cm have the same error, about 1%. It is the number of significant figures which decides the accuracy. Let the distance between two cities, as measured, be 312 km. Its expression in metres as 312000 m can create confusion regarding the significance of zeros. Therefore, it should be either written as 312×10^3 m, or as 3.12×10^5 m. The usual practice is to place the decimal point after the first significant figure, and take the appropriate power of ten. The first factor, here 3.12, which represents fully the significant figures/accuracy of the measurement is called the *number sequence* or *mantissa*, and the second factor the order of magnitude*.

In some cases even single digit accuracy measurements are not available. In such a case the result is expressed in order of magnitude only. As an example, age of the earth is of the order of 10^{16} s (10^9 yrs) or, in brief, $\sim 10^{16}$ s.

Error in the measurement of a physical quantity arising from the limitations of the measuring devices can be estimated as follows. Supose the density of a cylindrical solid rod is to be determined. Let the mass, length and radius of the rod

*When the number sequence is equal to or greater than 5, the order of magnitude is quite often taken to be ten times the second factor. The order of magnitude of, say, 6.1×10^{10} is 10^{11}.

Table 1.4 *Dimensions of Some Physical Quantities*

Quantity	*Symbol*	*Dimensional formula*	*Derived units in SI system*	*Quantity*	*Symbol*	*Dimensional formula*	*Derived units in SI system*
				Temperature	T	$[K]$	K
Length	l	$[L]$	metre	Entropy	S, s	$[ML^2T^{-2}K^{-1}]$	J/K
Mass	m, M	$[M]$	kilogram	Specific heat	c	$[ML^2T^{-2}K^{-1}]$	J/kg K
Time	t	$[T]$	second	Latent heat	L	$[L^2T^{-2}]$	J/kg
Displacement	r, d, s	$[L]$	metre	Current	I, i	$[I]$	ampere
Area	A, S	$[L^2]$	metre^2	Charge	Q, q	$[IT]$	coulomb
Volume	V, v	$[L^3]$	metre^3	Electric field strength	E	$[I^{-1}MLT^{-3}]$	volt/metre
Frequency	ν	$[T^{-1}]$	hertz (cycles/sec)				
Density	ρ	$[ML^{-3}]$	kg/m^3	Electric flux	Φ_E	$[I^{-1}ML^3T^{-3}]$	volt-metre
Velocity	v	$[LT^{-1}]$	metres/sec	Electric potential, emf	V, ε	$[I^{-1}ML^2T^{-3}]$	volt
Acceleration	a	$[LT^{-2}]$	metres/sec^2				
Force	F	$[MLT^{-2}]$	newton	Electric displacement	D	$[ITL^{-2}]$	C/metre^2
Work, Energy	W, E, T, K, U	$[ML^2T^{-2}]$	joule	Electric polarisation	P	$[ITL^{-2}]$	C/metre^2
Power	P	$[ML^2T^{-3}]$	watt	Electric dipole moment	p	$[ITL]$	C m
Momentum	p	$[MLT^{-1}]$	kg m/s	Resistance	R	$[I^{-2}ML^2T^{-3}]$	ohm
Angular displacement	θ	—	radian	Resistivity	ρ	$[I^{-2}ML^3T^{-3}]$	ohm-metre
Angular velocity	ω	$[T^{-1}]$	radians/sec	Conductivity	σ	$[I^2M^{-1}L^{-3}T^3]$	(ohm-metre)$^{-1}$
Angular acceleration	α	$[T^{-2}]$	radians/sec^2	Capacitance	C	$[I^2M^{-1}L^{-2}T^4]$	farad
Torque	τ	$[ML^2T^{-2}]$	N m	Inductance	L	$[I^{-2}ML^2T^{-2}]$	henry
				Magnetic field-B	B	$[I^{-1}MT^{-2}]$	tesla = weber/m^2
Angular momentum	L	$[ML^2T^{-1}]$	kg m^2/s	Magnetic flux	Φ_B	$[I^{-1}ML^2T^{-2}]$	weber = volt-sec
Rotational inertia	I	$[ML^2]$	kg m^2	Magnetic dipole moment	m	$[IL^2]$	ampere-metre^2
Gravitational constant	G	$[M^{-1}L^3T^{-2}]$	Nm^2/kg^2	Magnetic field-H	H	$[IL^{-1}]$	ampere-turns/metre
Gravitational field strength	g	$[LT^{-2}]$	N/kg	Magnetisation	M	$[IL^{-1}]$	ampere/metre
Pressure	p, P	$[ML^{-1}T^{-2}]$	N/m^2	Permeability	μ	$[I^{-2}MLT^{-2}]$	henry/metre
Coefficient of viscosity	η	$[ML^{-1}T^{-1}]$	Ns m^{-2}	Permittivity	ε	$[I^2M^{-1}L^{-3}T^4]$	farad/metre
Surface tension	σ	$[MT^{-2}]$	Nm^{-1}	Planck's constant	h	$[ML^2T^{-1}]$	joule-sec

be measured as M, l and r, respectively. The density ρ of the material is then given by,

$$\rho = \frac{M}{\pi r^2 l}$$

The variation of ρ in terms of variations of the other quantities can be found by taking logs of both the sides, and then differentiating. Thus,

$$\frac{\delta\rho}{\rho} = \frac{\delta M}{M} - 2\frac{\delta r}{r} - \frac{\delta l}{l}$$

The values δM, δr and δl represent the errors in the measurement of mass, radius and length, respectively, and may be positive or negative. Consider the worst case, which is that in which all the errors tend to add up. The error in density in such a case is given by,

$$\frac{\delta\rho}{\rho} = \frac{\delta M}{M} + 2\frac{\delta r}{r} + \frac{\delta l}{l}$$

The percentage error in density, $\frac{\delta\rho}{\rho} \times 100$, thus depends on the least counts of the measuring instruments, and also on the magnitudes of the various quantities measured. Further, error in the measurement of radius gets magnified two times. Thus accurate measurement of small quantities and quantities appearing in higher powers in the formula is more desirable.

When the errors arising from instruments, external conditions such as temperature variation, or from personal mistakes are eliminated, it is found that there is still a margin of error. This is due to accidental errors. Though such errors cannot be eliminated, their estimation is possible. Take a very large number of observations, say n, and take their mean value. Let it be X. This arithmetic mean is now taken as the most accurate value or the best estimate of the quantity required. Any particular measurement x will differ from the mean value X by an error, or deviation,

$$\varepsilon = x - X$$

The mean square deviation is then given by,

$$\sigma^2 = \frac{\Sigma\,\varepsilon^2}{n-1}$$

The root-mean-square deviation

$$\sigma = \sqrt{\frac{\Sigma\,\varepsilon^2}{n-1}}$$

is also called the *standard deviation*. Appearance of $(n-1)$, and not n, in the denominator can be explained on the ground that when there is only one observation, σ should be indeterminate, and not zero. No one number can ever give two pieces of information, here average value and standard deviation.

The final result is then expressed as

$$X \pm \sigma$$

In case the standard result (S) is already available in the literature, the error (E) in the observed value (O) is given by,

$$E = O - S,$$

and the percentage error is given by,

$$\%E = \frac{O-S}{S} \times 100$$

For the study of the theory of errors and various methods of their estimation reference may be made elsewhere*.

1.7 Vectors

Every physical quantity has a magnitude. One bucket contains 5 litres of water and another contains 3 litres of water, the total volume of water is just 8 litres. The quantities which can be specified completely by giving a magnitude only, and which can be combined by ordinary addition, are called scalars. Mass, length, time, pressure and temperature are some examples of scalar quantities.

A quantity which requires for its specification both magnitude and direction is called a vector, *e.g.*, displacement, velocity, acceleration, force and momentum. Apart from this, a vector has to obey certain other conditions explained later in this chapter.

***N.C. Barford, Experimental Measurements: Precision, Error and Truth (Addison-Wesley, 1967).**

Example 1.1 A man says that he has walked 6 km. He may be at B, B', B'', or B''' after his walk [Fig. 1.3(a)]. The distances travelled from A to B, B', B'' and B''' are the same, but the displacements (change of position in a specified direction) AB, AB', AB'' and AB''' are all different because each of them is specified by different magnitude and direction.

If a man goes 4 km from A to B due east and then 3 km from B to C due north [Fig. 1.3(b)], the distance covered by the man is 7 km, but the displacement AC is only 5 km in a direction making an angle of $\tan^{-1}\frac{3}{4}$ with the direction AB. We see that the total distance travelled is calculated by just adding the magnitudes of the distances, while the total displacement takes the direction also into consideration.

Graphically, a vector is represented by an arrow which has a foot (or origin, or starting point) and a head (or terminal point). The length of the arrow corresponds to the magnitude of the vector, and the direction is represented by the direction of the arrow. A vector whose origin and terminus coincide is a *zero*, or *null* vector.

Fig. 1.3 (a) *Persons starting from the same point may not have the same displacement even after travelling the same distance.* (b) *Total displacement may not be equal to the arithmetic sum of the individual displacements. Here three and four add up to five and not seven*

Analytically, a vector is represented by a letter with an arrow over it, as $\vec{A}$ or as $\underset{\sim}{A}$, and its magnitude is denoted by | **A** | or A.

Two vectors are *collinear* if they lie on the same straight line or on parallel straight lines. Collinear vectors may be of the same direction or of opposite directions.

Two vectors are said to be equal if they have the same magnitude and the same direction (*i.e.* they are parallel to each other) regardless of their origin, as vectors **A** and **B** in Fig. 1.4. A vector having the same magnitude as vector **A** but having a direction opposite to it is denoted by −**A** (Fig. 1.4).

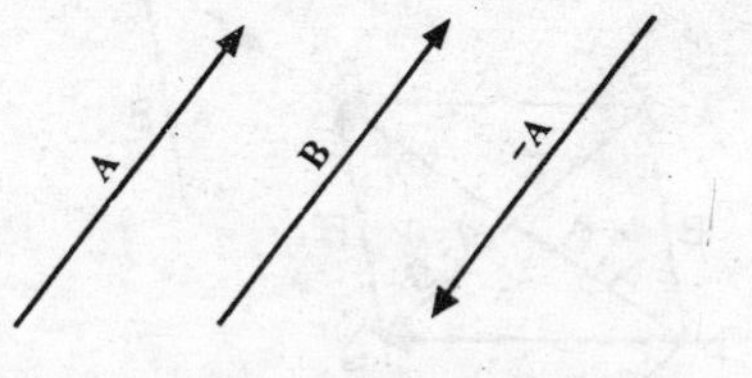

Fig. 1.4 *Equal and opposite vectors*

A unit vector **a** in the direction of **A** is a vector parallel to **A** having a unit magnitude, *i.e.*, **a** = **A**/A. We can represent vectors parallel to **A** in terms of the unit vector **a** (Fig. 1.5). A zero vector is considered to be of the same direction as any vector, its length is zero, i.e., | **0** | = 0.

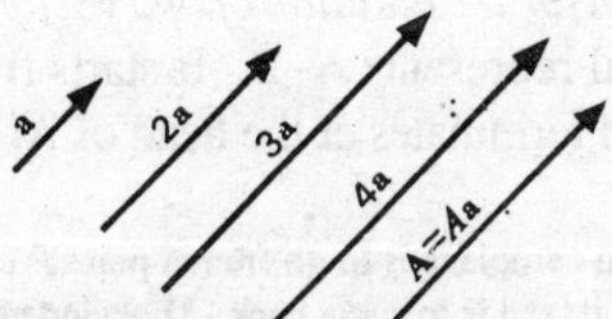

Fig. 1.5 *Vectors parallel to a unit vector*

1.8 Sum of Vectors

In Fig. 1.3(b), we have seen that the resultant displacement AC is the sum of the displacements AB and BC. Two vectors can be added just like two displacements. Sum of the vectors **A** and **B** is a vector formed by placing the origin of **B** on the terminal point of **A**, and then joining the origin of

A to the terminal point of **B** (Fig. 1.6). This is written as **A**+**B**. Similarly we can find the difference of two vectors, *e.g.* **A**−**B**. Now :

$$\mathbf{A}-\mathbf{B}=\mathbf{A}+(-\mathbf{B}) \quad (1.1)$$

Thus the problem reduces to finding the sum of **A** and −**B**. This has been done in Fig. 1.6.

Let us now complete parallelogram *PQRS*. Now, in Fig. 1.6, *PS* is equal to *QR* in magnitude and is parallel to it; and *SQ* is equal to *PT* in magnitude and is parallel to it. Therefore, *PS* = **B**, and *SQ* = **A**−**B**. Hence we get the rule : *Join the origins of* **A** *and* **B** *and complete the parallelogram with these two as adjacent sides. The diagonal with*

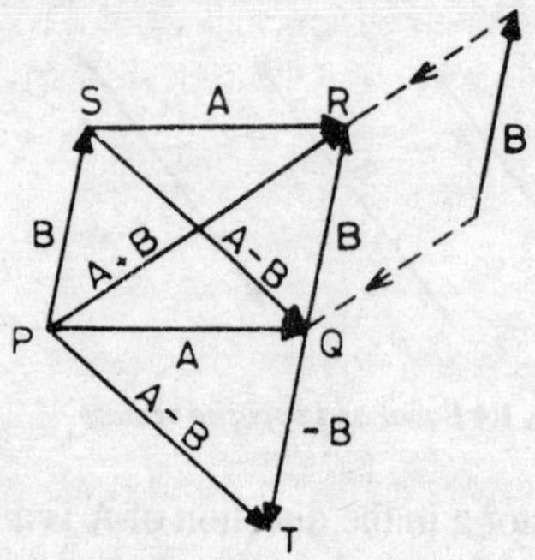

Fig. 1.6 *Sum and difference of two vectors. Two and two can be anything between zero and four*

the starting point at the common origin of **A** *and* **B** *represents the sum* **A**+**B**. *This is the law of parallelogram for the addition of two vectors.* The other diagonal represents **A**−**B**. It starts from the head of **B** and terminates at the head of **A**.

Example 1.2 An aeroplane is to go from a point *P* to a point *Q* to the north of it, and is to come back. The wind velocity is v' from west to east, and the speed of the plane with respect to air is v. Find the time taken for the return trip in terms of v, v' and l, the distance between *P* and *Q*.

As the plane is to go to the north, the vector sum of wind velocity v′ and air velocity v of the plane should be in this direction, as shown in Fig. 1.7(a). Therefore, the magnitude of the resultant velocity is $\sqrt{v^2-v'^2}$. Thus the time taken from *P* to *Q* is: $l/\sqrt{v^2-v'^2}$. As can be seen from Fig. 1.7(b), the plane will take the same time while coming back from *Q* to *P*. Hence, time t taken for the return trip is given by :

$$t=\frac{2l}{\sqrt{v^2-v'^2}}=\frac{2l/v}{\sqrt{1-v'^2/v^2}}$$

Since, in Fig. 1.6, *SR* is equal to *PQ* in magnitude and is parallel to it, *SR* = **A**. Therefore, *PR* represents the sum **B**+**A** also. Hence, we get:

$$\mathbf{A}+\mathbf{B}=\mathbf{B}+\mathbf{A} \quad (1.2)$$

Commutation relation [Eq. (1.2)] is, in fact, a property which the vectors must obey.

When there are more than two vectors we can add them successively to get the resultant vector. Let **A**, **B**, **C** and **D** be four vectors. Then we can sum them as (Fig. 1.8) :

$$\left.\begin{aligned}\mathbf{R}&=(\mathbf{A}+\mathbf{B})+(\mathbf{C}+\mathbf{D})\\&=\mathbf{R}_1+\mathbf{R}_2,\\ \text{or}\quad \mathbf{R}&=\mathbf{A}+\mathbf{B}+(\mathbf{C}+\mathbf{D})\\&=\mathbf{A}+(\mathbf{B}+\mathbf{C}+\mathbf{D})\end{aligned}\right\} \quad (1.3)$$

Question : Can two or three vectors of different magnitudes combine to give a zero resultant?

1.9 Components of a Vector

(a) Two-dimensional space : On a plane surface, or a two-dimensional space, we have seen that a vector can be specified completely by its magnitude and the direction with respect to a fixed axis. For example, a vector *PQ* is completely specified by the length of the arrow *PQ*, and the angle θ which *PQ* makes with a fixed axis *xx′* (Fig. 1.9).

Drop perpendiculars from *P* and *Q* on the axis *xx′*. The distance P_xQ_x is called the projection of *PQ* on *xx′*, or the component of *PQ* along *xx′*. It is very useful and convenient to express a vector in terms of its components along two mutually perpendicular axes. The magnitudes of these components in Fig. 1.10 are:

$$\left.\begin{aligned}A_x&=P_xQ_x=|PR|=A\cos\theta,\\ \text{and}\quad A_y&=P_yQ_y=|RQ|=A\sin\theta\end{aligned}\right\} \quad (1.4)$$

$$\therefore \quad |\mathbf{A}^2|=A^2=|PR^2|+|RQ|^2$$

$$=A_x^2+A_y^2 \quad (1.5)$$

Thus, the magnitude of a vector is given by the square-root of the sum of the squares of its components; and :

$$\tan\theta = \frac{|RQ|}{|PR|} = \frac{P_yQ_y}{P_xQ_x} = \frac{A_y}{A_x}$$

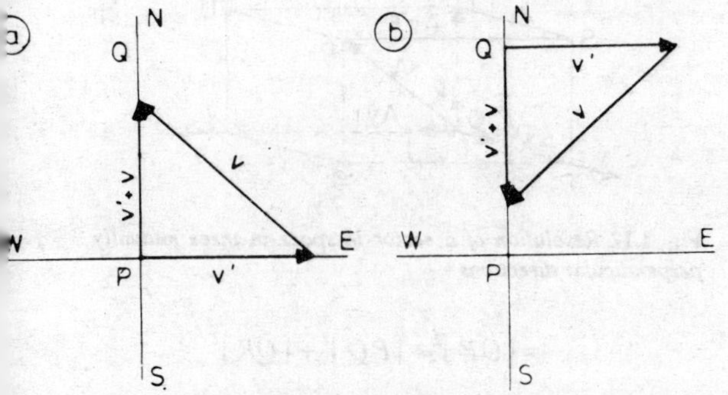

Fig. 1.7 *For Example 1.2*

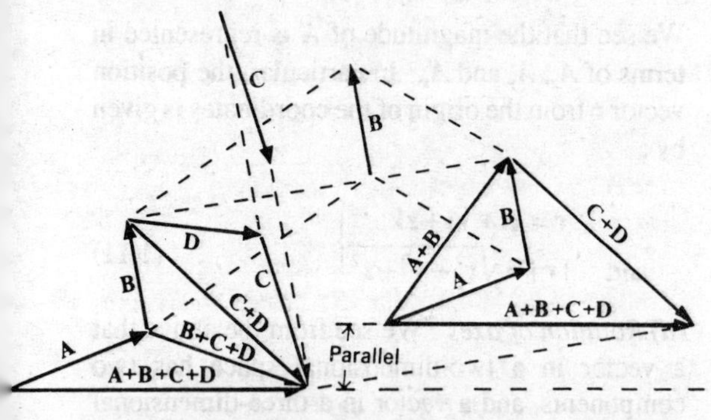

Fig. 1.8 *Summation of more than two vectors*

gives the direction which the vector makes with the x-axis.

If we define a unit vector, a vector having unit magnitude, along x-axis by **i**, and a unit vector along y-axis by **j**, then we can write :

$$\mathbf{A} = A_x\mathbf{i} + A_y\mathbf{j} \qquad (1.6)$$

$$= P_xQ_x\mathbf{i} + P_yQ_y\mathbf{j}$$

$$= PR + RQ = PQ \quad \text{(Fig. 1.10)}$$

(b) Position vector and displacement vector : Position of any point P can now be given by a vector joining the origin and the point P. We can denote the position vector by :

$$\mathbf{r} = x\mathbf{i} + y\mathbf{j}$$

where (x,y) are the coordinates of the point P in the x-y plane. If P_1 and P_2 are two points given by position vectors $\mathbf{r}_1$ and $\mathbf{r}_2$, then the displacement P_1P_2 (Fig. 1.11) is given by :

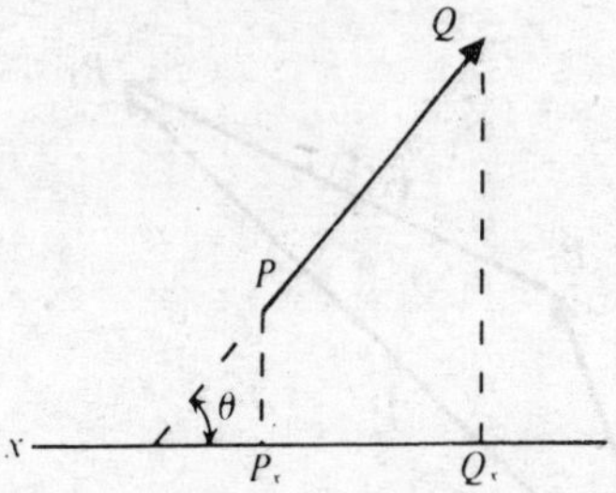

Fig. 1.9 *A vector in a two-dimensional space is completely specified if its length and inclination with respect to a reference direction are known*

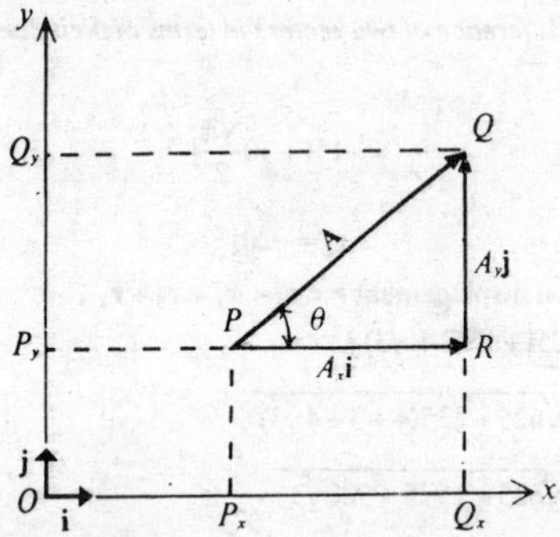

Fig. 1.10 *Resolution of a vector in a plane in two mutually perpendicular directions*

$$P_1P_2 = OP_2 - OP_1$$

$$= \mathbf{r}_2 - \mathbf{r}_1 \qquad (1.7)$$

$$= (x_2\mathbf{i} + y_2\mathbf{j}) - (x_1\mathbf{i} + y_1\mathbf{j})$$

$$= (x_2 - x_1)\mathbf{i} + (y_2 - y_1)\mathbf{j}$$

$$\therefore \quad |P_1P_2|^2 = |\mathbf{r}_2 - \mathbf{r}_1|^2$$

$$= (x_2 - x_1)^2 + (y_2 - y_1)^2 \qquad (1.8)$$

Example 1.3. A man moves 40 m east, then goes north an additional 50 m, then goes N-30°W for 30 m, and finally goes 20 m south. Find the resultant displacement.

Let us take the x- and y-axes in east and north directions respectively.

$$\therefore \quad \mathbf{r}_1 = 40\mathbf{i}$$

$$\mathbf{r}_2 = 50\mathbf{j}$$

$$\mathbf{r}_3 = 30\cos 120°\mathbf{i} + 30\sin 120°\mathbf{j}$$

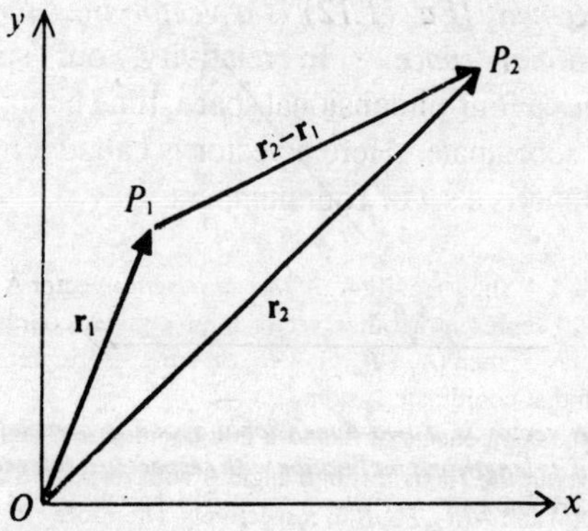

Fig. 1.11 *Difference of two vectors in terms of their components*

$$= -15\mathbf{i} + 30\frac{\sqrt{3}}{2}\mathbf{j}$$

$$\mathbf{r}_4 = -20\mathbf{j}$$

Hence, total displacement $\mathbf{r} = \mathbf{r}_1 + \mathbf{r}_2 + \mathbf{r}_3 + \mathbf{r}_4$

$$= 25\mathbf{i} + 15(2+\sqrt{3})\,\mathbf{j}$$

$$\therefore \quad |\mathbf{r}| = \sqrt{625 + 225(4+3+4\sqrt{3})}$$

$$= \sqrt{625 + 1575 + 900\sqrt{3}}$$

$$= 61\text{m}$$

If θ is the angle which this vector makes with eastern direction :

$$\tan\theta = \frac{15(2+\sqrt{3})}{25}, \quad \text{or } \theta = 66°$$

(c) Three-dimensional space : Let us take a unit cube. At any corner, three edges of the cube are perpendicular to one another. If we consider the three edges to be vectors, then they constitute a set of rectangular unit vectors denoted by **i, j** and **k** pointing towards, say, x-, y- and z- axes respectively.

Any vector **A** can, then, be represented by :

$$\mathbf{A} = A_x\mathbf{i} + A_y\mathbf{j} + A_z\mathbf{k} \tag{1.9}$$

where A_x, A_y and A_z are the components of **A** along x-, y- and z- axes, respectively. From Fig. 1.12 we can see that :

$$|OP| = A_x, |PQ| = A_y = |OV|,$$

$$|QR| = |OT| = A_z. \text{ Now, } |\mathbf{A}|^2 = |OR|^2 = |OQ|^2 + |QR|^2$$

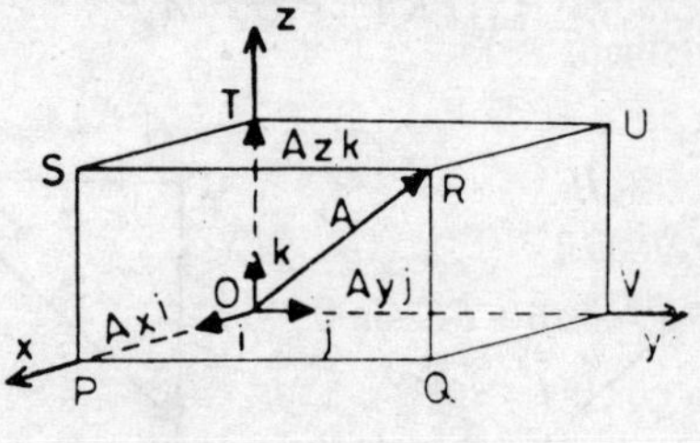

Fig. 1.12 *Resolution of a vector in space in three mutually perpendicular directions*

$$= |OP|^2 + |PQ|^2 + |QR|^2$$

$$= A_x^2 + A_y^2 + A_z^2,$$

$$\text{or} \quad |\mathbf{A}| = \sqrt{A_x^2 + A_y^2 + A_z^2} \tag{1.10}$$

We see that the magnitude of **A** is represented in terms of A_x, A_y and A_z. In particular, the position vector **r** from the origin of the coordinates is given by :

$$\left.\begin{aligned} \mathbf{r} &= x\mathbf{i} + y\mathbf{j} + z\mathbf{k}, \\ \text{and} \quad |\mathbf{r}| &= \sqrt{x^2+y^2+z^2} \end{aligned}\right\} \tag{1.11}$$

(d) Rotation of axes : We see from the above that a vector in a two-dimensional space has two components, and a vector in a three-dimensional space has three components. Once the components in a given coordinate system are known, the vector is specified completely as in Eqs. (1.6) and (1.9). Thus we can say that in a two-dimensional space a vector is two numbers (components along the two axes); in a three-dimensional space it is three numbers, and so on for higher order spaces.

Let us take an ordinary three-dimensional space. Take two coordinate systems S and S' with common origin as shown in Fig. 1.13. The z- and z'-axes are parallel and are perpendicular to the plane of the paper. Thus system S' has been obtained from system S after a rotation of angle θ about the z-axis. Consider a displacement vector

OP. From P drop perpendiculars on various axes as shown in Fig. 1.13. Also drop perpendiculars: CG on FP and Ox', DH on Oy', DI on FP. Let the components of OP in system S be (x, y, z), and in system S' be (x', y', z'). Therefore: $OC = DP = x$, $OD = CP = y$, $OE = FP = x'$, $OF = EP = y'$. We now see that:

$$\begin{aligned} x' &= OE \\ &= OJ + JE \\ &= OJ + GP \\ &= OC \cos\theta + PC \sin\theta \\ &= x\cos\theta + y\sin\theta \end{aligned}$$

$$\begin{aligned} y' &= OF \\ &= OH - HF \\ &= OH - ID \\ &= OD\cos\theta - DP\sin\theta \\ &= y\cos\theta - x\sin\theta \end{aligned}$$

Obviously, $z' = z$

Hence, putting all the above transformations at one place :

$$\left.\begin{aligned} x' &= x\cos\theta + y\sin\theta \\ y' &= -x\sin\theta + y\cos\theta \\ z' &= z \end{aligned}\right\} \quad (1.12)$$

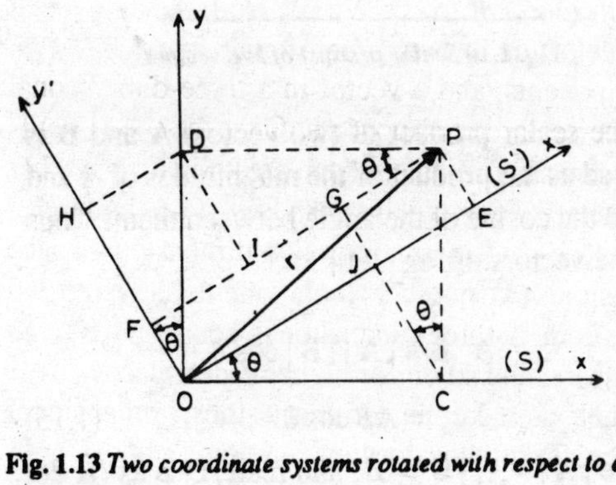

Fig. 1.13 ***Two coordinate systems rotated with respect to each other***

Equation (1.12) represents the law of transformation of displacement components, or the three numbers defining displacement vector, under rotation of axes. Components of other vectors will also transform in the same way. Velocity is an example.* Hence, *any directed physical quantity which can be represented by three numbers and these three numbers change to three other numbers under rotation of axes like the components of displacement [Eq. (1.12) is a vector in a three-dimensional space.* In relativity, our space becomes a four-dimensional space, time being the fourth coordinate. Here a vector is called a four-vector and is a set of four numbers.

Example 1.4 Show that if (A_x, A_y, A_z) represent a vector **A**, and (B_x, B_y, B_z) represent another vector **B** in a given coordinate system, say S, then $(A_x + B_x, A_y + B_y, A_z + B_z)$ also represent a vector in that coordinate system.

Let the components of **A** and **B** in a coordinate system S', rotated about the z-axis through angle θ with respect to S, be $(A_{x'}, A_{y'}, A_{z'})$ and $(B_{x'}, B_{y'}, B_{z'})$, respectively Therefore :

---

*In system S:

$$\text{velocity} = \frac{d\mathbf{r}}{dt} = \frac{d}{dt}(x\mathbf{i} + y\mathbf{j} + z\mathbf{k})$$

$$= \frac{dx}{dt}\mathbf{i} + \frac{dy}{dt}\mathbf{j} + \frac{dz}{dt}\mathbf{k}$$

$$= v_x\mathbf{i} + v_y\mathbf{j} + v_z\mathbf{k}$$

In system S':

$$\text{velocity} = \frac{d}{dt}(x'\mathbf{i}' + y'\mathbf{j}' + z'\mathbf{k}')$$

$$= \frac{dx'}{dt}\mathbf{i}' + \frac{dy'}{dt}\mathbf{j}' + \frac{dz'}{dt}\mathbf{k}'$$

$$= v_{x'}\mathbf{i}' + v_{y'}\mathbf{j}' + v_{z'}\mathbf{k}'$$

Therefore:

$$v_{x'} = \frac{dx'}{dt} = \frac{d}{dt}(x\cos\theta + y\sin\theta)$$

$$= \frac{dx}{dt}\cos\theta + \frac{dy}{dt}\sin\theta$$

$$= v_x\cos\theta + v_y\sin\theta$$

And similarly for other components.

$$A_{x'} = A_x \cos\theta + A_y \sin\theta$$

$$A_{y'} = -A_x \sin\theta + A_y \cos\theta$$

$$A_{z'} = A_z$$

and $$B_{x'} = B_x \cos\theta + B_y \sin\theta$$

$$B_{y'} = -B_x \sin\theta + B_y \cos\theta$$

$$B_{z'} = B_z$$

From the above equations we get :

$$(A_{x'} + B_{x'}) = (A_x + B_x)\cos\theta + (A_y + B_y)\sin\theta$$

$$(A_{y'} + B_{y'}) = -(A_x + B_x)\sin\theta + (A_y + B_y)\cos\theta$$

$$A_{z'} + B_{z'} = A_z + B_z$$

Thus the set of three numbers $(A_x+B_x, A_y+B_y, A_z+B_z)$ transforms under rotation of coordinates to the set of three numbers $(A_{x'}+B_{x'}, A_{y'}+B_{y'}, A_{z'}+B_{z'})$ according to the transformation rule for the components of a vector [Eq. (1.12)]. Therefore, the numbers $(A_x+B_x, A_y+B_y, A_z+B_z)$ represent a vector. This vector is denoted by **A**+**B**.

Similarly, it can be easily shown that $(A_x-B_x, A_y-B_y, A_z-B_z)$ is also a vector, and is represented by **A**−**B**.

Under the rotation (and also the translation) of coordinate axes, the respective components of all vectors change in the same fashion. Thus, if a vector relation holds good in one coordinate system, it will also hold good in coordinate systems which have been rotated (or translated) with respect to it. It may be recalled that the law of the addition of vectors was stated without reference to any coordinate system. Now, we know from experience that *the laws of physics are invariant with respect to rotation (or translation) of coordinates*. Hence the mathematical equation representing a physical law written in terms of vectors will not depend on a particular coordinate system. Let us take Newton's second law as an example. In system S it can be written as:

$$\left.\begin{aligned} F_x &= ma_x \\ F_y &= ma_y \\ F_z &= ma_z \end{aligned}\right\} \qquad (1.13)$$

where mass m has been subjected to a force having components (F_x, F_y, F_z) in S, resulting in an acceleration (of m) having components (a_x, a_y, a_z) in S. To express this law in S', we will have to rewrite Eq. (1.13) after putting primes on the force and acceleration components. But if we write this law in the form :

$$\mathbf{F} = m\mathbf{a} \qquad (1.14)$$

where **F** is a vector representing force and **a** is a vector representing acceleration, then this equation does not depend on a particular coordinate system. The technique of vector analysis has thus been developed to reduce mathematical details to a minimum.

1.10 Scalar Product

When we multiply a vector **A** by a scalar c the resultant vector is $c\mathbf{A}$. The direction of the vector remains unchanged, but the magnitude becomes c times that of **A**. When we come to the product of two vectors, we have to define the nature of the product. We can multiply the two vectors to yield a scalar, or another vector. We will discuss here only two types of products.

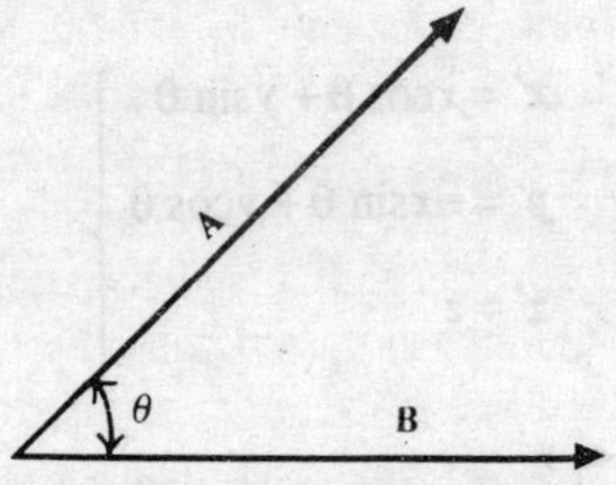

Fig. 1.14 *Scalar product of two vectors*

The scalar product of two vectors **A** and **B** is defined as the product of the magnitudes of **A** and **B** and the cosine of the angle between them. Thus for the vectors of Fig. 1.14 :

$$\mathbf{A}\cdot\mathbf{B} \equiv |\mathbf{A}|\,|\mathbf{B}|\cos\theta$$

$$= AB\cos\theta \qquad (1.15)$$

where $|\mathbf{A}| = A$, $|\mathbf{B}| = B$, and read $\mathbf{A}\cdot\mathbf{B}$ as (**A** dot **B**). We emphasize here that $\mathbf{A}\cdot\mathbf{B}$ is a scalar. If $\theta = 0$, $\mathbf{A}\cdot\mathbf{B} = AB$, If $\mathbf{A} = \mathbf{B}$, then : $\mathbf{A}\cdot\mathbf{A} = A^2$. Thus the dot product of a vector with itself gives the square of the magnitude of the vector.

If $\theta = 90°$,

$$\mathbf{A} \cdot \mathbf{B} = 0 \quad (1.16)$$

Therefore, two vectors are perpendicular to each other (orthogonal), if their scalar product is zero.

Example 1.5 For a set of rectangular unit vectors (**i**, **j**, **k**), we have :

$$\mathbf{i} \cdot \mathbf{i} = \cos 0° \, |\mathbf{i}| \, |\mathbf{i}| = 1.$$

Similarly, $\mathbf{j} \cdot \mathbf{i} = 0$, $\mathbf{j} \cdot \mathbf{j} = 1$, $\mathbf{j} \cdot \mathbf{k} = 0$,

$$\mathbf{k} \cdot \mathbf{i} = 0, \text{ and } \mathbf{k} \cdot \mathbf{k} = 1$$

Example 1.6 Find the scalar product of two vectors in terms of their components.

Let the two vectors be **A** and **B** having components (A_x, A_y, A_z) and (B_x, B_y, B_z), respectively, in a given coordinate system. Then:

$$\mathbf{A} \cdot \mathbf{B} = (A_x\mathbf{i} + A_y\mathbf{j} + A_z\mathbf{k}) \cdot (B_x\mathbf{i} + B_y\mathbf{j} + B_z\mathbf{k})$$

After opening the brackets and using the results of Example 1.5 for the dot product of unit vectors **i**, **j** and **k**, we get:

$$\mathbf{A} \cdot \mathbf{B} = A_xB_x + A_yB_y + A_zB_z$$

Thus the dot product of two vectors is equal to the sum of the products of respective components. This product is invariant under the rotation of the coordinate axes [Problem 1.11 (ii)], and hence, is a scalar.

1.11 Vector Product

The vector product of **A** and **B** is a vector $\mathbf{C} = \mathbf{A} \times \mathbf{B}$ (read as A cross B) having a magnitude equal to the product of the magnitudes of **A** and **B** and the sine of the angle between them. Thus in Fig. 1.15 (c) :

$$|\mathbf{C}| = |\mathbf{A} \times \mathbf{B}| \equiv |\mathbf{A}| \, |\mathbf{B}| \sin\theta \quad (1.17)$$

The direction of $\mathbf{A} \times \mathbf{B}$ is perpendicular to the plane of **A** and **B** in such a manner that **A**, **B** and **C** form a right-handed system*. Or the direction of **C** is that in which a right-handed screw advances when

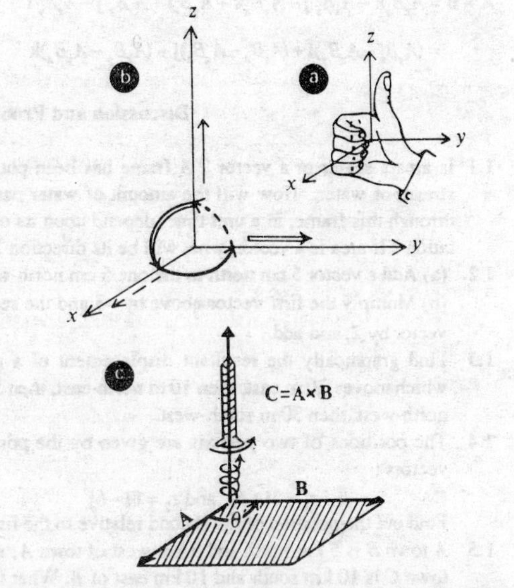

Fig. 1.15 **(a)** *Righi-handed coordinate system* **(b)** *Cyclic order of x-, y- and z-axes—rotation of x towards y through smaller angle takes the right-handed screw in positive z direction, rotation of y towards z...* **(c)** *Vector product of two vectors*

turned through the smaller angle from **A** to **B**. The magnitude of $\mathbf{A} \times \mathbf{B}$ is equal to the area of the parallelogram formed by the vectors **A** and **B**. (Area of a surface is represented by a vector whose magnitude is equal to the area and the direction is perpendicular to the surface.)

$$\text{If} \quad \theta = 0, \mathbf{A} \times \mathbf{B} = \mathbf{0}$$

Two vectors are said to be parallel to each other if their cross product is zero.

We can easily see that for a right-handed system :

$$\left.\begin{aligned} &\mathbf{i} \times \mathbf{i} = 0, \ \mathbf{i} \times \mathbf{j} = \mathbf{k}, \ \mathbf{i} \times \mathbf{k} = -\mathbf{j}, \\ &\mathbf{j} \times \mathbf{i} = -\mathbf{k}, \mathbf{j} \times \mathbf{j} = 0, \ \mathbf{j} \times \mathbf{k} = \mathbf{i}, \\ &\mathbf{k} \times \mathbf{i} = \mathbf{j}, \ \mathbf{k} \times \mathbf{j} = -\mathbf{i}, \ \mathbf{k} \times \mathbf{k} = 0 \end{aligned}\right\} \quad (1.18)$$

Example 1.7 Find the cross product of two vectors in terms of their components.

Let the two vectors be $A(A_x, A_y, A_z)$ and $\mathbf{B}(B_x, B_y, B_z)$. Then:

$$\mathbf{A} \times \mathbf{B} = (A_x\mathbf{i} + A_y\mathbf{j} + A_z\mathbf{k}) \times (B_x\mathbf{i} + B_y\mathbf{j} + B_z\mathbf{k})$$

After opening the brackets and using the results of Eq. (1.18), we get :

*The motion of a right-handed screw is given by the right hand with the thumb stretched and fingers curled [Fig. 1.15(a)]. The rotation of the screw is represented by the curled fingers and the advancement of the screw is in the direction of the stretched thumb. If we choose a coordinate system (*x*, *y*, *z*) in such a way that the rotation of *x* into *y* is given by the direction of *z*, that of *y* into *z* by the direction of *x*, and that of *z* into *x* by the direction of *y*, according to the right-hand rule [Fig. 1.15(b)] the coordinate systesm is a right handed coordinate system.

$$\mathbf{A}\times\mathbf{B} = A_xB_y\mathbf{k} - A_xB_z\mathbf{j} - A_yB_x\mathbf{k} + A_yB_z\mathbf{i} + A_zB_x\mathbf{j} - A_zB_y\mathbf{i}$$

$$= (A_yB_z - A_zB_y)\mathbf{i} + (A_zB_x - A_xB_z)\mathbf{j} + (A_xB_y - A_yB_x)\mathbf{k}$$

Discussion and Problems

1.1 Is area a scalar or a vector ? A frame has been put in a stream of water. How will the amount of water passing through this frame, in a unit time, depend upon its orientation ? If area is a vector, what will be its direction ?

1.2 (a) Add a vector 5 cm north to the one 5 cm north-west.
(b) Multiply the first vector above by –3 and the second vector by 2, and add.

1.3 Find graphically the resultant displacement of a point which moves 20 m east, then 10 m north-east, then 50 m north-west, then 30 m south-west.

1.4 The positions of two persons are given by the position vectors :

$$\mathbf{r}_1 = 3\mathbf{i} + 4\mathbf{j}, \text{ and } \mathbf{r}_2 = 8\mathbf{i} - 6\mathbf{j}$$

Find out the position of the second relative to the first.

1.5 *A* town *B* is 5 km north and 5 km west of town *A*; and a town *C* is 10 km south and 10 km east of *B*. What is the position of *C* with respect to *A*, and what is the distance between *A* and *C* ?

1.6 Two vectors are : $\mathbf{A} = 3\mathbf{i} + 4\mathbf{j}$, and $\mathbf{B} = -2\mathbf{i} + \mathbf{j}$. Find : (i) the length of each vector and its inclination with *x-axis* ; (ii) the sum and difference of the two vectors; (iii) the scalar and vector products; and (iv) the angle between the two vectors.

1.7 Show that :
(a) $|\mathbf{A}+\mathbf{B}| = (A^2+B^2+2AB\cos\theta)^{1/2}$
where θ is the angle between vectors **A** and **B**. [Hint : Let $\mathbf{R} = \mathbf{A} + \mathbf{B}$. Find $\mathbf{R}\cdot\mathbf{R}$. Also do it by the graphical method.]
(b) $(A+B) \geq |\mathbf{A}+\mathbf{B}| \geq (A-B)$

1.8 Write the displacement vector in three-dimensional space.

1.9 Using components of a vector prove that :
(i) $\mathbf{A}+(\mathbf{B}+\mathbf{C}) = (\mathbf{A}+\mathbf{B})+\mathbf{C}$
(ii) $m\mathbf{A} = \mathbf{A}m$
(iii) $(m+n)(\mathbf{A}+\mathbf{B}) = m\mathbf{A}+m\mathbf{B}+n\mathbf{A}+n\mathbf{B}$

1.10 Given:
$\mathbf{r}_1 = -\mathbf{i}+2\mathbf{j}+2\mathbf{k}$
$\mathbf{r}_2 = 3\mathbf{i}+4\mathbf{j}-\mathbf{k}$
$\mathbf{r}_3 = -2\mathbf{i}+2\mathbf{j}+5\mathbf{k}$
(i) Find the magnitude of $\mathbf{r}_1$, and $\mathbf{r}_1+\mathbf{r}_2+\mathbf{r}_3$
(ii) Write the unit vector in the direction $\mathbf{r}_2$, and $\mathbf{r}_1+\mathbf{r}_2+\mathbf{r}_3$

1.11 Prove that :
(i) $\mathbf{A}\cdot\mathbf{B} = \mathbf{B}\cdot\mathbf{A}$
(ii) The dot product of two vectors is invariant under rotation of coordinates.
[Hint : Write the dot product in primed coordinate system and then change the primed components to the unprimed ones using Eq. (1.12).]
(iii) The unit vector **a** in the direction **A** is :

$$\mathbf{a} = \frac{\mathbf{A}}{\sqrt{\mathbf{A}\cdot\mathbf{A}}}, \text{ or } \frac{|\mathbf{A}|\mathbf{A}}{\mathbf{A}\cdot\mathbf{A}}$$

(iv) $|\mathbf{A}|^2 = (\mathbf{A}\cdot\mathbf{i})^2 + (\mathbf{A}\cdot\mathbf{j})^2 + (\mathbf{A}\cdot\mathbf{k})^2$

1.12 Prove that :
(i) $\mathbf{B}\times\mathbf{A} = -\mathbf{A}\times\mathbf{B}$
(ii) $\mathbf{A}\times(\mathbf{B}+\mathbf{C}) = \mathbf{A}\times\mathbf{B}+\mathbf{A}\times\mathbf{C}$

1.13 Show that $(\mathbf{A}\times\mathbf{B})$ is a vector.
[Hint : Find the components of this product in the primed coordinates, and then express them in terms of the components in the unprimed system. Check whether two sets of components obey the transformation rule (1.12).]

1.14 Show that the component of vector **A** in the direction of a unit vector **u** is $\mathbf{A}\cdot\mathbf{u}$.

1.15 Prove that if the sum and difference of two vectors are perpendicular, the vectors have equal magnitudes.
[Hint : Put the relevant dot product equal to zero.]

1.16 Prove that if the magnitudes of the sum and difference of two vectors are equal, the vectors are perpendicular.

1.17 Using Eq.(1.12), or Fig. 1.13, express *x*, *y* and *z* in terms of *x′*, *y′* and *z′*.

1.18 (a) If $\mathbf{A} = \mathbf{B} + \mathbf{C}$, show that the projection of **A** along any direction is equal to the sum of the projections of **B** and **C** along that direction.
(b) Derive Eq. (1.12) making use of the above property.
[Hint: Take $\mathbf{B} = x\mathbf{i}$ and $\mathbf{C} = y\mathbf{j}$.]

Chapter 2

Force

2.1 Motion along a Straight Line

In order to know the position of a particle moving along a straight line as the time passes, one must know the initial position, the direction of motion and the quantity which tells us how fast the particle is moving, the latter being known as the velocity of the particle. Mathematically, the velocity **v** is defined as:

$$\mathbf{v} \equiv \frac{\text{displacement}}{\text{time}} = \frac{\mathbf{s}}{t} \qquad (2.1)$$

Since time is not considered to have any direction, the direction of velocity is the same as that of displacement. Conversely, we can also say that : *displacement = velocity × time*

= area under velocity-time graph

Example 2.1 The motion of a cyclist for certain time is given in Table 2.1.

Table 2.1 *Velocity of a Cyclist at Various Time Intervals*

Time interval	*Duration of interval (s)*	*Velocity during interval (m/s)*
1	60	2
2	40	3
3	30	4
4	20	5
5	10	4
6	10	3

Draw the velocity-time (v-t) graph, and find the final displacement of the cyclist.

The v-t graph is shown in Fig. 2.1. Adding the areas in various rectangular segments, we find that the total area under the curve is 530 m. Therefore, the total displacement is 530 m.

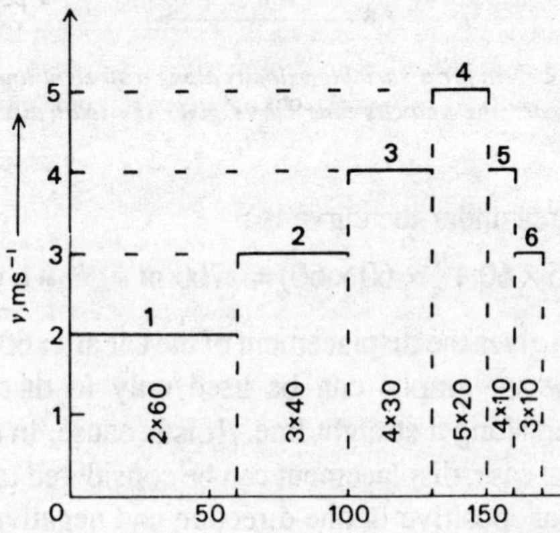

Fig. 2.1 *In a staight line motion the areas of various velocity-time rectangles add up to give the total distance travelled.*

Now, a velocity-time curve of any shape can be divided into infinitesimally small time intervals, and velocity can be treated as constant during a particular interval. The areas of these very thin rectangles will add up to the area under the actual curve. Thus, the above conclusion that *the area under a v-t curve gives displacement*, is a general one. For example the velocity of a car is recorded in Table 2.2, and is shown in Fig. 2.2.

Table 2.2 *Velocity of a Car as a Function of Time*

Time (s)	*Velocity (m/s)*	*Change in velocity during the time (m/s/s)*
0	15	
10	25	1
20	35	1
50	65	1
60	75	1

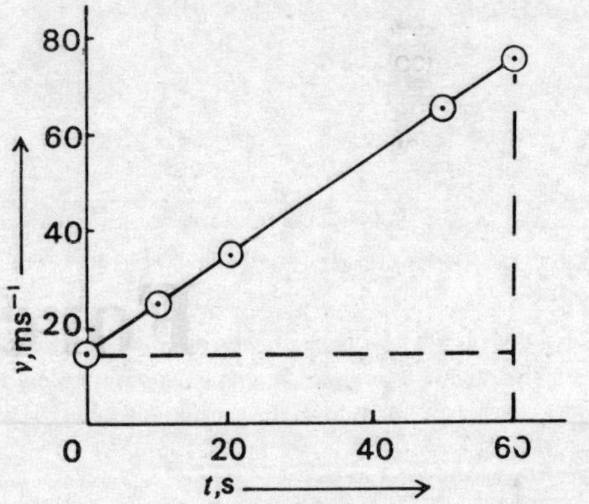

Fig. 2.2 *Even for a variable velocity along a straight line the area under the velocity-time curve gives the total distance covered.*

The area under the curve is :

$$(15 \times 60 + \tfrac{1}{2} \times 60 \times 60) = 2700 \text{ m}$$

which gives the displacement of the car after 60 s.*

The *v-t* graphs can be used only to discuss motion along a straight line. It is because, in this special case, displacement can be considered to be a scalar—positive in one direction and negative in the opposite direction. This method will not help us to study motion in space in general.

The slope of the time-velocity curve is a quantity which tells us about the *change of velocity per unit time*, and is termed as *acceleration*, **a**, of the particle. Or:

*Using calculus:

$$v = \frac{ds}{dt}$$

where *ds* is the distance covered in time *dt*.

Therefore, the distance covered

$$s = \int v\,dt = \int_0^{60} (15 + t)\,dt,$$

because speed at any time *t*, from Table 2.2 is (15+*t*). Or:

$$s = \left[15t + \frac{t^2}{2}\right]_0^{60}$$

$$= \left[15 \times 60 + \frac{60 \times 60}{2}\right]$$

$$= 900 + 1800$$

$$= 2700 \text{ m}$$

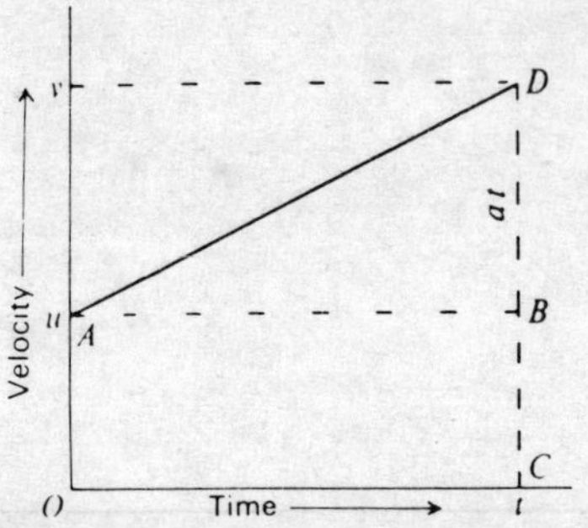

Fig. 2.3 *Velocity-time graph for motion along a straight line with constant acceleration*

$$\mathbf{a} = \frac{\mathbf{v}_2 - \mathbf{v}_1}{t_2 - t_1}$$

$$= \frac{\textit{change in velocity}}{\textit{interval of time}} \qquad (2.2)$$

Let us look at it in another way. Initially, let the velocity be *u*, and after a time *t* let it be *v*. Let the acceleration be *a*. From velocity-time graph (Fig. 2.3) we see that :

$$v - u = at,$$

or $$v = u + at \qquad (2.3)$$

The area under the curve is :

$$s = \text{rectangle } OABC + \Delta\, DAB$$

$$= ut + \frac{1}{2} t \times at,$$

or $$s = ut + \frac{1}{2} at^2 \qquad (2.4)$$

Eqs. (2.3) and (2.4) give:

$$v^2 = u^2 + 2as \qquad (2.5)$$

Eqs. (2.3) to (2.5) are the equations of motion under constant acceleration.

Example 2.2 How do you differentiate between average and instantaneous values of velocity and acceleration ?

If, when a particle moves from *A* to *B*, the distance is given by $\Delta x = x_2 - x_1$, and the elapsed time for motion between these two points is $\Delta t = t_2 - t_1$ (Fig. 2.4), the average velocity is given by :

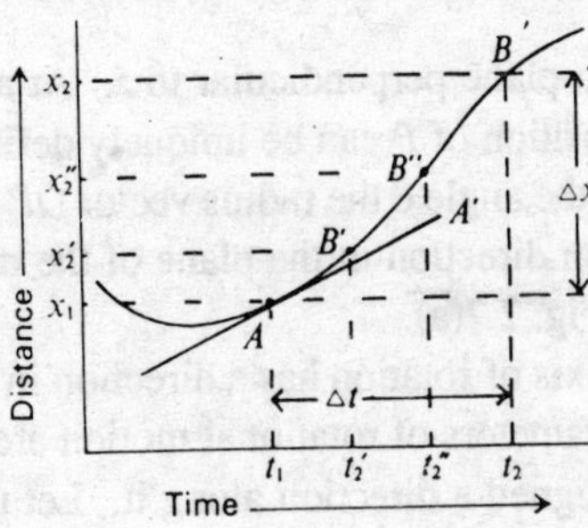

Fig. 2.4 *The average velocity between A and B, in rectilinear motion, equals the slope of the line joining A and B. The instantaneous velocity at A equals the slope of the tangent at A.*

$$< v > = \Delta x / \Delta t$$

$\Delta x/\Delta t$ is called the average velocity, because we know only the positions of A and B and the time taken to go from A to B. We do not know anything about how the motion took place and what path was followed. The velocity of a particle at a given instant of time is called the instantaneous velocity. The different average velocities during various time intervals, *e.g.* $\dfrac{\Delta x'}{t_2' - t_1}$, $\dfrac{\Delta x''}{t_2'' - t_1}$, etc., may differ. In Fig. 2.4, the time interval from A to B has been divided into three equal parts. It can be easily seen that the averge velocity in the first interval (from t_1 to t_2') is the least. In any interval the average velocity is equal to the slope of the line joining the two end points. As we let B approach A, the ratio $\Delta x/\Delta t$ approaches a definite limiting value. In the limiting value this slope becomes equal to the slope of the tangent AA' at A. This limiting value of $\Delta x/\Delta t$ is called the instantaneous velocity at the point A, or at the instant t_1. Or, the velocity at time t is the limiting value approached by $\Delta x/\Delta t$ as both Δx and Δt approach zero. Mathematically:

$$v = \lim_{\Delta t \to 0} \frac{\Delta x}{\Delta t} = \text{the derivative of } x \text{ with respect to } t$$

$$= \frac{dx}{dt}$$

Similarly, instantaneous acceleration is given by:

$$a = \lim_{\Delta t \to 0} \frac{\Delta v}{\Delta t} = \frac{dv}{dt}$$

2.2 Motion in Space

We have so far considered only the change in magnitude of velocity. But the acceleration is defined as the rate of change of velocity. Hence any change in the direction of velocity must also involve acceleration. (As velocity is a vector even a change in direction amounts to change in velocity.) Let us consider the case when the magnitude of velocity remains constant and only the direction changes.

A particle is moving with a constant speed (magnitude of velocity). The direction of velocity is different at A and B (Fig. 2.5). Let us consider two perpendicular unit vectors **i** and **j** along x- and y- axes respectively. The component of velocity in the direction of **j** is zero at A. But as soon as the direction changes, it gets a component along **j**. Thus, there is a change in the velocity in the y-direction. There is a corresponding change in x-direction also such that the speed is same as before. Mathematically:

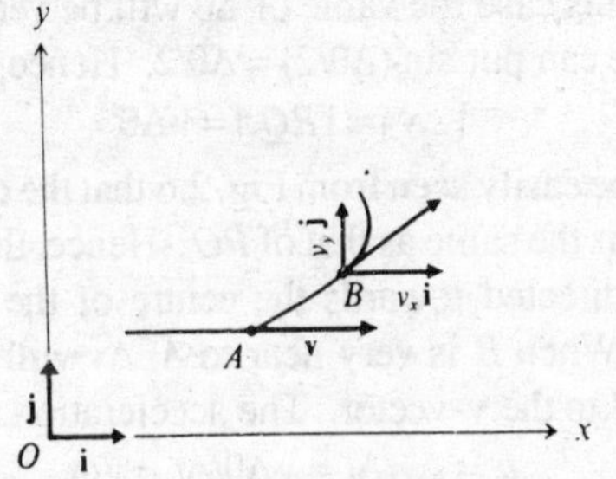

Fig. 2.5 *For motion in space, the same speed but different directions also amounts to different velocities.*

Velocity at A is: $\mathbf{v}_A = \mathbf{i}v + 0\mathbf{j}$

Velocity at B is: $\mathbf{v}_B = \mathbf{i}v_x + \mathbf{j}v_y$

Therefore, change in velocity $= \mathbf{i}(v_x - v) + \mathbf{j}v_y$. If the time taken is Δt sec, the acceleration is given by:

$$\mathbf{a} = \mathbf{i}\frac{v_x - v}{\Delta t} + \mathbf{j}\frac{v_y}{\Delta t} \quad (2.6)$$

Consider the motion of a particle along a circle with a constant speed v (Fig. 2.6). The direction of velocity changes constantly. Let the particle move from A to B in a small time interval Δt. Let the angle AOB subtended by arc AB at the centre be $\Delta\theta$. Extend $\mathbf{v}_A$ and $\mathbf{v}_B$ to meet at a point P. Let $|PQ| = |PR| = v$. It is clear that the magnitude of change in velocity is:

$$|\Delta \mathbf{v}| = |\mathbf{v}_B - \mathbf{v}_A| = |PQ - PR| = |RQ|$$

$$= 2v \sin(\Delta\theta/2)$$

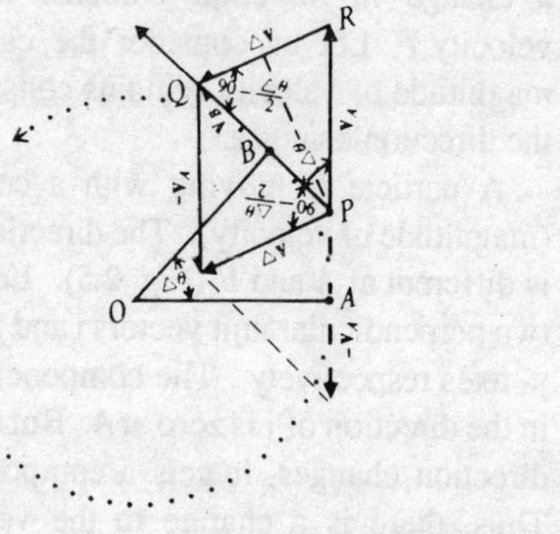

Fig. 2.6 *Motion along a circle with constant speed*

To find acceleration at *A*, *B* has to be very near to it. In this case the value of $\Delta\theta$ will be very small, and we can put $\sin(\Delta\theta/2) = \Delta\theta/2$. Hence, at *A*:

$$|\Delta \mathbf{v}| = |RQ| = v\Delta\theta \quad (2.7)$$

It can be easily seen from Fig. 2.6 that the direction of RQ is the same as that of PO. Hence, the vector $\Delta\mathbf{v}$ is directed towards the centre of the circular path. When *B* is very near to *A*, $\Delta\mathbf{v}$ will become normal to the **v**-vector. The acceleration is, thus :

$$a = \Delta v/\Delta t = v\Delta\theta/\Delta t = v\omega \quad (2.8)$$

where ω, the *angular velocity, is the rate at which the angle changes.* In this case where speed is constant, ω will also be constant. Hence, the acceleration is equal to $v\omega$ in magnitude, and is directed towards the centre.

This acceleration is known as centripetal acceleration. All along the circular motion, this acceleration is acting perpendicular to the velocity vector. Conversely, whenever a particle in motion is acted upon by a constant acceleration always perpendicular to its direction of motion, the particle will execute a circular motion with uniform speed.

Question : Is the centripetal acceleration a constant vector ?

2.3 Angular Vectors

Take a particle *P* rotating about an axis-*z* fixed in space [Fig. 2.7(a)]. Draw a perpendicular *PO* on *z*. Let $OP = R$. The path of the particle will be a circle of radius *R* with *O* as its centre. This circle will lie in a plane perpendicular to *z*. At a given time the position of *P* can be uniquely defined by specifying the angle θ the radius vector *OP* makes with a given direction in the plane of the motion, say *Ox* in Fig. 2.7(a).

As the axis of rotation has a direction in space, various parameters of rotational motion around it may be assigned a direction along it. Let us first take the angular displacement.

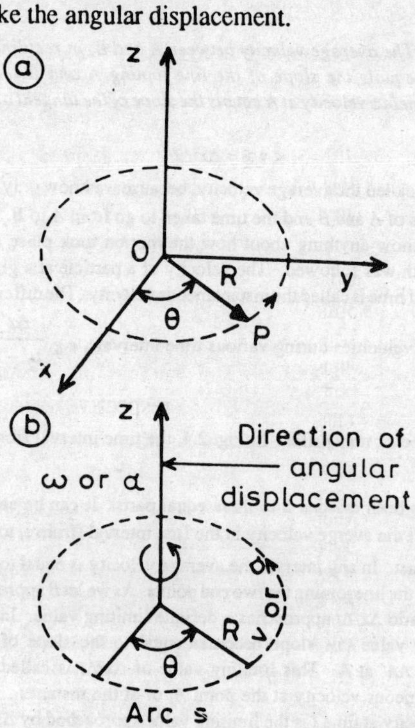

Fig. 2.7 (a) *Rotation of a particle about a fixed axis.* **(b)** *Angular displacement, velocity and acceleration are represented as vectors with directions given by the right-hand screw rule.*

An angle can be assigned a direction along the axis about which the angular displacement (rotation) takes place. If we rotate a right-handed screw in the manner in which the angular displacement is taking place, then the direction in which the screw advances is the direction of rotation [Fig. 2.7 (b)]*. In a right-handed coordinate system if

---

*The vectors to which we assign directions like this are called *axial vectors*. The vectors like **r** which have their natural directions are called *polar vectors*. Vector product of two polar vectors is an axial vector.

the rotation is from x to y, the direction of rotation is in the z-direction. Thus, the angular displacement from y to z is in the x-direction, and from z to x in the y-direction. (Also see footnote on p.19.) The magnitude of rotation by an angle θ is given by (Fig 2.7) :

Angle θ (in radians) = arc s/radius R

$\therefore$ *Angular displacement*

= *displacement along the arc/radius* (2.9)

It has been stated in Section 1.8 that for two vectors **A** and **B** we must have **A+B** = **B+A**. But in the case of finite angular displacements this type of relation is not always true, and in general we have :

$$\theta_1 + \theta_2 \neq \theta_2 + \theta_1$$

Therefore, though finite angular displacement has magnitude and direction, it is not a vector. Angular displacements about z- and x-axes have been shown in Figs. 2.8(a) and 2.8(b) respectively. Fig. 2.8(c) gives the total angular displacement $\theta_x + \theta_z$ *i.e.* first a rotation about x-axis is taken and then Fig. 2.8(b) is rotated about z-axis. Angular displacement $\theta_z + \theta_x$ is shown in Fig. 2.8(d). Here, first a rotation is taken about the z-axis, and then Fig. 2.8(a) is rotated about the x-axis. From Figs. 2.8(c) and 2.8(d) it is obvious that the orientations of the shaded planes are different from each other. Hence, the directions of angular displacements $\theta_x + \theta_z$ and $\theta_z + \theta_x$ are not same. Verify this with the help of a book. Take each rotation to be 90°. We say that the addition of finite angular displacements is not commutative. But when angular displacements are infinitesimally small, we can take :

$$\Delta\theta_x + \Delta\theta_z = \Delta\theta_z + \Delta\theta_x$$

because the difference between both the sides of this equation will be negligible.

If we consider motion along a circular path of radius R with a speed v, we can find angular velocity. The angular velocity ω is a vector in the direction of angular displacement.

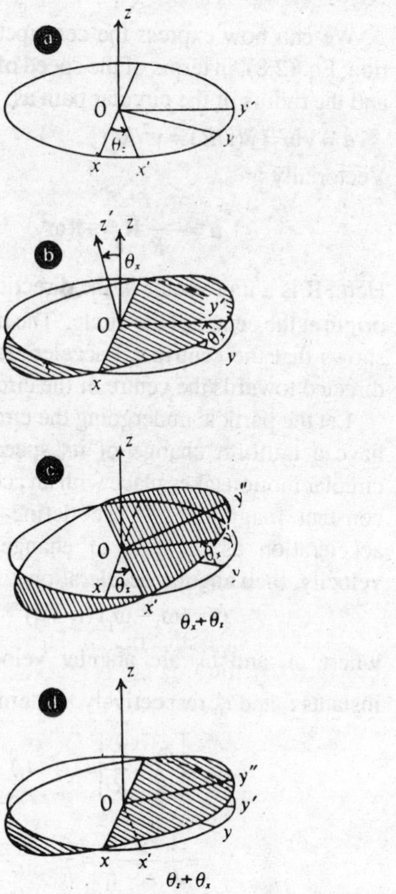

Fig. 2.8 A+B *may not equal* B+A. **(a)** *Rotation about z-axis* **(b)** *rotation about x-axis* **(c)** *rotation of (b) about z-axis* **(d)** *rotation of (a) about x-axis*

$$\omega = \frac{\textit{angular displacement}}{\textit{time taken to sweep the angle}} = \frac{\Delta\theta}{\Delta t}$$

$$= \frac{\textit{displacement along the arc}}{\textit{radius} \times \textit{time taken to traverse the arc}}$$

$$= \frac{\textit{speed along the arc}}{\textit{radius}} = \frac{v}{R},$$

or $$v = \omega R \qquad (2.10)$$

In right-handed coordinate system, ω is positive if the rotation is counterclockwise about the axis of rotation.

We can now express the centripetal acceleration, Eq. (2.8), in terms of the speed of the particle and the radius of the circular path as:

$a = v\omega = v(v/R) = v^2/R$

Vectorially :

$$\mathbf{a} = -\frac{v^2}{R}\hat{\mathbf{R}} = -\mathbf{R}\omega^2 \qquad (2.11)$$

Here, $\hat{\mathbf{R}}$ is a unit vector in the direction of **R** with origin at the centre of the circle. The negative sign shows that the centripetal acceleration is always directed towards the centre of the circle.

Let the particle undergoing the circular motion have a uniform change of its speed v, *i.e.* the circular motion takes place with an acceleration of constant magnitude. If we define the angular acceleration as the rate of change of angular velocity, then angular acceleration :

$$\alpha \equiv (\omega_2 - \omega_1)/(t_2 - t_1)$$

where ω_2 and ω_1 are angular velocities at the instants t_2 and t_1, respectively. Alternatively :

$$\alpha = \left[\frac{v_2}{R} - \frac{v_1}{R}\right] / (t_2 - t_1)$$

$$= \frac{1}{R}\frac{v_2 - v_1}{t_2 - t_1} = \frac{a}{R} \qquad (2.12)$$

$$\therefore \quad \alpha = \frac{\textit{acceleration along the circular path}}{\textit{radius of the circle}}$$

Question : Addition of finite angular displacements is, in general, non-commutative. But $\omega_1 + \omega_2 = \omega_2 + \omega_1$. Explain.

Example 2.3 Show that at any time the transverse velocity of a particle rotating about an axis is given by :

$$\mathbf{v} = \omega \times \mathbf{r}$$

where **r** is the position vector as measured from a point on the axis.

Let us consider a point P with position vector OP (**r**), undergoing a rotation about the z-axis with an angular velocity ω (Fig. 2.9). Let O be the reference point. From Fig. 2.9 it is clear that the transverse velocity vector **v** is perpendicular to the plane in which the vectors **r** and ω lie. Also we see that the magnitude of **v** is given by :

$$|\mathbf{v}| = |\mathbf{r}| \sin\theta\, \omega$$

where θ is the angle between **r** and ω. Hence, from Fig. 2.9 :

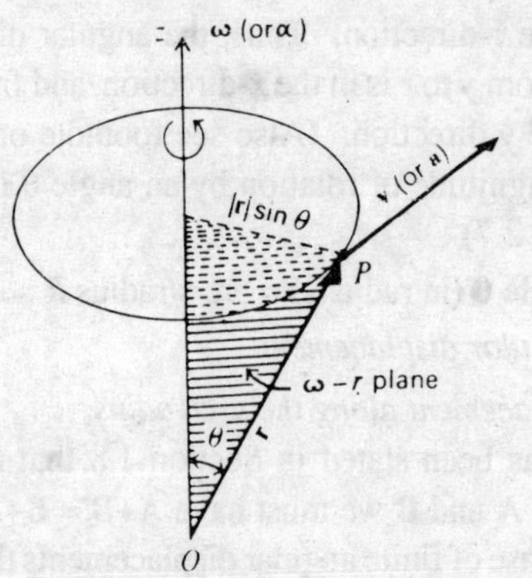

Fig. 2.9 *Transverse velocity of a point rotating about an axis in terms of the radius vector and the angular velocity*

$$\mathbf{v} = \omega \times \mathbf{r}$$

Similarly, for transverse acceleration we can prove that :

$\mathbf{a} = \alpha \times \mathbf{r}$, where α is angular acceleration.

Question : What are the angular speeds of the seconds needle and the minutes needle of a watch ? Could you conveniently express the motion of these needles in terms of linear variables ?

2.4 Force

From experience we know that if we want to make an object move we will have to exert a push or a pull on it. Also, if we want to stop a body in motion we will have to push it in the direction opposite to its motion. Any such push or pull by which we change the state of motion of a body is termed as *force*. Similarly, if we want to change the direction of motion of a body moving in a straight line we have to apply a sidewise force on it.

What we see is that *the application of a force either changes the magnitude or the direction of velocity, or both.* In other words the force exerted produces an acceleration in the body.

It has been seen experimentally that if we apply a force in a certain direction the body will be accelerated in that direction*. Also the acceleration produced will be large if a large force is applied,

*Physical Science Study Committee, *Physics* (D.C. Heath & Co.), 1965. At relativistic velocities the force is not in the direction of the acceleration.

and will be less if the force applied is of small magnitude. Mathematically :

$$\mathbf{F} \propto \mathbf{a} \quad (2.13)$$

The proportionality constant is known as the inertial mass of the body, which is a scalar and is denoted by m_i. Therefore :

$$\mathbf{F} = m_i\mathbf{a} \quad (2.14)$$

If the same force acts on two different bodies with respective inertial masses m_{i1} and m_{i2}, we have :

$$m_{i1}\mathbf{a}_1 = m_{i2}\mathbf{a}_2,$$

or

$$\frac{m_{i1}}{m_{i2}} = \frac{\mathbf{a}_2}{\mathbf{a}_1} \quad (2.15)$$

If mass of one of the bodies is known the other can be determined. The standard of mass is kilogram and masses of all other bodies are compared with this standard placed at International Bureau of Weights and Measures near Paris.

If a number of forces act on a body simultaneously, then we can write:

$$\Sigma\mathbf{F} = m_i\mathbf{a} \quad (2.16)$$

where $\Sigma\mathbf{F}$ is the vector sum of all external forces acting on the body. Eq. (2.16) is the mathematical statement of Newton's *second law of motion* which he stated as : *The change of motion is proportional to the motive force impressed, and is made in the direction of the straight line in which that force is impressed. This, in other words, can be stated as : The acceleration is proportional to the resultant force and is in the direction of the force.*

If **a** is constant, $\Sigma\mathbf{F}$ must also be constant. The motion *with a constant acceleration is the motion* under the action of a constant force. If **a** is zero, no resultant force is acting on the body. Or, in other words, *if no external force acts on a body, it will continue to move with a constant velocity without any change in the magnitude and direction of the velocity; if it is at rest it will continue to be at rest.* This is what is known as Newton's *first law of motion.* We see that the first law of motion is just a special case of the second law.

When the unit of mass is kilogram the unit of force is given by :

$$\text{unit of force} = \text{kilogram} \times \frac{\text{metre}}{\text{second}^2} = \text{newton}$$

In cgs units the force is expressed as dyne = 10^{-5} newton. In fps system the unit of force is pound, and 1 N = 0.2248 1b.

Question : Does Newton's second law define force ?

Example 2.4 A 16 kg block is placed on a horizontal plane. What constant horizontal force is required to give it a velocity of 5 ms^{-1} in 2 seconds, starting from rest (neglect friction) ?

As the force is constant and the block starts from rest, the motion will be along a straight line.

We are given that :

Initial velocity $u = 0$, velocity after 2 seconds, $v = 5$ m/s

Acceleration required, $a = \dfrac{v-u}{t}\,\dfrac{\text{ms}^{-1}}{\text{s}}$

$$= \frac{5-0}{2}\text{m/s}^2 = \frac{5}{2}\text{m/s}^2$$

Hence, the force required is : $F = 16 \times \frac{5}{2}\ \text{kg}\ \frac{\text{m}}{\text{s}^2}$

$$= 40\ \text{N}$$

2.5 Centripetal Force

In Section 2.3 we saw that for a particle revolving in a circle of radius R with a speed v, an acceleration v^2/R acts towards the centre of the circle. The magnitude of the resultant force acting on a particle of mass m_i undergoing circular motion is, then :

$$F = m_i v^2/R \quad (2.17)$$

The direction of this force is towards the centre and it is called the *centripetal force.* Centripetal force is just like any other force exerting push or pull. The only special significance attached to it is that while other forces cause change in the magnitude of the velocity, centripetal force changes the direction of the velocity. Being always perpendicular to the velocity, the centripetal force cannot change the magnitude of the velocity. This is a purely deflecting force

If an object is tied to a cord and whirled in a circle, it will be necessary to exert an inward force to keep the circular motion going. If the cord breaks, the circular motion ceases and the object flies off along a tangent to the circle.

Example 2.5 A small body of mass 100 g revolves in a circle on a horizontal frictionless surface, attached by a cord 25 cm long to a pin set in the surface. If the body makes 7 revolutions in 2 seconds, find the force exerted on the body by the cord.

Radius of circle formed = R = 25 cm

$$\text{Angular velocity} = = \frac{7}{2} \times 2\pi$$

$$= \frac{7}{2} \times 2 \times \frac{22}{7} = 22 \text{ rad/s}$$

Centripetal acceleration = $25 \times 22 \times 22$ cm/s^2

= 12100 cm/s^2

Therefore, the centripetal force is :

$$F = m_i a = 100 \text{ g} \times 12100 \text{ cm/s}^2$$

$$= 12.1 \times 10^5 \text{ dyn} = 12.1 \text{ N}$$

On a curved road, you must have observed that the road is provided with a slope across its breadth. This is to supply the necessary centripetal force. When a car moves on a curved path, the forces acting are :

(i) The weight of the vehicle mg acting vertically downward.

(ii) The normal force **N**.

(iii) The centripetal force **P**.

In horizontal tracks [Fig. 2.10(a)] the force **P** is to be provided by friction in the case of a motor car, and by a force exerted by the rails against the flanges on the wheels in the case of a railway car.

In order not to have to rely on friction, or to reduce wear on the rails and flanges, the road, or the track, may be banked by an angle [Fig. 2.10(b)]. If the car is to negotiate the curve without friction and without slipping, the force between the road and the wheels is to be perpendicular to the road. The compression of road must provide force sufficient to : *(i)* balance the weight of the car, and *(ii)* exert a centripetal force required for turning.

If v is the speed of the car and R is the radius of the track, then from Fig. 2.10(b) :

$$N \sin\theta = mv^2/R, \text{ and, } N\cos\theta = mg$$

$$\therefore \qquad \tan\theta = v^2/Rg$$

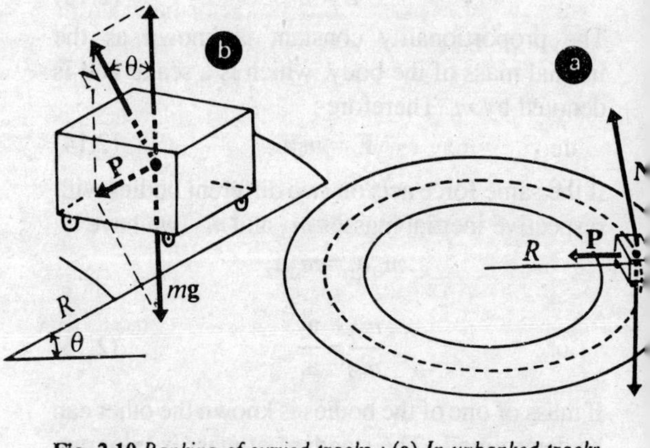

Fig. 2.10 *Banking of curved tracks :* **(a)** *In unbanked tracks, negotiating a curve requires either friction or some other inward push.* **(b)** *In banked tracks the necessary inward push is provided by the reaction force.* **P** *is the resultant of* **N** *and* **mg**.

Thus, the banking angle depends on the speed of the vehicle. The banking of a road or track is provided for some average speed. If a vehicle moves with a very large speed on a curved track, there is a possibility of the vehicle turning turtle.

2.6 Weight of a Body

It is a common experience that every thing on the surface of the earth is pulled towards the earth's centre even when it is at rest. The magnitude of this force on an object is called the *weight* of the object. If we take a standard kilogram mass and measure its weight, it will be nearly 9.8 N; if we take standard 2 kg mass, it will weigh 2 × 9.8 N. The weight of 2 kg mass will be exactly twice that of 1 kg mass anywhere on the surface of the earth. As the mass here is connected with the gravitational pull on the body, and it has nothing to do with its inertial state (*i.e.* the state of motion), we call it gravitational mass, m_g. Even in the absence of any motion of the body the gravitational force is experienced by such a mass. We see that :

$$\mathbf{F}_g \propto m_g,$$

or $$\mathbf{F}_g = m_g \mathbf{g} \qquad (2.18)$$

where g is the proportionality factor between the gravitational force and the gravitational mass. If two bodies balance each other at one place on the earth, they will balance at all other places. The value of g may vary at different places but at any given place it will be the same for all bodies. The factor g is the gravitational force per unit mass and is called the *gravitational field intensity.*

Suppose we allow a body to fall freely under the influence of gravitational force $\mathbf{F}_g$ (considering at present the resistance offered by air to be negligible). Then according to Newton's second law:

$$\mathbf{F}_g = m_i \mathbf{a} \qquad (2.19)$$

As soon as the body starts moving the inertial mass appears in Eq. (2.19). Eqs. (2.18) and (2.19) show that:

$$m_i \mathbf{a} = m_g \mathbf{g},$$

or $$\mathbf{a} = m_g \mathbf{g}/m_i \qquad (2.20a)$$

It has been experimentally observed that the ratio (m_i/m_g) is the same for all objects. Or:

$$\frac{m_{i1}}{m_{g1}} = \frac{m_{i2}}{m_{g2}} = \frac{m_{i3}}{m_{g3}} = \ldots\ldots \qquad (2.20b)$$

Since inertial and gravitational masses are proportional to each other, we take them to be equal. In future we will not differentiate between them, and will write $m_g = m_i = m$. From Eq. (2.20) it follows that:

$$\mathbf{a} = \mathbf{g} \qquad (2.21)$$

Hence, all objects under the influence of gravitational force are accelerated in the same manner irrespective of their masses. This experimental evidence itself suggests that $m_i \propto m_g$.

Question: Imagine a situation in which only protons inside a body oppose the change in its velocity, but the gravitational force on it depends on protons, neutrons and electrons. Will Eq. 2.20(b) hold good under this situation?

Slug: In British engineering system the unit of mass is slug. Previous measure for mass in Britain was a standard avoirdupois pound which is equal to 0.4536 kg.

The unit of force is given by:
Pound = avoirdupois pound × g = 1 lb
where g = 32.17 ft/sec^2
One slug is the mass which when acted upon by a force of 1 lb gets accelerated by 1 ft/sec^2. Thus:

$$1 \text{ slug} = \frac{1 \text{ lb}}{1 \text{ ft/s}^2} = \frac{1 \text{ avoirdupois pound} \times 32.17 \text{ ft/s}^2}{1 \text{ ft/s}^2}$$

$$= 32.17 \times \text{avoirdupois pound}$$

$$= 32.17 \times 0.4536 \text{ kg} = 14.59 \text{ kg}$$

2.7 Gravitational Force

The four basic interactions upon which all known forces and interactions are based are given

Table 2.3 The Four Basic Interactions

Type	*Source*	*Relative strength*	*Range*
1. Gravitational	Mass	~10^{-38}	Long
2. Weak	All elementary particles	~10^{-15}	Short (~10^{-15} m)
3. Electro-magnetic	Electric charge	~10^{-2}	Long
4. Nuclear	Hadrons (protons, neutrons, mesons)	1	Short (~10^{-15} m)

in Table 2.3. The magnitude of gravitational (or electrostatic) force between two point particles at rest and separated by a distance r, is given by:

$$F = C/r^2 \qquad (2.22)$$

where C is a constant. Such forces are called inverse square central forces. A central force is directed along the line joining the two particles. If one particle is at the origin and the second particle is at the position **r**, the force on the second particle is given by:

$$\mathbf{F} = \frac{C}{r^2}\hat{\mathbf{r}} \qquad (2.23)$$

where $\hat{\mathbf{r}}$ is the unit vector along **r** pointing away from the origin. The gravitational force between two point masses m_1 and m_2 is attractive and is proportional to the product $m_1 m_2$. Therefore:

$$\mathbf{F}_g = -\frac{Gm_1m_2}{r^2}\hat{\mathbf{r}} \tag{2.24}$$

Here G is another constant called the *gravitational constant,* and its value is 6.673×10^{-11} N m^2 kg^{-2}. Eq. (2.24) is a representation of Newton's law of gravitation.

If one of the masses, say m_1, is taken to be the mass of the earth M, and $m_2 = m$ is the mass of any other particle, the gravitational force between the earth and the particle is :

$$\mathbf{F}_g = -\frac{GM}{r^2}m\hat{\mathbf{r}} \tag{2.25}$$

From Eqs. (2.18) and (2.25) we can find the value of the gravitational field intensity of the earth's field, or the acceleration due to gravity, as :

$$\mathbf{g} = -\frac{GM}{r^2}\hat{\mathbf{r}} \tag{2.26}$$

Question : How will you measure the gravitational field intensity at a point ?

Example 2.6 Find out the gravitational field due to a uniform spherical shell at a point outside it, and at a point inside it.

Let the geometry be as shown in Fig. 2.11. Let us first find out the gravitational field at a point P outside the shell. Take a very thin ring on the shell, with its centre N on line OP, as shown. Gravitational fields due to two elements of equal areas on the ring and opposite to each other are shown. If we resolve these fields in two perpendicular directions—along PO and normal to it—then components along PO will add, and components perpendicular to it will cancel out. The same will hold good for other area elements on the ring. Now :

width of the ring $= R\Delta\theta$,

radius of the ring $= R\sin\theta$,

and therefore,

surface area of the ring $= (2\pi R \sin\theta)R\Delta\theta$.

If σ is the mass per unit surface area of the shell, then:

mass of the ring $= 2\pi R^2 \sin\theta\Delta\theta \times \sigma$

As each mass element on the ring is at the same distance x from the point P, the gravitational field at P due to the whole ring will be :

$$\Delta g = G.\frac{2\pi\sigma R^2 \sin\theta\Delta\theta}{x^2}\cos\alpha \tag{2.27}$$

in the direction PO. The gravitational field due the complete shell can be found out by summing the right hand side of Eq. (2.27) over the whole shell. The result is *:

$$g = \frac{MG}{r^2} \tag{2.28}$$

where M is the total mass of the shell.

If a point mass M is kept at O in Fig. 2.11, the gravitational field at P due to this mass will be the same as given by Eq. (2.28). Hence, *for the purpose of calculating the gravitational field* g *at a point outside a uniform spherical shell the mass of the shell can be treated to be kept at its centre.*

At a point inside the shell, the gravitational field is zero.**

To find out the gravitational field due to a uniform solid sphere at a point outside it, we can divide the sphere into a large number of concentric shells. As the point is outside all the shells, we can assume the mass of each shell to be kept at the common centre. Hence, Eq. (2.28) holds good for a uniform solid sphere also, with M being the mass of the sphere.

Velocity of a satellite : Let us consider the motion of a satellite moving with speed v along a circular path around the earth at a height h from its surface (Fig. 2.12). The circular motion of the satellite must be due to a centripetal force. This centripetal force is provided by the earth's gravitational force. The centripetal force required for the circular motion of the satellite is:

$$\mathbf{F} = -\frac{mv^2}{r}\hat{\mathbf{r}} \tag{2.29}$$

---

*In triangle PQN:

$$\cos\alpha = \frac{PN}{PQ} = \frac{PO - ON}{PQ} = \frac{r - R\cos\theta}{x}$$

In triangle OQP,

$$x^2 = R^2 + r^2 - 2Rr\cos\theta$$

Differentiating,

$$2xdx = 2Rr\sin\theta d\theta,\ \sin\theta d\theta = \frac{xdx}{Rr}$$

Eq. (2.27) then becomes :

$$dg = G\frac{2\pi\sigma R^2}{x^2}\frac{xdx}{Rr}\frac{r - R\cos\theta}{x}$$

$$= \frac{2\pi\sigma GR}{x^2}\frac{dx}{r}\left[r - \frac{R(R^2+r^2-x^2)}{2Rr}\right]$$

$$= \frac{\pi\sigma GR}{r^2}\left[\frac{r^2 - R^2}{x^2} + 1\right]dx$$

Integration of this expression over the range $(r-R)$ at A to $(r+R)$ at B for x gives the result of Eq. (2.28).

** x now ranges from $(R-r)$ at A to $(R+r)$ at B.

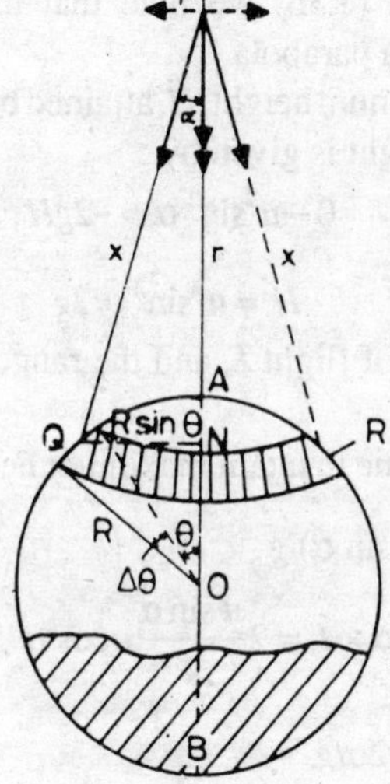

Fig. 2.11 *For Example 2.6*

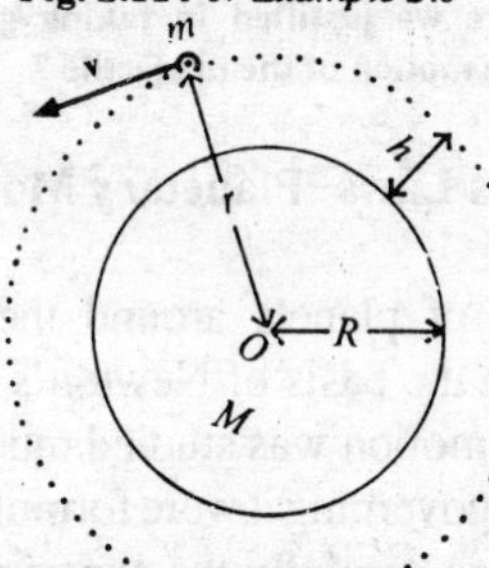

Fig. 2.12 *Motion of a satellite around the earth*

The gravitational force of the earth acting on the satellite is :

$$\mathbf{F}_g = -\frac{GMm}{r^2}\hat{\mathbf{r}} \qquad (2.30)$$

Equating (2.29) and (2.30) we get :

$$v^2 = GM/r, \qquad (2|31)$$

or
$$v^2 = GM/(R+h), \qquad (2.32)$$

or
$$v = \sqrt{\frac{GM}{R+h}} \qquad (2.33)$$

The velocity of the satellite depends only on the mass of the earth and its distance from the earth. As the mass of the earth is constant, the velocity of the satellite depends only on its distance from the earth. The velocity of a satellite, taking h as negligible compared to the radius of the earth, works out to be about 8 km/s.

Example 2.7 What is the time-period of revolution of a satellite around the earth, considering the earth to be a sphere of uniform density ρ?

From relation (2.31) :

$$v^2 = GM/r, \text{ or } r^2\omega^2 = GM/r, \text{ or } \omega^2 = GM/r^3$$

This gives,
$$\frac{4\pi^2}{T^2} = \frac{GM}{r^3},$$

or
$$T^2 = \frac{4\pi^2}{GM}r^3, \text{ or } T^2 \propto r^3$$

Thus, the square of the period of revolution of a satellite is directly proportional to the cube of the radius of the orbit.

In terms of density :

$$\omega^2 = G\frac{4\pi}{3}\rho\frac{R^3}{r^3} = G\frac{4\pi}{3}\rho\frac{R^3}{(R+h)^3}$$

$$= G\frac{4\pi}{3}\rho\left(1+\frac{h}{R}\right)^{-3}$$

$$= G\frac{4\pi}{3}\rho\left(1-3\frac{h}{R}\right)$$ by binomial expansion where we have neglected higher powers of $\frac{h}{R}$. Hence :

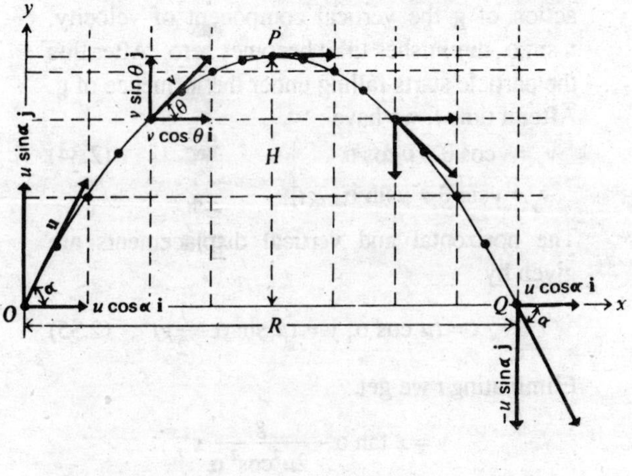

Fig. 2.13 ***Projectile motion. Heavy dots indicate the positions of the projectile after equal time intervals. It is a very good example of the fact that the motions in mutually perpendicular directions are independent.***

$$\omega = \sqrt{\frac{4\pi}{3}\rho G\left(1-\frac{3h}{R}\right)},$$

or $$\frac{2\pi}{T} = \sqrt{\frac{4\pi}{3}\rho G}\left(1-\frac{3h}{R}\right)^{\frac{1}{2}}$$

$$= \sqrt{\frac{4\pi\rho G}{3}}\left(1-\frac{3h}{2R}\right)$$

or $$T = \sqrt{\frac{3\pi}{\rho G}}\left(1+\frac{3h}{2R}\right)$$

If $h \ll R$, we see that time-period of revolution depends only on the avergae density of the earth, and is about 84 minutes.

Question: Why does not a satellite fall towards the earth ?

Projectile motion : The presence of gravitational field gives rise to projectile motion. Let a particle be projected with a velocity **u** making an angle α with the horizontal surface at time $t = 0$ (Fig. 2.13). Let the velocity after a time t be **v** at an inclination θ to the horizontal. Then initially at $t = 0$, the horizontal component of **u** = $u_x = u \cos \alpha$, and the vertical component of **u** = $u_y = u \sin \alpha$

Since no force is acting in the horizontal direction, the horizontal component of velocity remains unchanged throughout the motion. The vertical component is acted upon by an acceleration **g** acting vertically downward. Under the action of **g** the vertical component of velocity, $u \sin \alpha$, diminishes till it becomes zero. After this the particle starts falling under the influence of **g**. After a time t, we have :

$$\left.\begin{aligned} v_x &= v\cos\theta = u\cos\alpha \\ v_y &= v\sin\theta = u\sin\alpha - gt \end{aligned}\right\} \quad (2.34)$$

The horizontal and vertical displacements are given by:

$$x = tu \cos \alpha;\ y = tu \sin \alpha - \frac{1}{2}gt^2 \quad (2.35)$$

Eliminating t we get :

$$y = x \tan \alpha - \frac{g}{2u^2\cos^2\alpha}x^2,$$

or $$y = ax - bx^2 \quad (2.36)$$

Here a and b are constants since for a given trajectory u, α and g are constant.

From Eq. (2.36) we note that the path of a projectile is a parabola.

The maximum height, H, attained by the particle during its flight is given by :

$$0 - u^2\sin^2\alpha = -2gH,$$

or $$H = u^2\sin^2\alpha/2g \quad (2.37)$$

The time of flight T, and the range R are given by :

$T = 2 \times$ time to attain maximum height

$$= (2 \times u \sin \alpha)/g \quad (2.38)$$

$$R = u\cos\alpha \times T = 2\frac{u\sin\alpha}{g}u\cos\alpha$$

$$= u^2 \sin 2\alpha/g \quad (2.39)$$

Question : Are we justified in taking g as a constant throughout the motion of the projectile ?

2.8 Kepler's Laws—Planetary Motion

The motion of planets around the sun can be explained on the basis of Newton's law of gravitation. This motion was studied much earlier and precise laws governing it were formulated. Kepler, after analysing carefully the experimental data of Brahe, gave the following three laws:

1. The planets move in elliptical orbits with sun at one focus.
2. The position vector of a given planet relative to the sun sweeps out equal areas in equal times. In other words, the areal velocity of the planet remains constant. Thus the orbital speed of the planet will be maximum at perihelion and minimum at aphelion.*
3. The squares of the periods of revolution are proportional to the cubes of the average distances of the planets from the sun.

*The distance between the planet and the sun when the planet is closest to the sun is called the perihelion distance; and when the planet is farthest from the sun, it is called the aphelion distance. The corresponding distances for the orbit of the moon around the earth—and for the orbits of the earth's artificial satellites—are called the perigee and apogee distances, respectively.

The first and the third laws are a consequence of the inverse square force (force varying inversely as square of the distance) between a planet and the sun. The third law has been proved in Example 2.7 for the special case of a circular orbit. The second law holds good for all central forces* as a consequence of the law of conservation of angular momentum. This has been proved in Section 6.3.

The important data about the solar system have been given in Table 2.3.

Example 2.8 Orbital velocity of mercury at perihelion is 58.9 $\times 10^3$ m/s. What is its velocity at aphelion ? The perihelion and aphelion distances of mercury are 4.60×10^{10} m and 6.98×10^{10} m, respectively.

At perihelion and aphelion planets move perpendicular to the lines joining them with the sun. Therefore,

Areal velocity of mercury at perihelion

$$= \frac{1}{2} \times 58.9 \times 10^3 \times 4.60 \times 10^{10} \text{ m}^2/\text{s}$$

If the orbital velocity of mercury at aphelion is v, then its areal velocity here will be $\frac{1}{2} \times v \times 6.98 \times 10^{10}$ m. Therefore, from Kepler's second law :

$$\frac{1}{2} \times v \times 6.98 \times 10^{10} \text{ m} = \frac{1}{2} \times 58.9 \times 10^3 \times 4.60 \times 10^{10} \text{ m}^2/\text{s}$$

$$v = 38.8 \times 10^3 \text{ m/s}$$

2.9 Newton's Third Law of Motion

Newton's third law is stated as: *To every action there is always an opposite and equal reaction, or the mutual actions of two bodies upon each other are always equal.*

Suppose that a man pulls on one end of a rope attached to a block (Fig. 2.14). $\mathbf{F}_1$ is the force exerted by the man on the rope and is called *action*. Its *reaction* $\mathbf{F}_1' = -\mathbf{F}_1$, is the force exerted by the rope on the man. $\mathbf{F}_2$ represents the force (action) exerted on the block by the rope, its reaction $\mathbf{F}_2' = -\mathbf{F}_2$ is the force exerted on the rope by the block. The total force acting on the rope is $\mathbf{F}_1 + \mathbf{F}_2'$. $|\mathbf{F}_1| > |\mathbf{F}_2'|$ means that the rope has an accelerated motion in the direction of $\mathbf{F}_1$ with an acceleration $(F_1 - F_2)$ divided by the mass of the rope. If $F_1 = F_2'$, the rope is in equilibrium, i.e. either the rope is at rest or is in a state of motion with constant velocity.

*A central force has a direction which always passes through a fixed point.

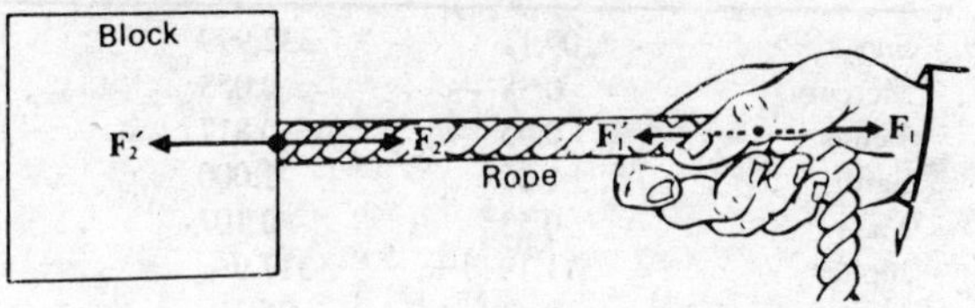

Fig. 2.14 *Action of A on B is equal to the reaction of B on A*

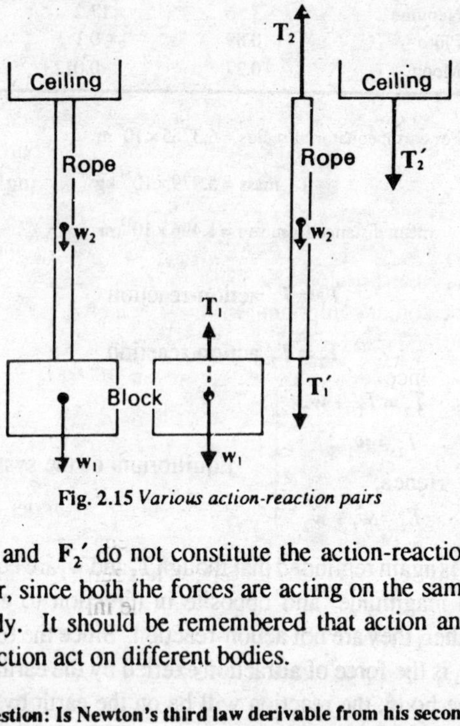

Fig. 2.15 *Various action-reaction pairs*

$\mathbf{F}_1$ and $\mathbf{F}_2'$ do not constitute the action-reaction pair, since both the forces are acting on the same body. It should be remembered that action and reaction act on different bodies.

Question: Is Newton's third law derivable from his second law ?

The rope, which is subjected to pulls at its ends, is said to be in *tension.* The magnitude of the tension at the right-hand end is F_1, and that at the left-hand end is F_2. When the rope is in equilibrium, the tension is same at both the ends. We can further understand the tension and action-reaction by the following example. Consider a weight w_1 hanging from a ceiling by a rope of weight w_2 (Fig. 2.15). The various forces acting on different parts are shown separately. We can easily see that the magnitudal relations are:

Table 2.3 *Important Data about the Solar System*

Body	*Equatorial radius*	*Mass* (On scale: Earth = 1)**	*Mean orbital radius*	*Period of rotation*	*Period of orbital motion (year)*
Sun	109.11	332,999	—	25^d09^h	—
Mercury	0.38	0.053	0.387	58^d16^h	0.241
Venus	0.95	0.817	0.723	242^d23^h	0.615
Earth	1.00	1.000	1.000	23^h56^m	1.000
Mars	0.53	0.107	1.523	24^h37^m	1.881
Jupiter	11.19	317.9	5.203	9^h50^m	11.862
Saturn	9.46	95.1	9.541	10^h14^m	29.46
Uranus	3.77	14.5	19.190	10^h48^m	84.01
Neptune	3.56	17.2	30.086	15^h40^m	164.79
Pluto	0.89	~ 0.1	39.5	6^d?	248.43
Moon	0.27	0.0123	0.00257	27^d07^h	27^d07^h

**For earth: equatorial radius $= 6.3785 \times 10^6$ m

mass $= 5.979 \times 10^{24}$ kg

mean distance from sun $= 1.496 \times 10^{11}$ m

$$T_1 = T_1' \text{ action-reaction}$$

$$T_2 = T_2' \text{ action-reaction}$$

$$\left.\begin{aligned} T_2 &= T_1' + w_2 \\ T_1 &= w_1 \\ \text{Hence,}& \\ T_2 &= w_1 + w_2 \end{aligned}\right\} \text{Equilibrium of the system}$$

It is again reminded that though T_1 and w_1 are equal in magnitude* and opposite in direction to each other, they are not action-reaction. Since the force w_1 is the force of attraction exerted by the earth on the body, the reaction will be on the earth by the body.

Finally, in the motion of a body we have to consider only the forces acting on the body, and not the forces exerted by it on other bodies. Only then the motion of a body will be possible.

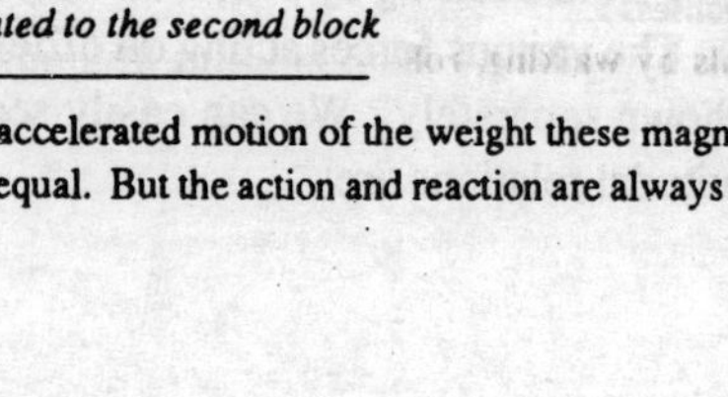

Fig. 2.16 ***The pull on one block may not be completely communicated to the second block***

*In the accelerated motion of the weight these magnitudes will not be equal. But the action and reaction are always equal.

Example 2.9 Two blocks of 1 kg and 2 kg are tied together with an inextensible string of negligible mass (Fig. 2.16). These are pulled on a frictionless horizontal surface with a force of 9 N. (i) what is the acceleration ? (ii) What is the force acting on the 1 kg mass ?

Different pairs of forces acting on the two blocks are shown in Fig. 2.16.

(i) As the string is inextensible, the acceleration of every part of the system will be the same. Let this be a.

As the net force on the 2 kg block is $(F_1 - F_2')$, we get :

$$F_1 - F_2' = 2 \text{ kg} \times a.$$

As the thread is of negligible mass : $F_2 - F_3' = 0$. Therefore,

$$F_2 = F_2' = F_3 = F_3'.$$

The force on 1 kg block is $F_2 = 1 \text{ kg} \times a$. Thus, $F_1 = 2 \text{ kg } a + 1 \text{ kg } a = 3 \text{ kg } a$. Therefore, $a = 9 \text{ N}/3 \text{ kg} = 3 \text{ m s}^{-2}$.

This acceleration could also be found by considering the blocks and the string as one system closed in a black box. As the mass of this system is 3 kg and the force acting on it is 9 N, the acceleration will be 3 ms^{-2}.

(ii) Force on 1 kg block $= F_3 = 1 \text{ kg} \times 3 \text{ m s}^{-2} = 3$ N.

2.10 Frictional Force

If we push a block on a horizontal surface (or roll a ball) we observe that after some time the motion ceases, though no retarding force is exerted on the block (or the ball). From where is the retarding force acting ? Whenever one surface moves in contact with another, a force is generated that tends to retard the motion. This force is called *friction*. It is well known that it is easier to roll a cylindrical body than to slide it. Thus, rolling friction is less than the sliding friction. We can also see that it is harder to start a slide than to keep it going, *i.e.* we require more force to initiate the motion of a body than to keep it in uniform motion. Thus starting friction (or static friction) is larger than the sliding friction (or kinetic friction).

Consider a block at rest on a horizontal table. Pull the block with the help of a string. When we apply a small force through the string the block does not move. When we increase the pull uniformly, we observe that at some definite pull the block just begins to move. This pull is just equal to the limiting frictional force exerted on the block by the table acting along the surface of contact. Once motion is started, this same force produces accelerated motion, because the kinetic friction is less than the static friction. If we want to keep the block moving with a constant velocity we will have to reduce the pull on the string till the pull is just equal to the kinetic frictional force.

One factor that does affect the friction strongly is the normal force, perpendicular to the surface of contact, that pushes the two objects together. For a block resting on a horizontal table or sliding along it, the normal force is equal in magnitude to the weight of the block. It has been seen that the maximum value of static frictional force is nearly proportional to the normal force. The ratio of this limiting force and the normal force is called the coefficient of static friction, μ_s. Thus, if the magnitude of the frictional force at a particular moment is F_f, then

$$F_f \leq \mu_s N \qquad (2.40)$$

where N is the magnitude of the normal force. The equality sign holds only when F_f has its maximum value.

When the block is in motion the kinetic frictional force $\mathbf{F}_k$ is also nearly proportional to the normal force. Or,

$$F_k = \mu_k N$$

where μ_k is called the coefficient of kinetic friction.

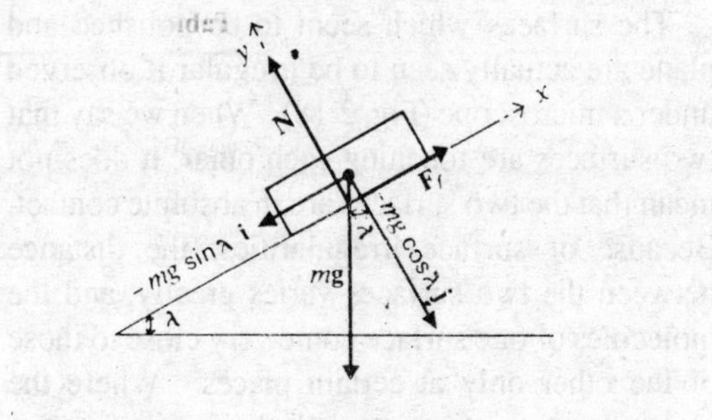

Fig. 2.17 *Sliding of an object along a rough incline*

Both μ_s and μ_k are dimensionless quantities and depend on the nature of the surfaces in contact. For a given pair of surfaces $\mu_s > \mu_k$.

Example 2.10 A good example demonstrating the frictional force is given by observing a block of mass *m* on a surface which can be tilted (Fig. 2.17). On a plane inclined at an angle λ, the forces acting on the block can be resolved as: (i) the weight *mg* of the block downwards, and (ii) the force *N* normal to the plane. We can see that $N = mg \cos\lambda$.

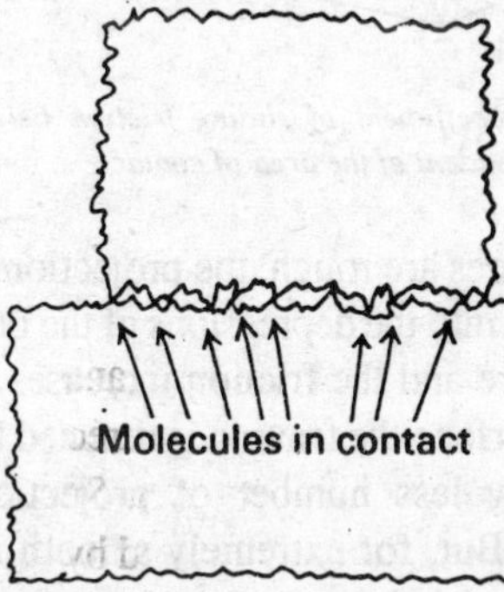

Fig. 2.18 *Mechanism of frictional forces*

But, the component $mg \sin\lambda$ acts on the block and tries to pull down the block along the plane. At small λ even when the force $mg \sin\lambda$ is acting on the block, the block does not move. This is because the frictional force is acting up the plane. The inclination is increased till a position is attained when the block just slides. At this position:

$$F_f = mg \sin\lambda = \mu_s N = \mu_s mg \cos\lambda, \quad \text{or} \quad \tan\lambda = \mu_s$$

Question: How could a person, who is at rest on a completely frictionless ice covering a pond, reach the shore? Could he do this by walking, rolling, swinging his arms, or kicking his feet?

The surfaces which seem to be polished and plane are actually seen to be irregular if observed under a microscope (Fig. 2.18). When we say that two surfaces are touching each other, it does not mean that the two surfaces are in absolute contact. Because of surface irregularities, the distance between the two surfaces varies greatly, and the molecules of one surface come very close to those of the other only at certain places. Where the molecular separation is small, the force of attraction is large. When one object slides over the other, a force must be exerted to separate the molecules at these contact points. The total molecular attraction that must be overcome to maintain the motion is the friction.

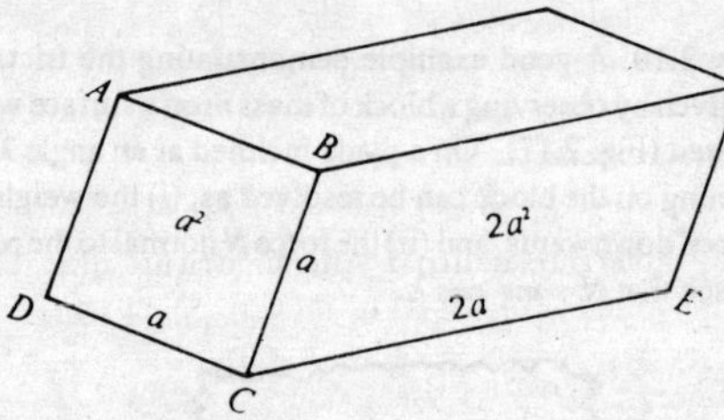

Fig. 2.19 *The coefficient of sliding friction between two surfaces is independent of the area of contact.*

If the surfaces are rough, the projections of one surface can fit into the depressions of the other, the contact is more and the friction increases. When we polish a surface the friction is reduced because there are now less number of projections and depressions. But, for extremely smooth and flat surfaces the molecules are mostly very close together over the entire surface. The friction is very large under such conditions. In rolling the surfaces are primarily peeled off, while in sliding they mainly tear off each other.

Example 2.11 If the frictional force depends on the number of contact points, why do we consider it independent of the area of contact?

Let us consider a block (Fig. 2.19). The frictional force to slide the block on the face *ABCD* is the same as that to slide it on the face *BCEF*, though the area of *ABCD* = $\frac{1}{2}$ area of *BCEF*, and the contact points in *ABCD* are half of those in *BCEF*. This is because when the block is resting on the face *ABCD* the weight per unit area is double than that when the block is resting on the face *BCEF*. The increased compression on the surface *ABCD* brings more points into contact, and the total number of points of contact remains unchanged, keeping the same frictional force.

The coefficient of friction depends on many factors, such as the nature of the materials, their surface finish, surface contamination, presence of surface films, temperature, etc. Frictional force between two dry surfaces can be reduced considerably by injecting a layer of a lubricant in between the two. If a layer of oil or compressed air is introduced between two surfaces, the frictional force reduces. The layer of oil separates the two surfaces and the chief frictional force is now the viscous drag of the fluid (see Section 8.4).

2.11 Equilibrium of Concurrent Forces

Concurrent forces are the forces which run (current) together (con-), i.e., their lines of action all pass through a common point. Such forces can produce translational motion in a body on which they act, but not rotational motion (see Chapter 6 for rotational motion). If under the action of concurrent forces the body remains in translational equilibrium, then the resultant of all these forces must be zero, otherwise the body will accelerate in the direction of their resultant following Newton's second law. At equilibrium vector sum of all these forces being zero, the vector diagram will be a closed polygon. Since the forces may be applied in any direction in three-dimensional space, the force polygon will not necessarily be confined to a plane. In the case of three forces, it will be a triangle.

Let the concurrent forces be resolved along three perpendicular directions, say *x*-, *y*- and *z*-axes, and let F_x, F_y and F_z, respectively, be the algebraic sums of these components. If *R* represents the resultant of the concurrent forces, then

$$R^2 = F_x^2 + F_y^2 + F_z^2$$

When the body is in equilibrium, $R = 0$, i.e., $F_x = 0$, $F_y = 0$ and $F_z = 0$. Hence the algebraic sums of the resolved components are also zero.

It may be mentioned that any one of the concurrent forces in equilibrium can be said to be the balancing force for all the others. This is because their resultant will be equal and opposite to this force. The reader can easily convince himself about it with the help of the vector polygon. In other words, the force labelled by us can be regarded as the equilibrant (i.e., the balancer) of all the others.

If the concurrent forces are also *coplanar*, the polygon will also be in their plane, say, the *xy*-plane. Forces can now be resolved into two directions *x*- and *y*-directions.

Discussion and Problems

2.1 What does the area under acceleration-time graph indicate?

2.2 Is the gravitational field of the earth really constant?

2.3 Would all objects accelerate at the same rate in the gravitational field if the ratios of their inertial and gravitational masses are different?

2.4 When a horse pulls a cart then, according to Newton's third law, force exerted by the horse on the cart is equal to the force exerted by the cart on the horse. How, then, the cart moves?

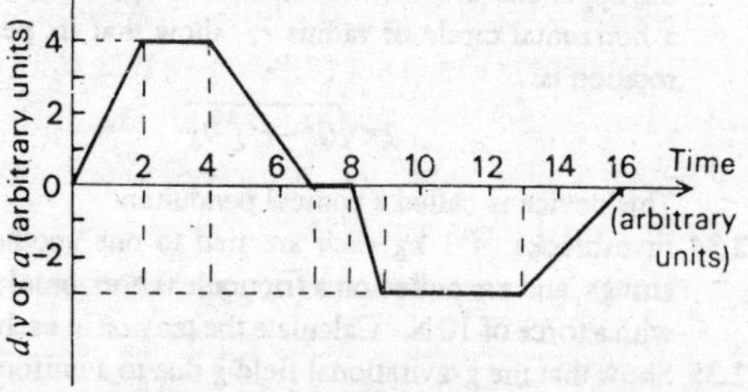

Fig. 2.20 *For Problem 2.5*

2.5 Motion of an object of mass 10 units moving along a straight line has been recorded in Fig. 2.20 (idealized, of course).
(a) If the record is the distance as a function of time: (i) when is the object farthest away from the starting point? and (ii) when is its speed maximum?
(b) If the record is a velocity–time graph: (i) when is the object farthest away from the starting point? and (ii) when is its speed maximum?
(c) If the graph represents acceleration as a function of time: (i) when is the object farthest away from the starting point? and (ii) when is its speed maximum?
(d) Plot the force-time graph in each of the above cases.

2.6 An automobile changes its velocity from 40 km/hr to 50 km/hr in 10 sec. What is its acceleration?

2.7 An automobile moves with a velocity of 80 km/hr due north. After 5 minutes it is moving at the same speed but due east. What is its average acceleration?

2.8 A car and a truck start from rest at the same time, with the car initially some distance behind the truck. The truck has a constant acceleration of 3 ft/sec^2, and the car has an acceleration of 5 ft/sec^2. The car overtakes the truck after the truck has moved 100 ft. (i) How long did it take the car to overtake the truck? (ii) Initially how far was the car behind the truck? (iii) What is the velocity of each when they are abreast?

2.9 A truck is going at a speed of 15 m/s and a car starts at a point ahead of it by 100 m with an acceleration of 1 m/sec^2. When will they meet? What is the distance covered between the two meetings?

2.10 In order to get into orbit, an astronaut has to accelerate from zero velocity to the orbital velocity of 8 km/s. Let us assume that his acceleration during the take off period is 3 *g*. How long does it take the rocket to reach the orbital velocity and how far does it travel during this time? Assume *g* to remain constant.

2.11 An object moving with angular velocity ω_1 at certain time has an angular acceleration α. After time t it has an angular velocity ω_2 and has covered an angle θ. Show that:

(i) $\omega_2 = \omega_1 + \alpha t$; (ii) $\theta = \omega_1 t + \frac{1}{2}\alpha t^2$; and

(iii) $\omega_2^2 - \omega_1^2 = 2\alpha\theta$

2.12 The angular speed of an automobile engine is increased from 1000 rpm to 2500 rpm in 10 seconds. (i) What is its angular acceleration assuming it to be uniform? (ii) How many revolutions have been made during this time?

2.13 An automobile travelling at 50 m.p.h. has wheels of 30 in diameter. (i) What is the angular speed of the wheels about the axle? (ii) If the wheels are brought to a stop uniformly in 30 turns, what is the angular acceleration? (iii) How far does the automobile advance during the braking period?

2.14 A car (weighing one metric ton) moving with a speed of 10 m/s, comes to rest, when brakes are applied, within a distance of 20 m. What is the retarding force? How long a time was required for the car to come to rest.

2.15 A body is released from rest and slides down a frictionless inclined plane, requiring 4 seconds to cover a distance of 100 cm. (i) What was its acceleration? (ii) How many centimetres would it have fallen in the same time? (iii) What is the inclination of the plane?

2.16 An object of mass 100 kg is being pulled by two forces as in Fig. 2.21. (i) Find out the acceleration and its direction. (ii) We want the object to move along the direction *xox'*. Find out the magnitude and the direction of the minimum force to be applied to enable us to do so.

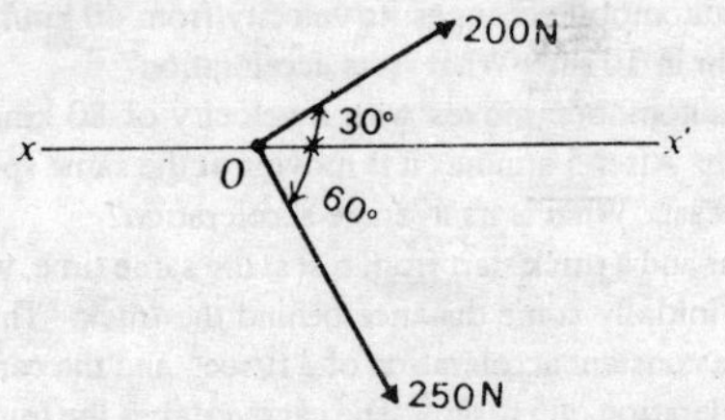

Fig. 2.21 *For Problem 2.16*

2.17 What is the period of revolution for a moon satellite if mass of the moon is 1/81 that of the earth and the radius is 1/3.7 that of the earth? Time period for an earth satellite is about 84 minutes.

2.18 Suppose we wish to establish a space platform in a circle in the earth's equatorial plane and at such a height above the earth's surface that it always remains above the same point, i.e., it is a geostationary satellite. Find the height of the space platform.

2.19 An object, kept on a pillar, is being pulled down by two persons with the help of ropes of length 5 m each. They stand 5 m apart. What is the force exerted by the ropes on the object when each person exerts a force of 250 N?

2.20 For a given initial speed of a projectile, plot a graph between the angle of projection and the range. From this graph find out the angle for which the range is maximum.

2.21 Consider a projectile fired at 30° from a horizontal surface. The vertical component of the initial velocity is 49 m/s. Ignoring air resistance, find the: (i) time of flight of the projectile; (ii) range; (iii) velocity after 7 sec; and (iv) the initial velocity.

2.22 A bomb is released from an airplane flying horizontally at an elevation of 1600 ft with a velocity of 600 m.p.h. (i) How far does the bomb travel horizontally before striking the earth? (ii) What will be the magnitude and the direction of its velocity just before striking? (iii) For how long is it in air? (iv) Trace the path of the bomb.

2.23 A hunter aims and shoots a bullet at a monkey on a tree. At the instant the bullet leaves the gun, the monkey falls down from the branch on which he was sitting. Will the monkey be hit by the bullet regardless of its speed?

2.24 A block of 20 kg moves on a level frictionless surface, connected by a light inextensible cord passing over a small frictionless pulley to a second hanging block of 8 kg. What is the acceleration of the system and what is the tension in the cord connecting the two blocks?

2.25 A body of 10 kg attached to a spring balance has been allowed to fall down along with the balance with an acceleration of 5 m/s^2. What is the reading of the balance? Assume $g = 10\ m/s^2$.

2.26 A body of 10 kg has been hung from the roof of an earth satellite with the help of a spring balance. What would be the reading of this balance?

2.27 Find the speed of an aeroplane moving in a vertical circle of diameter 1.0 km, so that the pilot feels weightless at the top of this circle. What will be his weight at the bottom?

2.28 We have to push with a force of 100 N to slide an object across the floor with a constant speed of 2 km/hr. Calculate the frictional force on the object.

2.29 A body of mass 10 kg is moving with a constant velocity of 6 m/s on a horizontal surface. The coefficient of sliding friction between the body and the surface is 0.20. (i) What horizontal force is required to maintain the motion? (ii) If the force is removed, how soon will the body come to rest?

2.30 A block is pushed with a velocity of 5 m/s up an inclined plane having a sliding coefficient of friction equal to 0.25, and an inclination of $\tan^{-1}\frac{3}{4}$. (i) How far will the block move up the plane before coming to rest? (ii) After how much time will the block return from top to its starting position? (iii) What will be the velocity of the block when it returns to the starting position?

2.31 Two blocks weighing 8 kg and 16 kg respectively are connected by a string and slide down a 30° inclined plane (with 8 kg weight ahead of 16 kg weight). The coefficient of sliding friction between 8 kg block and the plane is 0.25, and that between 16 kg block and the plane is 0.50. (i) Calculate the acceleration of each block. (ii) Calculate the tension in the string.

2.32 A curve of 200 m radius on a road is banked at a correct angle for a velocity of 15 m/s. If an automobile negotiates this curve at 30 m/s, what should be the minimum coefficient of friction between the tyres and the road so that the automobile does not skid?

2.33 A particle of mass m is supported by a thread of length l, the upper end of which is fixed. If the particle rotates in a horizontal circle of radius r, show that its period of rotation is:

$$2\pi\sqrt{(l^2-r^2)^{1/2}/g}$$

This device is called a conical pendulum.

2.34 Five blocks of 1 kg each are tied to one another with strings, and are pulled on a frictionless horizontal surface with a force of 10 N. Calculate the tension in each string.

2.35 Show that the gravitational field g due to a uniform solid sphere of radius R and mass M at a point P distant r from its centre is given by:

(*i*) $g = \dfrac{GM}{r^2}$ for $r > R$

(*ii*) $g = \dfrac{GM}{R^3}r$ for $r < R$.

[*Hint:* Divide the sphere into concentric shells.]

2.36 Show that the gravitational field due to an infinite and uniform sheet at a point distant r from it is given by:

$$g = 2\pi\sigma G,$$

where σ is the mass per unit area of the sheet.

[*Hint:* Taking foot of the normal on the sheet from the point at which field is needed draw concentric rings on the sheet.]

2.37 Calculate the period of revolution of mars around the sun. Its mean distance from the sun is 1.523 AU.

Chapter 3

Frames of Reference

3.1 Introduction

Description of all physical processes involves correct observation of their position, velocity and forces acting on them, as also the place from where the observation is made and what the observer is doing. Let there be two persons A and B living in the same town. A lives due north of the railway station and B lives due south of it. A will say that the railway station is due south of his residence, while B will say that it is due north of his residence. Both these statements about the position of the railway station, though different, are correct because the observers are situated at different places. Similarly, let A sit in a moving train and let another train pass by him from opposite direction. The second train will appear to move faster to A than to B, if B is standing on the ground. Thus, we see that the position and the state of the observer are important for description of a physical phenomenon. The frame in which the observer sits and makes his observations is known as his *frame of reference*. In this frame he will fix a point, called *origin;* the position, velocity, etc. of an object will be described in terms of a coordinate system chosen by him.

The simplest way to describe a process in a given frame of reference is with the help of Cartesian coordinates (see Section 1.9). The position vector is given by:

$$\mathbf{r} = x\mathbf{i} + y\mathbf{j} + z\mathbf{k} \tag{3.1}$$

where the unit vectors **i**, **j** and **k** form an orthonormal set of vectors and x, y, z are the components along **i**, **j** and **k**, respectively.

In this frame of reference we can write the distances between two points given by position vectors $\mathbf{r}_1$ and $\mathbf{r}_2$, in terms of their components along **i**, **j** and **k**. We have:

$$\mathbf{r}_2 - \mathbf{r}_1 = (x_2 - x_1)\mathbf{i} + (y_2 - y_1)\mathbf{j} + (z_2 - z_1)\mathbf{k},$$

$$\text{or} \quad |\mathbf{r}_2 - \mathbf{r}_1|^2 = (x_2 - x_1)^2 + (y_2 - y_1)^2 + (z_2 - z_1)^2 \tag{3.2}$$

The distance between any two points is an invariant in all stationary frames of reference. If we change the origin (which is nothing but choosing another frame of reference stationary with respect to the earlier frame of reference) the position vectors of different points will change, but the displacement vector $(\mathbf{r}_2 - \mathbf{r}_1)$ between any two points remains unaltered. Also there will be no change in velocity and acceleration by a mere change of origin. All such frames (where only origin in different) are equivalent to one another. The kinematics of a body, as observed from all these equivalent frames, will be identical.

The position vector of point P_1 in a frame of reference with origin at O is OP_1 (Fig. 3.1), and the position vector of P_1 in a reference frame with origin at O' is $O'P_1$. Then:

$$O'P_1 = OP_1 - OO' \tag{3.3}$$

Let us consider the displacement of a particle from P_1 to P_2 in time t. The velocity of the particle is given by:

$$v = P_1P_2/t, \tag{3.4}$$

where $$P_1P_2 = OP_2 - OP_1 \tag{3.5}$$

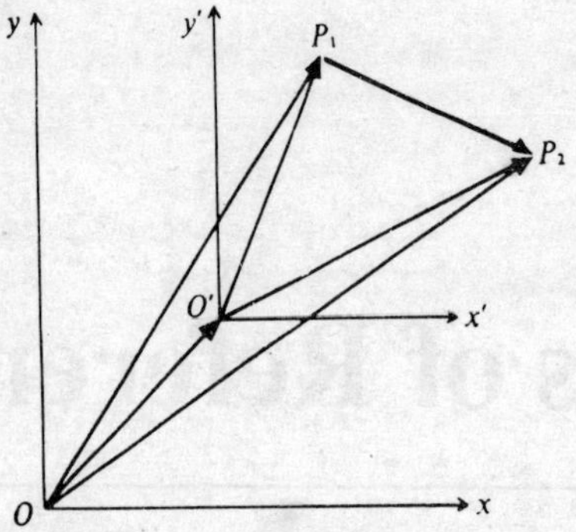

Fig. 3.1 *Two frames of reference at rest with respect to each otner*

Now, the value of P_1P_2 as observed in the frame with origin at O' is:

$$P_1P_2 = O'P_2 - O'P_1$$

$$= (OP_2 - OO') - (OP_1 - OO')$$

$$= OP_2 - OP_1 \qquad (3.6)$$

From Eqs. (3.5) and (3.6) it is obvious that P_1P_2 is same whether we take the origin at O or at O'. Hence, Eq. (3.4) shows that the velocity vector remains unaltered by changing the origin.

3.2 Frames of Reference Moving with Uniform Velocity with respect to Each Other

In the previous section we considered only the shift of origin. Now let us assume that the frame of reference with origin at O' moves with a velocity **V** with respect to the frame with origin at O (Fig. 3.2). The observer at O sees that the frame O' is moving with a velocity **V** with respect to him, whereas he may also conclude that the frame O' is at rest and his own frame is moving with a velocity –**V** with respect to this. There is no method to check the truth of either of the two conclusions. Both the statements are equivalent. Similar conclusions will be drawn by the observer at O'. This situation leads us to a basic statement: *No frame of reference has an absolute velocity.*

Fig. 3.2 *Uniform relative motion*

Under these circumstances we stop talking about absolute velocities of frames of reference and only talk of relative velocities. We simply say that the frame O' is moving with a velocity **V** relative to the frame O. It is the presence of the frame O' which makes an observer at O conclude that he is moving with velocity –**V** with respect to the frame O'. Had there been no other frame of reference except the frame O, the observer at O cannot know whether his frame is moving with a constant velocity or not moving at all.

Question: You are sitting in a closed compartment of a train which does not give any jerk while in motion. Can you suggest some experiment to prove that the train is moving (after the train has attained a uniform velocity)?

In other words we cannot differentiate between a frame moving with a constant velocity and a frame at rest. A frame at rest is equivalent to any other frame which is moving with a constant velocity. All frames of reference moving with constant velocities are known as *inertial frames of reference*. All inertial frames of reference are equivalent.

Since there is no physical meaning of absolute velocity, we are led to the fundamental hypothesis: *The basic laws of physics are identical in all reference systems which move with uniform (unaccelerated) velocities with respect to one another.* This hypothesis is known as the Galilean invariance principle.

Newton's second law of motion is the fundamental law of Newtonian mechanics. We must then have the law:

force = rate of change of momentum

to be valid for an observer stationed in any inertial frame of reference.

3.3 Galilean Transformations

Now we come to the important point–How are the measurements made by the observer in the frame O related to the observations made by the observer in the frame O'? If we know the relationship between the length and the time-intervals measured by the two observers, we can find the relationships for other physical quantities.

Here we will assume that the time t of an event as observed by the observer in frame O coincides with the time t' for the same event as observed by the observer at O', i.e. :

$$t = t' \tag{3.7}$$

Let O be the origin of the coordinate system (x, y, z), and O' be the origin of the coordinate system (x', y', z') (Fig. 3.3). The axes (x', y', z') are taken parallel to the axes (x, y, z). Let the inertial coordinate system O' move with a velocity **V** with respect to the inertial system O. For simplicity, we choose **V** to be in the direction of x- (or x'-) axis.

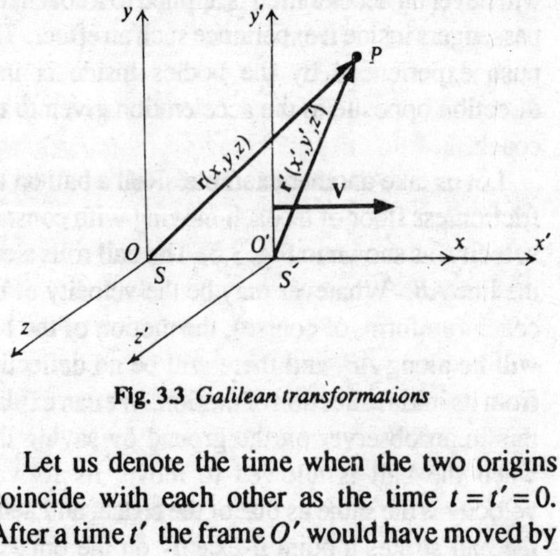

Fig. 3.3 *Galilean transformations*

Let us denote the time when the two origins coincide with each other as the time $t = t' = 0$. After a time t' the frame O' would have moved by a distance Vt' along the x-direction. At this instant let the position vectors of a point P be $\mathbf{r}(x, y, z)$ in the frame O and $\mathbf{r}'(x', y', z')$ in the frame O'. Then we can see that:

$x = x' +$ distance traversed by frame O' along the x-direction,

or $$x = x' + Vt'$$

We can thus write the complete set of transformation equations as:

$$\left.\begin{aligned} x &= x' + Vt' \\ y &= y' \\ z &= z' \\ t &= t' \end{aligned}\right\} \tag{3.8}$$

This set is known as Galilean transformations.

Immediately we see that the x-component of velocity:

$$v_x = \frac{x_2 - x_1}{t_2 - t_1} = \frac{(x_2' - x_1') + V(t_2' - t_1')}{t_2' - t_1'}$$

$$= v_x' + V, \text{ and } v_y = v_y';\ v_z = v_z'$$

In vector notation we can write:

$$\mathbf{v} = \mathbf{i}v_x + \mathbf{j}v_y + \mathbf{k}v_z$$

$$= \mathbf{i}(v_x' + V) + \mathbf{j}v_y' + \mathbf{k}v_z'$$

$$= (\mathbf{i}v_x' + \mathbf{j}v_y' + \mathbf{k}v_z') + \mathbf{i}V,$$

or $$\mathbf{v} = \mathbf{v}' + \mathbf{V} \tag{3.9}$$

where $\mathbf{v}'$ is the velocity measured in frame O' and $\mathbf{v}$ is the velocity measured in frame O. The inverse transformation of Eq. (3.9) is simply:

$$\mathbf{v}' = \mathbf{v} - \mathbf{V} \tag{3.10}$$

Eq. (3.9) [or Eq. (3.10)] is the law of velocity addition.

Question: Rain is falling vertically downward when a man stands on the ground. But as soon as the man starts running to take shelter, the showers start falling in inclined direction. Explain.

Example 3.1 Water in a river moves east at 9 km/hr and a boat heads north at 12 km/hr with respect to water. Find the velocity of the boat with respect to the ground.

Let us take the ground as reference frame O and water as reference frame O'. Let the coordinate axes be as in Fig. 3.4. Velocity of boat relative to water : $\mathbf{v}' = 12\mathbf{j}$; and velocity of water relative to the ground : $\mathbf{V} = 9\mathbf{i}$.

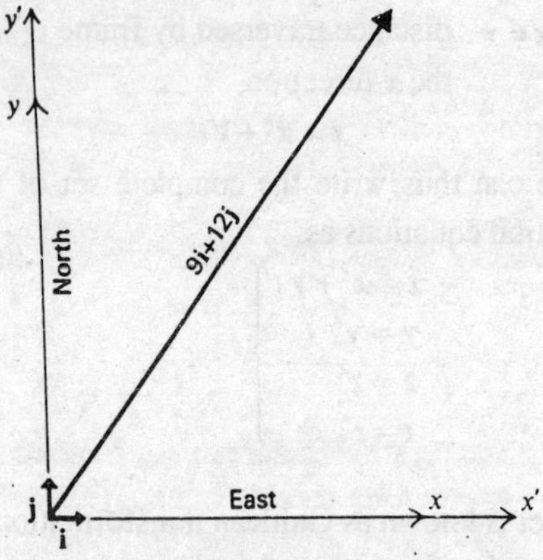

Fig. 3.4 *For Example 3.1*

∴ velocity of boat relative to the ground:

$$\mathbf{v} = \mathbf{v}' + \mathbf{V} = 12\mathbf{j} + 9\mathbf{i}$$

Therefore, $|\mathbf{v}| = 15$ km/hr

and $\tan\theta = 4/3$, or $\theta = \tan^{-1} 4/3 = 53°$

where θ is the angle the boat makes with the east.

A particle at a point P has position vector $\mathbf{r}(x, y, z)$ in O, and $\mathbf{r}'(x', y', z') = \mathbf{r}'(x - Vt, y, z)$ in O'. The velocity of the particle is $\mathbf{v}$ in O and $\mathbf{v}'$ in O'. Let the particle move from P to P' under the influence of some force, so that the velocity changes to $\mathbf{u}$ in O and $\mathbf{u}'$ in O'. We then have

$$\mathbf{u} = \mathbf{u}' + \mathbf{V} \tag{3.11}$$

From Eqs. (3.11) and (3.9) we get:

$$\mathbf{u} - \mathbf{v} = \mathbf{u}' - \mathbf{v}'$$

Therefore, change in velocity in frame O = change in velocity in frame O',

or $\Delta\mathbf{v} = \Delta\mathbf{v}'$,

or $$\mathbf{a} = \Delta\mathbf{v}/\Delta t = \Delta\mathbf{v}'/\Delta t' = \mathbf{a}' \tag{3.12}$$

The acceleration **a** as observed in frame O is the same as the acceleration **a**′ observed by an observer in the frame O'.

How does the force **F** transform from the frame O to O'? If the mass of the particle at P is M, the force observed in the frame O is $\mathbf{F} = M\mathbf{a}$, and the force observed in the frame O' is $\mathbf{F}' = M\mathbf{a}'$. From Eq. (3.12), as $\mathbf{a} = \mathbf{a}'$, we get:

$$\mathbf{F} = \mathbf{F}' \tag{3.13}$$

We therefore conclude that under Galilean transformation the observers in all inertial reference frames would agree on the magnitude and direction of force **F**, independent of the relative velocities of the reference frames.

Question: You are provided with a very accurate weighing machine. Will you observe any difference in your weight when you are: (i) standing on the ground, (ii) travelling in a train moving with a constant velocity of 60 km/hr, (iii) travelling in a supersonic plane flying with a uniform velocity of 2000 km/hr?

3.4 Accelerated Frames

If we sit in a closed coach moving with a constant velocity, we cannot know by any experiment the state of motion of the coach because all the physical processes appear similar in all inertial frames. We cannot say whether the coach is at rest or is in motion. But if the coach is suddenly stopped, we will be pushed in the direction in which the coach was moving. You must have felt such an experience in a train or a car. When the speed of a car is increased the passengers inside feel themselves to be stuck up to their backs, but when brakes are suddenly applied the passengers are pushed forward. No such feeling is experienced when the speed of a vehicle is constant. All this means that whenever an acceleration is applied to a coach, the passengers inside it experience such an effect. The push experienced by the bodies inside is in a direction opposite to the acceleration given to the coach.

Let us take another example. Roll a ball on the frictionless floor of a coach moving with constant velocity, as shown in Fig. 3.5. The ball rolls along the line AB. Whatever may be the velocity of the coach (uniform, of course), the motion of the ball will be along AB, and there will be no deflection from its initial direction of motion. We can explain this to an observer on the ground by saying that when the ball is allowed to move, its forward velocity is the same as that of the coach, and hence the ball strikes a point B exactly on the opposite side.

If the coach is moving with a constant acceleration $\mathbf{a}_o$, the ball rolls along the curved path AC.

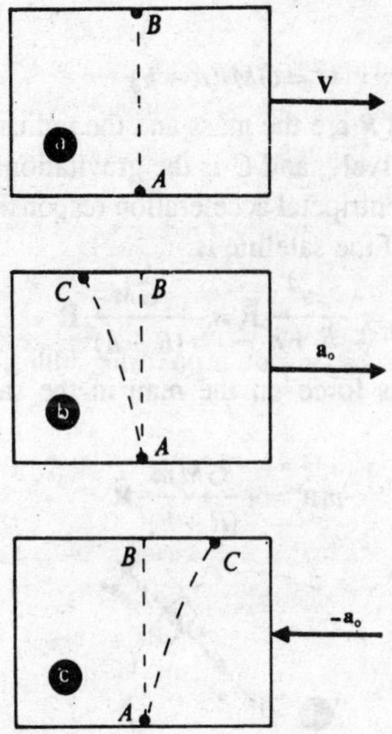

Fig. 3.5 *Accelerated frames. Path followed by a ball rolled across the floor of a coach moving with:* **(a)** *uniform velocity,* **(b)** *acceleration* $\mathbf{a}_o$, **(c)** *deceleration* $\mathbf{a}_o$.

The path of motion is that of a body which is under the influence of a constant acceleration $-\mathbf{a}_o$, i.e. a parabola. This happens because the point B moves a distance CB ahead of the ball during the time in which it crosses from A to B. When the ball is let go from A, its forward velocity remains constant at its initial value as time passes, but the velocity of the coach increases with time. This is very strange. No real force is applied to bend the path of the rolling ball, but to an observer sitting in the coach it appears to experience an acceleration equal and opposite to the acceleration applied to the coach (frame of reference).

If brakes are applied to the coach, i.e. the coach is given an acceleration $-\mathbf{a}_o$, the ball is pushed forward with an acceleration $\mathbf{a}_o$.

Example 3.2 **A coach is moving with an acceleration of $\mathbf{a}_o$, and a ball is rolled from off the coach with a velocity v perpendicular to the motion of the coach. Find the deviation of the path when the ball strikes the other side of the coach.**

Let b be the breadth of the coach. The time for the ball to reach other side of the coach is b/v. The distance BC *(Fig. 3.5) is given by :*

$$BC = 0 \times t + \frac{1}{2}(-\mathbf{a}_o)t^2 = -\frac{1}{2}\mathbf{a}_o(b/v)^2 = -\frac{1}{2}(\mathbf{a}_o b^2/v^2)$$

We see that to an observer in an accelerated frame the objects in it seem to experience an acceleration exactly equal and opposite to the acceleration of the frame. This gives rise to a force :

$$\mathbf{F}_0 = -M\mathbf{a}_o \qquad (3.14)$$

where M is the mass of an object in the frame, and $\mathbf{a}_o$ is the acceleration of the frame. This is called a fictitious force or a pseudo force. This is called fictitious because no real force is applied directly to the object. The force is a consequence of the acceleration of the reference frame.

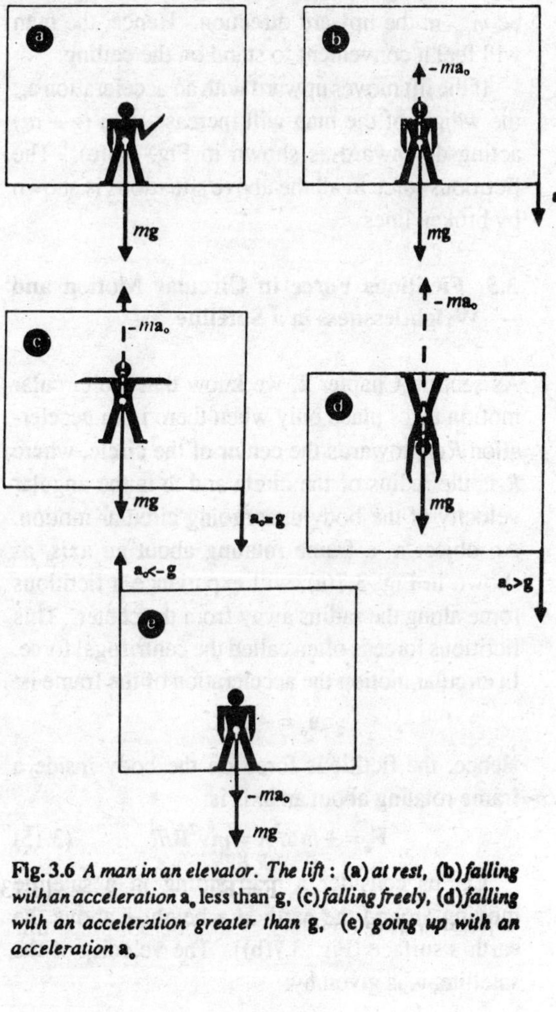

Fig. 3.6 *A man in an elevator. The lift :* **(a)** *at rest,* **(b)** *falling with an acceleration* $\mathbf{a}_o$ less than g, **(c)** *falling freely,* **(d)** *falling with an acceleration greater than* **g**, **(e)** *going up with an acceleration* $\mathbf{a}_o$

For better understanding of the problem let us discuss the force experienced by a man in an elevator, as shown in Fig. 3.6. When the lift is at rest the weight of the man is mg where m is the mass of the man and **g** is the acceleration due to gravity [Fig. 3.6(a)]. If the lift moves downward with an acceleration $\mathbf{a}_o$, a fictitious force $m\mathbf{a}_o$ acts on the man in the upward direction [Fig. 3.6(b)], and the weight of the man (as measured by a spring balance fixed in the elevator) reduces to $m(\mathbf{g} - \mathbf{a}_o)$. If $\mathbf{a}_o$ equals **g**, i.e. the lift is falling freely [Fig. 3.6(c)], the man will feel himself weightless. If $\mathbf{a}_o > \mathbf{g}$, the force on the man in the upward direction will be more than the force in the downward direction [Fig. 3.6(d)]. If $\mathbf{a}_o = 2\mathbf{g}$, the resultant will be mg in the upward direction. Hence, the man will feel it convenient to stand on the ceiling.

If the lift moves upward with an acceleration $\mathbf{a}_o$, the weight of the man will increase to $m(g + a_o)$ acting downward as shown in Fig. 3.6(e). The fictitious force in all the above situations is shown by broken lines.

3.5 Fictitious Force in Circular Motion and Weightlessness in a Satellite

As seen in Chapter 2, we know that the circular motion takes place only when there is an acceleration $R\omega^2$ towards the center of the circle, where R is the radius of the circle and ω is the angular velocity of the body undergoing circular motion. An object in a frame rotating about an axis, as shown in Fig. 3.7(a), will experience a fictitious force along the radius away from the center. This fictitious force is often called the centrifugal force. In circular motion the acceleration of the frame is:

$$\mathbf{a}_o = -\omega^2\mathbf{R}$$

Hence, the fictitious force on the body inside a frame rotating about an axis is:

$$\mathbf{F}_o = +m\omega^2\mathbf{R} = mv^2\hat{\mathbf{R}}/R \qquad (3.15)$$

Let us consider a man sitting in a satellite moving around the earth at a height h above the earth's surface [Fig. 3.7(b)]. The velocity of the satellite, v, is given by:

$$v^2 = GM/(R+h)$$

where M and R are the mass and the radius of the earth respectively, and G is the gravitational constant. The centripetal acceleration responsible for the motion of the satellite is:

$$\mathbf{a}_o = -\frac{v^2}{R+h}\hat{\mathbf{R}} = -\frac{GM}{(R+h)^2}\hat{\mathbf{R}}$$

The fictitious force on the man in the satellite, therefore, is:

$$-m\mathbf{a}_o = +\frac{GMm.}{(R+h)^2}\hat{\mathbf{R}}$$

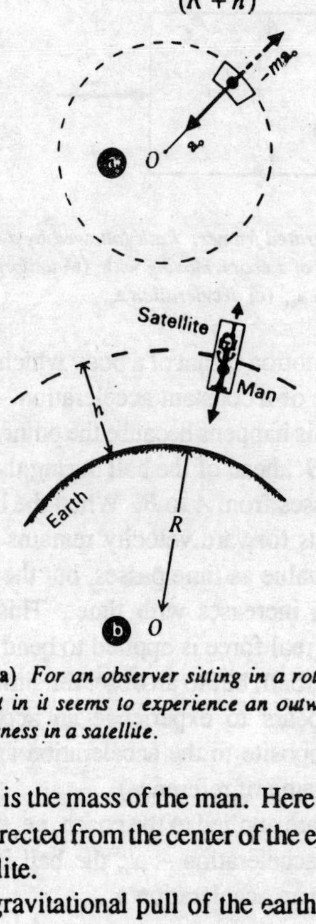

Fig. 3.7 **(a)** *For an observer sitting in a rotating frame an object kept in it seems to experience an outward force.* **(b)** *Weightlessness in a satellite.*

where m is the mass of the man. Here $\hat{\mathbf{R}}$ is the unit vector directed from the center of the earth towards the satellite.

The gravitational pull of the earth on the man is:

$$\mathbf{F} = -\frac{GMm}{(R+h)^2}\hat{\mathbf{R}}$$

Hence, the total force acting on the man (as observed in the frame of reference of the satellite) is:

$$\mathbf{F} - m\mathbf{a}_o = 0$$

Thus, the man in the satellite feels as if he is not experiencing any force. This is what is called the condition of weightlessness. Any object can hang in a satellite without any support.

Question: How will an observer in an inertial frame explain the above situation? And the situation in a freely falling lift?

3.6 Constancy of Speed of Light

We have seen in Sections 3.2 and 3.3 that in mechanics the relations between various kinematical and dynamical parameters look similar in all inertial frames, provided we transform these parameters using Eqs. (3.8). This principle of relativity was first stated by Newton as: "The motions of all bodies included in a given space are the same among themselves, whether that space is at rest or moves uniformly forward in a straight line."

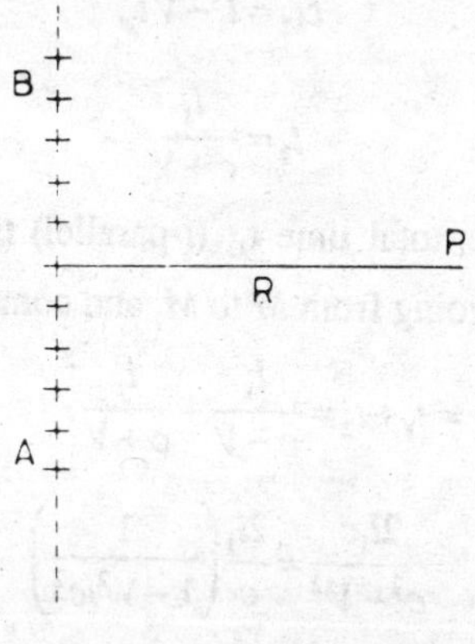

Fig. 3.8 *Electromagnetic field due to a line charge*

But this is not true for electromagnetic phenomena. Let us take a simple example of electromagnetic force of an infinitely long line of charge (Fig. 3.8). For an observer at rest with respect to the line charge, a test charge kept at P will experience an electric force only. An observer moving along the line charge will also see the same electric force because, according to the Galilean transformations, the length interval, and hence the linear charge density, has not changed. But to this observer the charge at P should also experience a magnetic force. Now total force remains invariant under Galilean transformations. Therefore, the magnetic force should be zero, which is not so.

This leads us to consider that either the nature is not consistent in that the Newton's principle of relativity holds for mechanics and not for electrodynamics, or we have to revise either the principle of relativity or electromagnetic equations, or classical mechanics. We will now discuss the following three possibilities:

(1) The relativity principle of Newton holds in mechanics but not in electromagnetism. The laws of electrodynamics hold only in one particular reference frame. Thus the speed of light (or electromagnetic waves) is c only in this frame and it is different for other frames.

(2) The principle of relativity is quite general but the laws of electricity need modifications.

(3) The principle of relativity is always true.

Let us first check the possibility that there is a preferred frame in which only the speed of light* is c. For an observer moving with respect to this frame, the speed of light will be different. This will happen if, like sound, something "physical" is required for the propagation of light. In that case we will be able to tell the absolute speed of the frame of the observer. The supposed physical medium required for the propagation of light was called ether.

This aspect was fully explored by Michelson and Morley during 1880's. The outline of the apparatus used by them has been given in Fig. 3.9. Here S is a source of light. Light starting from S is split into two coherent beams going perpendicular to each other by a partially silvered mirror M. These two beams after reflection from plane mirrors M_1 and M_2 go into some optical instrument T where they interfere. Whether they interfere constructively or destructively will depend upon the times they take in going from M to M_1 and M_2 and back. If the mirrors M_1 and M_2 are not quite at right angles, the interference pattern will consist of nearly parallel lines (Fig. 3.10), like the one we

*It will always be understood that we are talking about the speed of light in vacuum.

get from a thin wedge of air between two glass plates.

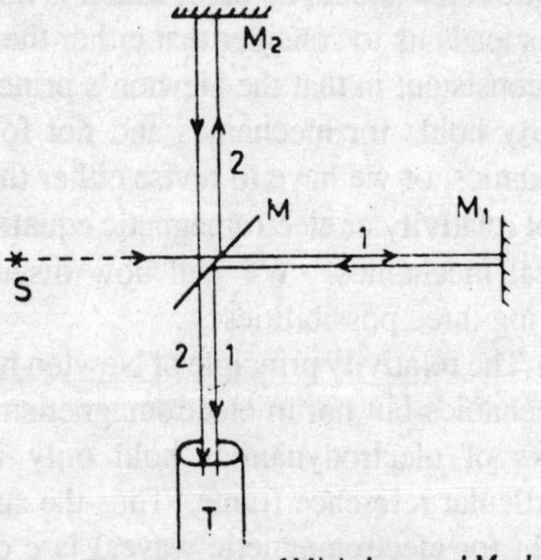

Fig. 3.9 *The apparatus of Michelson and Morley*

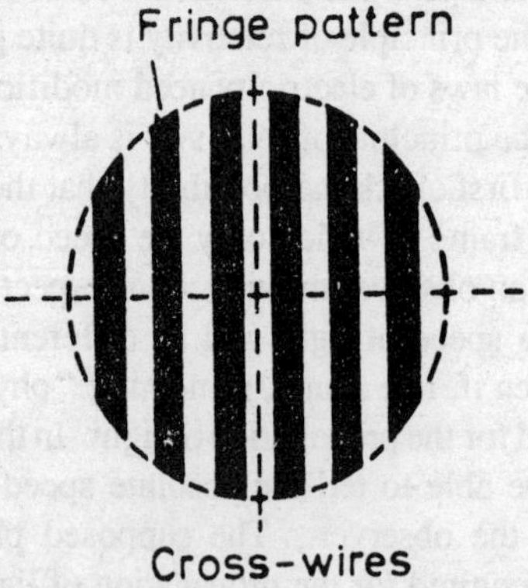

Fig. 3.10 *Interference pattern of the beams reflected from two plates inclined to each other*

Let arm MM_1 be aligned in the direction of motion of the earth. The hypothetical medium ether is supposed to be at rest. Then for an observer stationary with respect to ether the time t_1 taken by light in going from M to M_1 is given by [Fig. 3.11(a)]:

$$ct_1 = l_1 + Vt_1,$$

or
$$t_1 = \frac{l_1}{c - V} \tag{3.16}$$

Here l_1 is the length of the arm MM_1, c is the velocity of light in ether, and V is the velocity of earth in ether. Time t_2 taken by the beam in coming back from M_1 to M is given by [Fig. 3.11(b)]:

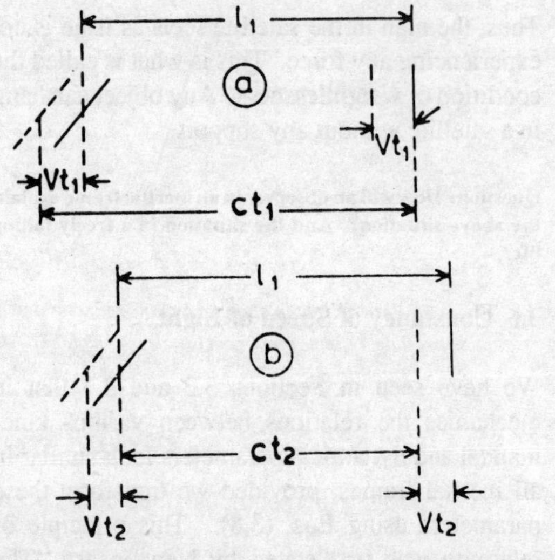

Fig. 3.11 (a) *Path covered by light in going to a plane reflector when the reflector moves in the direction of propagation of light and is kept perpendicular to the direction of motion,* **(b)** *Path covered when coming back*

$$ct_2 = l_1 - Vt_2,$$

or
$$t_2 = \frac{l_1}{c + V} \tag{3.17}$$

Therefore, total time t_{11} (t-parallel) taken by the beam in going from M to M_1 and coming back is:

$$t_{11} = t_1 + t_2 = \frac{l_1}{c - V} + \frac{l_1}{c + V}$$

$$= \frac{2l_1 c}{c^2 - V^2} = \frac{2l_1}{c}\left(\frac{1}{1 - V^2/c^2}\right)$$

$$= \frac{2l_1}{c(1 - \beta^2)} \tag{3.18}$$

where for brevity V/c is written as β.

For this observer, the path of beam 2 will look like that in Fig. 3.12. Therefore, if t_3 is the time in going from M to M_2,

$$(ct_3)^2 = (Vt_3)^2 + l_2^2$$

$$t_3 = \frac{l_2}{\sqrt{c^2 - V^2}}$$

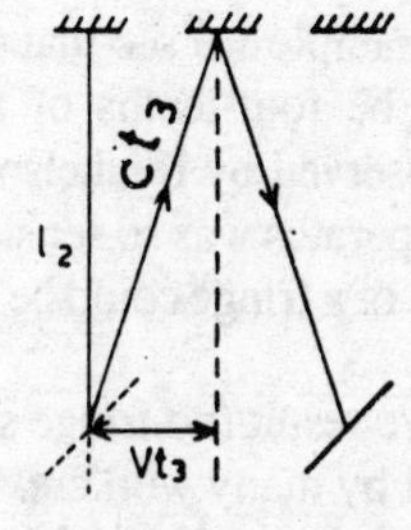

Fig. 3.12 *Path of light in going to a plane reflector and coming back when the reflector moves in the direction of the source and is kept parallel to it*

where l_2 is the length of the arm MM_2. Therefore, total time $t_\perp$ (t-perpendicular), taken by beam 2 in going from M to M_2 and coming back is:

$$t_\perp = 2t_3 = \frac{2l_2}{\sqrt{c^2 - V^2}}$$

$$= \frac{2l_2}{c} \frac{1}{\sqrt{1 - \frac{v^2}{c^2}}}$$

$$= \frac{2l_2}{c\sqrt{1-\beta^2}} \qquad (3.19)$$

It is apparent from Eqs. (3.18) and (3.19) that these two quantities of time are not equal even if $l_1 = l_2$. The same will be the conclusion of the observer sitting at the apparatus because time has been assumed to be "absolute" (see Problem 3.16). For this observer the ether wind will have a speed V in a direction opposite to that of the earth.

Example 3.3 The speed of water current in a river is 3 km/hr and the speed of a boat with respect to water is 5 km/hr. The river is 500 m wide. An observer is standing on a bank at M (Fig. 3.13). Calculate the time taken by the boat in going from M to M_1 and coming back. What is the time taken in going from M to M_2 and coming back? Here MM_1 = 500 m and M_2 is a point on the other bank directly opposite to M.

Here the situation is similar to the one discussed above. The observer is stationary with respect to the apparatus; water current corresponds to the ether wind; and the boat corresponds to a light signal.

Speed of the boat (with respect to the bank) up the stream is $(c_b - V_w)$, where c_b is the speed of the boat relative to water, and V_w is the speed of water current. Speed of the boat down the stream is $(c_b + V_w)$. Therefore, time taken from M to M_1 and back is

$$t_{\parallel} = \frac{0.5 \text{ km}}{2 \text{ km/hr}} + \frac{0.5 \text{ km}}{8 \text{ km/hr}}$$

$$= \left(\frac{0.5}{2} + \frac{0.5}{8}\right) \text{hr}$$

$$= 0.31 \text{ hr}$$

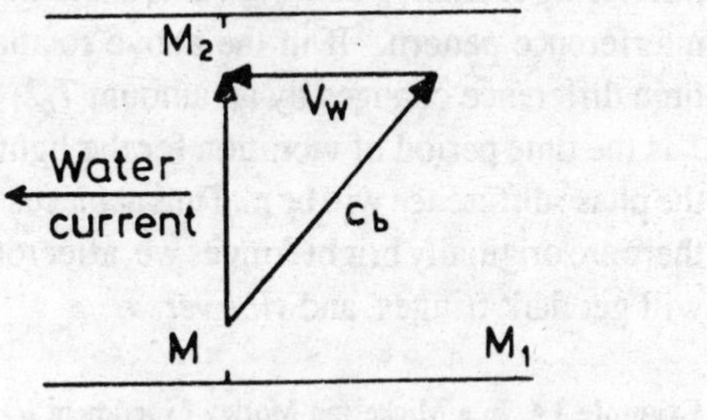

Fig. 3.13 *For Example 3.3*

Now, when boat is to go to M_2, it has to point in such a direction that its resultant velocity is in the direction MM_2. This direction has been shown in Fig. 3.13. Therefore, the required speed is 4 km/hr. Hence, the time taken round the trip is,

$$t_\perp = \left(\frac{0.5}{4} + \frac{0.5}{4}\right) \text{hr}$$

$$= 0.25 \text{ hr}$$

Thus $\qquad t_\perp < t_\parallel$.

Time taken for both these journeys, if water were still,

$$t = \left(\frac{0.5}{5} + \frac{0.5}{5}\right) \text{hr}$$

$$= 0.2 \text{ hr}$$

Thus, this time is shorter than both the times calculated above.

From expressions (3.18) and (3.19) we see that the difference of transit times of beams 1 and 2 is

$$\Delta t = \frac{2}{c}\left(\frac{l_1}{1-\beta^2} - \frac{l_2}{\sqrt{1-\beta^2}}\right) \qquad (3.20)$$

If we rotate the apparatus by 90°, the difference in transit times of beams 1 and 2 becomes:

$$\Delta t' = \frac{2}{c}\left(\frac{l_1}{\sqrt{1-\beta^2}} - \frac{l_2}{1-\beta^2}\right) \qquad (3.21)$$

Hence, this rotation changes the time differences by an amount:

$$\Delta t' \sim \Delta t = \frac{2}{c}\left(\frac{l_1 + l_2}{\sqrt{1-\beta^2}} - \frac{l_1 + l_2}{1-\beta^2}\right)$$

$$= \frac{2(l_1 + l_2)}{c}\left(\frac{1}{\sqrt{1-\beta^2}} - \frac{1}{1-\beta^2}\right) \qquad (3.22)$$

This change in time differences causes a corresponding change in phase differences between the interfering beams. This would amount to a shift in interference pattern. If in the above rotation the time difference changes by an amount $T/2$, where T is the time period of vibration for the light used, the phase difference will be π. Thus at places where there are originally bright fringes we, after rotation, will get dark fringes, and *vice versa*.

Example 3.4 In a Michelson-Morley experiment $l_1 = l_2 = 11$ m, and the light used has a wavelength of 5500 Å. Calculate the number of fringes which will move past the cross-wires in Fig. 3.10 when the apparatus is rotated through 90°. The orbital speed of earth about the sun is about 30 km/s.

Here, $$\beta = \frac{V}{c} = \frac{30 \times 10^3 \text{ ms}^{-1}}{3 \times 10^8 \text{ ms}^{-1}}$$

$$= 10^{-4}$$

Therefore, we can use binomial expansion in Eq. (3.22) and retain only first terms in β^2. The change in the time differences then becomes:

$$\Delta t' \sim \Delta t = \frac{2(l_1 + l_2)}{c}\left(1 + \frac{1}{2}\beta^2 - 1 - \beta^2\right)$$

$$= \frac{2(l_1 + l_2)}{c}\left(-\frac{\beta^2}{2}\right)$$

$$= \frac{22 \text{ m}}{3 \times 10^8 \text{ ms}^{-1}} \times 10^{-8}$$

$$= \frac{22}{3} \times 10^{-16} \text{ s}$$

Therefore, the number of fringes moving past the cross-wires is,

$$N = \frac{\Delta t' \sim \Delta t}{T} = (\Delta t' \sim \Delta t)\frac{c}{\lambda}$$

$$= \frac{22}{3} \times 10^{-16} \text{ s} \times \frac{3 \times 10^8 \text{ ms}^{-1}}{5500 \times 10^{-10} \text{ m}}$$

$$= \frac{22}{55} = \frac{2}{5} = 0.4$$

From this example we see that the shift in the pattern should be four-tenths of a fringe. But this was not observed by Michelson and Morley, though their apparatus was so sensitive that a shift as small as 0.01 of a fringe could be measured with it.

This negative result (no fringe shift at all) was later confirmed by many workers. Many alternative explanations were offered for this null result, but Michelson and Morley checked all the alternatives only to arrive at the conclusion that *the measured speed of light is the same, i.e. c, for all directions in every inertial frame.*

Let us now take up the second possibility. The laws of electricity can be modified in a way to give the result that light always travels at velocity c with respect to its source. This explanation was also ruled out by the results of Michelson-Morley experiment using extraterrestrial sources. More recent experiments using fast moving terrestrial sources (a measurement of the speed of electromagnetic radiation from the annihilation of rapidly moving positrons, and radiation from the decay of rapidly moving π° mesons) confirm that the velocity of light is independent of the velocity of the source.

We conclude from the above discussion that the hypothesis of all pervading ether is untenable. The experiment also forces us to conclude that the laws of electricity are correct. The speed of light is the same in all inertial frames, independent of the relative motion of the source and the observer.

The third possibility was pursued by Einstein and he gave his principle of relativity in 1905: *A physical law must have the same meaning in all inertial frames.* Further, the constancy of the speed of light is a physical law, otherwise we would have been able to detect the absolute motion of a frame of reference. The constancy of the speed of light (known as the second postulate of relativity) is a consequence of the fundamental (first) postulate.

3.7 Simultaneity

Einstein arrived at the above conclusion by doubting the absoluteness of time. He said that the observers in different inertial frames have, like space coordinates, their own time. He explained the 'strange' result of constancy of speed of light by attributing 'strange' properties to space and time. Events that are simultaneous for one observer may not be simultaneous for the other.

Let us take two events in a particular frame. An event is described by the place where it occurred and the time when it occurred. Before we go ahead, let us first discuss how time scale is set up in a particular frame of reference. If two events occur at the same place and we have a clock at that place, then if the reading of the clock is the same for each event, we regard these events as simultaneous. But what happens if these events are spatially separated? In this case the readings of the clocks as taken by an observer will depend upon the distance of the observer from the clocks. This is due to the finite time taken by the light signal in going from the clock (place of occurrence of the event) to the observer. Thus various observers at different distances from the event will record different times. But we want that observers in the same reference frame should agree in this respect. For this purpose we first synchronize the clocks. The clocks are fixed in position along with an observer attached to each of them. Now a third observer is stationed midway between the two. He notes that the pair of clocks reads the same. Or, we can let a light flash go at the mid-point and instruct the observers sitting at the clocks to set them at zero reading when light reaches them. This procedure takes care of the finite speed of light. This way we can synchronize as many clocks as we want. Then the time interval between two events at a distance is calculated from the times recorded by the observers sitting there with the clocks synchronized using the above procedure. Events occurring at two different places in that frame are called simultaneous when the clocks at the respective places record the same time for them.

Let us now take two inertial frames moving with velocity V relative to each other. Let one frame be the earth say S, and the other be a space ship (or a train) say S'. Let the observer in S' synchronize two clocks by placing one clock at each end of the ship and then sending light signals from a flash gun kept at the mid-point between the clocks. These light flashes will reach the clocks simultaneously and may be used to trigger off some mechanism which will put the clocks hands at the same common predetermined reading (may be zero). The observer in S will not agree that the two clocks are synchronous. He would reason that since the space ship is moving forward, the clock in the front end is running away from the light signal and, therefore, the light has to go more than half way in order to catch up. Similarly, he will argue, the rear clock is advancing to meet the light signal, so this distance will be shorter. Hence, to him, the signal reaches the rear clock first. We see from this that when an observer in S' thinks that the times at two locations are simultaneous, equal values of t' in his coordinate system must correspond to different values of t in the other coordinate system. Thus simultaneity is a relative concept.

3.8 Time Dilation

Let us consider a 'light-clock'. It consists of a rod with flash bulb F at one end and a plane mirror M at the other end (Fig. 3.14). A light pulse goes from F to M and comes back. When it comes back we have a 'tick' (or a flash) of the clock and a new pulse is emitted. Consider two such identical light clocks at rest.

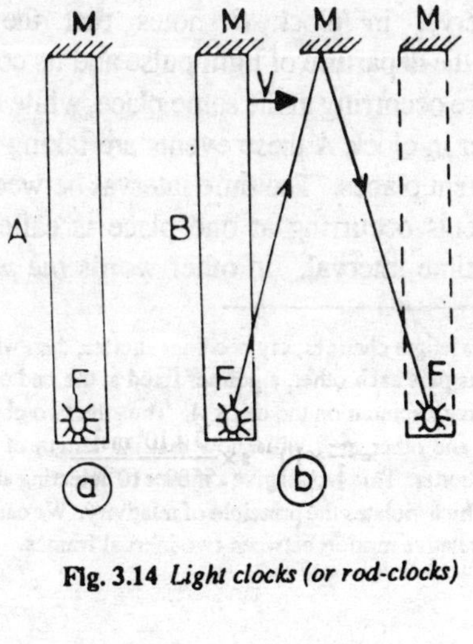

Fig. 3.14 *Light clocks (or rod-clocks)*

The time between ticks for both the clocks is $\tau = 2l/c$, where l is the length of the rod. Let clock B now move to the right with velocity **V**. Its length will remain unchanged.* For an observer fixed in clock A, the path of the light pulse in clock B will become longer as in Fig. 3.14(b). Remember we are now measuring the interval between two flashes of a moving clock. In this case the laboratory observers have to make sure that they are coincident with the moving clock at the time it flashed. Otherwise there will be an error due to the finite time taken by light in going from the clock to the observer. If the time between two ticks is now τ', then

$$\left(\frac{c\tau'}{2}\right)^2 = \left(\frac{V\tau'}{2}\right)^2 + l^2,$$

or
$$\tau' = \frac{2l/c}{\sqrt{1-\frac{v^2}{c^2}}}$$

or
$$\tau' = \frac{\tau}{\sqrt{1-\beta^2}} \qquad (3.23)$$

For an observer sitting at B the time between two ticks will still be $\tau = 2l/c$. Thus the time interval between two events in B (here the first event is 'light leaving F' and the second event is 'light coming back to F') appears to be longer to an observer in A than to an observer in B. Hence, the clocks in B will be slow to an observer in A. In other words, the moving clocks become slow.

Observer in clock B notes that the two events–the departure of light pulse and its coming back–are occurring at the same place, while for an observer in clock A these events are taking place at different places. The time interval between the two events occurring at one place is called the proper time interval. In other words *the proper time interval is the time interval recorded by a clock attached to the observed body. A non-proper (or improper) time interval would be a time interval measured by two different clocks at two different places.* Therefore, in Eq. (3.23), τ is the proper time interval and τ' is the improper time interval. Improper time interval is always greater than the proper time interval.

*If this length changes, say becomes shorter, then when the two clocks pass each other, a pointer fixed at the end of clock B will leave a scratch on the clock A. Thus the two observers one in A and other in B will know that the length of B has become shorter. This would give a means of detecting absolute motion which violates the principle of relativity. We can detect only the relative motion between two inertial frames.

Question: How will the time interval in clock A look to an observer sitting at clock B?

Einstein said that not only light clocks, but also other types like mechanical and radioactive ones, slow down by the same factor $1/\sqrt{1-\beta^2}$. He argued that it has nothing to do with the nature of the particular clock. It is due to an intrinsic property of time itself. If it were not so, then the mechanical clocks will differ with the light clocks in the moving frames, and this difference will give the absolute velocity of that frame. This violates the principle of relativity.

Question: Will chemical reaction rates also slow down in moving frames? What about life?

Example 3.5 The charged pions are unstable particles with a proper average life of 2.6×10^{-8} s. A pion beam has a speed of $V = 0.6\,c$. What is the average life? How far will a pion go in this time?

The average life for an observer with respect to whom the beam is moving, is given by:

$$\tau' = \frac{2.6\times10^{-8}\text{ s}}{\sqrt{1-0.36}} = \frac{2.6\times10^{-8}\text{ s}}{0.8}$$

$$= 3.25\times10^{-8}\text{ s}$$

Distance travelled in this time

$$d = 3.25\times10^{-8}\text{ s}\times0.6\times3\times10^{8}\text{ ms}^{-1}$$

$$= 5.85\text{ m}$$

The distance can be measured experimentally. Eq. (3.23) has been verified in this manner.

3.9 Length Contraction

Process of measurement of length requires two readings on the meter scale. If the object, the length of which we want to measure, is in motion along its length, then the two readings are to be taken *at the same time*, as it should be clear from Fig. 3.15. Remember, the observers equipped with clocks are to be coincident with the points of observation. Because simultaneity is a relative concept, length measurements will also depend on the frame of reference and will be relative.

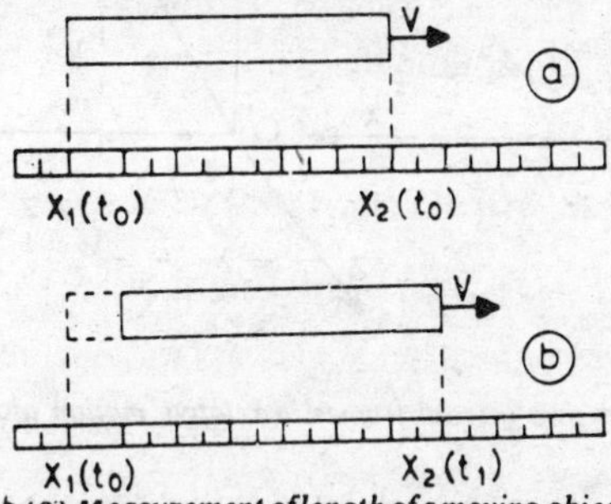

Fig.3.15 (**a**) *Measurement of length of a moving object requires two readings–one at each end–taken at the same time.* (**b**) *If the two readings are not taken at the same time the length could be anything depending upon the time difference*

Let us again go back to the rod clock of Fig. 3.14. Let its proper length be l and the proper time interval between two ticks be τ.* Therefore $\tau = 2l/c$. It will be the proper time interval because it is measured at the same place, i.e. the flash bulb, in the frame of the rod clock. Let it move with respect to a laboratory observer with velocity V *along its length*. Let the length of the rod for this observer be l'. The time recorded between two flashes will be

$$\tau' = \frac{l'}{c+V} + \frac{l'}{c-V}$$

$$= \frac{2cl'}{c^2 - V^2}$$

It will be the non-proper time interval because in the laboratory frame, the two ticks will take place at different space points. Therefore, using Eq. (3.23):

*In relativity we usually call the frame in which the observer is at rest the proper frame. The length of a rod in a frame in which it is at rest is called the proper length.

$$\frac{2l'}{c}\left(\frac{1}{1-\frac{V^2}{c^2}}\right) = \frac{1}{\sqrt{1-\frac{V^2}{c^2}}} \times \frac{2l}{c},$$

or
$$l' = l\sqrt{1-\beta^2} \tag{3.24}$$

Thus, for the laboratory observer, length of the moving rod gets contracted by a factor $\sqrt{1-\beta^2}$. This is called Lorentz contraction. We can conclude from relativity principle that lengths in the laboratory frame appear contracted for an observer in the moving frame.

Let us take one more example. Let a passenger in an express train pass through a station with uniform velocity V. For observers at the platform the passenger passes the platform in time $\Delta t = l/V$, where l is the length of the platform for them. Clearly this is the non-proper time because the events observed (passenger passes the back end of the platform, passenger passes the front end of the platform) occur at two different places in the ground frame. For the passenger the platform passes by him with speed V, and the two ends of the platform cross him at the same place in his frame. Therefore, the time measured by him $\Delta t'$ will be the proper time and will be $\Delta t' = l'/V$, where l' is the length of the platform for him. Therefore

$$\Delta t = \frac{\Delta t'}{\sqrt{1-\beta^2}},$$

or
$$l = \frac{l'}{\sqrt{1-\beta^2}}$$

Example 3.6 Suppose we observe a meter stick passing by (moving along its length) with a velocity 80% of the speed of light. How long will we measure it to be?

$$l' = 1\text{ m} \times \sqrt{1-0.64}$$
$$= 0.6\text{ m} = 60\text{ cm}$$

Question: How will the volume of a moving object change?

Example 3.7 Two clocks *A* and *B* are kept along the *x'*-axis a distance *l'* apart in *S'*-frame (Fig. 3.16), and are synchronized. The frame *S'* is moving with speed *V* relative to frame *S* along the *x-axis* (*x'-axis*). Find out the disagreement between the two clocks as seen by an observer in *S*.

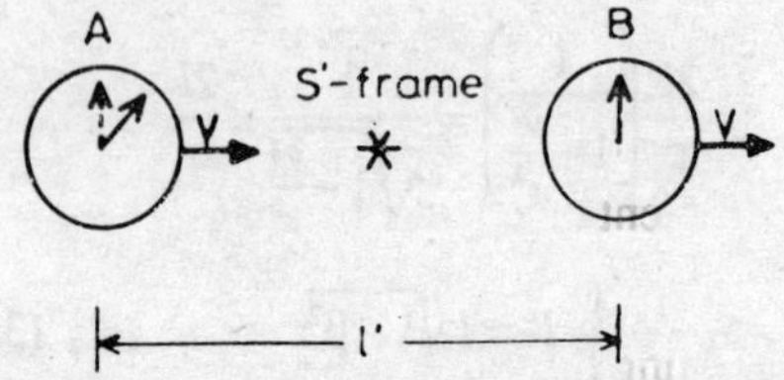

Fig. 3.16 *For Example 3.7*

Let the clocks in S' be synchronized with the aid of a light flash at their mid-point. Let in S-frame $t = 0$ when the flash goes out. When light pulse meets A, at $t = t_A$, we have

$$ct_A = \frac{l'}{2}\sqrt{1-\beta^2} - Vt_A \tag{3.25}$$

The factor $\sqrt{1-\beta^2}$ has been included because l' appears to be contracted to an observer in S. Similarly,

$$ct_B = \frac{l'}{2}\sqrt{1-\beta^2} + Vt_B \tag{3.26}$$

Therefore, the disagreement between the settings of the two clocks will be:

$$t_B - t_A = \frac{l'}{2}\sqrt{1-\beta^2}\left(\frac{1}{c-V} - \frac{1}{c+V}\right)$$

$$= \frac{l'V\sqrt{1-\beta^2}}{c^2-V^2}$$

with clock A running ahead. But during this interval, S observes clock A to run slow by a factor of $\sqrt{1-\beta^2}$. Thus, to an observer in S the clock A will read

$$\frac{l'V\sqrt{1-\beta^2}}{c^2-V^2} \times \sqrt{1-\beta^2} = \frac{l'V}{c^2} \tag{3.27}$$

when clock B is set to read zero.

3.10 Lorentz Transformations

Let us now find out relationships between space-time coordinate measurements of an event by an observer in one inertial frame and the space-time coordinate measurements of the same event by an observer in another inertial frame moving with velocity **V** relative to the first. With the help of these relations we should be able to explain all the physical phenomena discussed in the preceding sections. Let the two frames be S and S' as shown in Fig. (3.17), with respective axes being parallel. Let S' move with velocity V relative to S along Ox ($O'x'$). Let the clocks in both the frames start when O' coincides with O. Further, let at this instant a light signal be flashed at O (O'). Let in Fig. (3.17) P be a point on the wavefront of this signal.

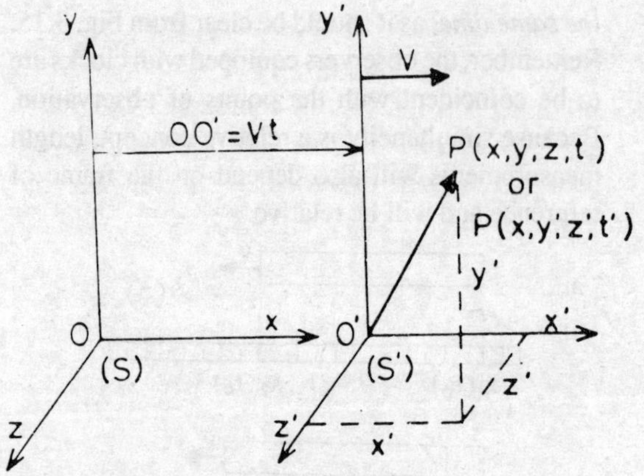

Fig. 3.17 *Two inertial frames in relative motion along x-x' direction*

Since, in frame S the speed of light is c, the wavefront will be spherical and spreading from O. The radius of this sphere will be ct at time t. Therefore:

$$x^2 + y^2 + z^2 = (ct)^2 \tag{3.28}$$

In frame S' also the speed of light is c, and the wavefront will be a sphere with center at O' and having a radius ct' at time t'. Hence

$$x'^2 + y'^2 + z'^2 = (ct')^2 \tag{3.29}$$

The coordinates in two frames are to be connected in such a way that we get Eq. (3.29) starting from Eq. (3.28), and *vice versa*. Galilean transformations give:

$$x' = x - Vt,\ y' = y,\ z' = z,\ t' = t \tag{3.30}$$

If we substitute Eq. (3.30) in Eq. (3.29), we get:

$$x^2 - 2xVt + V^2t^2 + y^2 + z^2 = c^2t^2 \tag{3.31}$$

This relation is not in agreement with Eq. (3.28). Therefore, the Galilean transformations are not adequate.

It has already been pointed out in Section 3.8 that the length of an object in a direction perpendicular to its velocity does not change. Therefore, $y = y'$ and $z = z'$. Thus we get from Eqs. (3.28) and (3.29)

$$x^2 - c^2t^2 \equiv x'^2 - c^2t'^2,$$

or $$(x^2 - x'^2) - c^2(t^2 - t'^2) \equiv 0 \tag{3.32}$$

For an observer in S, O' has the coordinate Vt. Therefore $x = Vt$ when $x' = 0$. Similarly for an observer in S', O has the coordinate $-Vt'$. Or, $x' = -Vt'$ when $x = 0$. The simplest relations which satisfy these conditions are:

$$x' = k(x - Vt), \tag{3.33a}$$

and $$x = k'(x' + Vt') \tag{3.33b}$$

Here k and k' are dimensionless quantities, and their values should be a function of V.

Linearity of the above equations can be justified on the following grounds:

1. There must be a one-to-one correspondence between what each observer sees. Any observation from frame S must have a unique interpretation in frame S'. If higher order equations are used then one set of (x, t) will correspond to more than one sets of (x', t'), and vice-versa. This is because a linear equation has one solution and a higher order equation has more than one solutions.
2. Eqs. (3.33a) and (3.33b) have a possibility of reducing to Galilean relations if k and k' reduce to unity as V approaches zero.
3. The wave fronts expand at a uniform rate in both the frames.

Eliminating x' from Eq. (3.33a) with the help of Eq. (3.33b),

$$t' = k\left[t - \frac{x}{V}\left(1 - \frac{1}{kk'}\right)\right] \tag{3.34}$$

Substituting the values of x' and t' from Eqs. (3.33a) and (3.34), respectively, in Eq. (3.32) :

$$x^2 - k^2(x - Vt)^2 - c^2\left[t^2 - k^2\left\{t - \frac{x}{V}\left(1 - \frac{1}{kk'}\right)\right\}^2\right] = 0$$

or

$$\left\{1 - k^2 + \frac{c^2k^2}{V^2}\left(1 - \frac{1}{kk'}\right)^2\right\}x^2 + \left\{2Vk^2 - \frac{2c^2k^2}{V}\left(1 - \frac{1}{kk'}\right)\right\}xt$$

$$+ (c^2k^2 - c^2 - k^2V^2)t^2 = 0$$

For this identity to hold good for all x and t the coefficients of x^2, xt and t^2 must vanish separately. The coefficient of t^2 gives

$$k = \frac{1}{\sqrt{1 - V^2/c^2}} \tag{3.35a}$$

The coefficient of xt requires that

$$(V^2 - c^2)kk' + c^2 = 0$$

After inserting the value of k,

$$k' = \frac{1}{\sqrt{1 - V^2/c^2}} \tag{3.35b}$$

The equality of k and k' is necessitated by the fact that there is no preferred inertial frame of reference.

Exercise : Verify that above values of k and k' also make the coefficient of x^2 vanish.

Hence the desired transformations are:

$$\left.\begin{aligned} x' &= \frac{x - Vt}{\sqrt{1 - \frac{V^2}{c^2}}} \\ y' &= y \\ z' &= z \\ t' &= \frac{t - \frac{Vx}{c^2}}{\sqrt{1 - \frac{V^2}{c^2}}} \end{aligned}\right\} \tag{3.36}$$

Exercise: Verify that substitution from Eq. (3.36) in Eq. (3.29) gives Eq. (3.28).

The set of Equations (3.36) is called the Lorentz transformations. It may be noted that when V/c approaches zero, the Lorentz transformations reduce to the Galilean transformations.

We can write (x, y, z, t) in terms of (x', y', z', t') by noting that the frame S moves with velocity $-V$ as observed from the frame S'. Thus:

$$\left.\begin{aligned} x &= \frac{x' + Vt'}{\sqrt{1 - \frac{V^2}{c^2}}} \\ y &= y' \\ z &= z' \\ t &= \frac{t' + \frac{Vx'}{c^2}}{\sqrt{1 - \frac{V^2}{c^2}}} \end{aligned}\right\} \tag{3.37}$$

The above relations can also be obtained by directly solving Eq. (3.36).

Exercise: Derive Eq. (3.36) by using the results of Eqs. (3.23), (3.24) and (3.27).

Let us note one point. Galilean transformation relations are inadequate not only in electromagnetism but also in mechanics at high speeds. If $c = \infty$, the Lorentz transformation equations go over to Galilean transformation equations. In that case all the clocks could be synchronized simultaneously and the time will become absolute. However, when $V \ll c$, the Galilean transformations become an approximation of Lorentz transformations.

3.11 Relativistic Addition of Velocities

We have seen in Section 3.3 that if a frame of reference S' (say train) moves with a velocity **V** with respect to another frame S (say ground), then a particle having a velocity **v′** in S' has a velocity **v** in S given by Eq. (3.9). But this is not so in relativistic kinematics. Let us first take a case in which all the velocities are along the common x-x' direction (Fig. 3.18). If the positions of the particle in S' at times t_1' and t_2' are x_1' and x_2', we get:

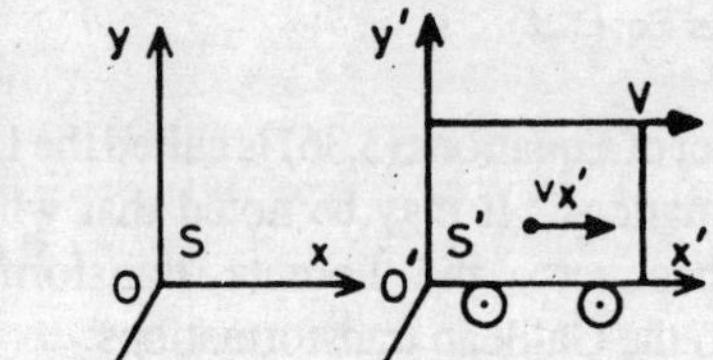

Fig. 3.18 *Relativistic addition of velocities*

$$(x_2' - x_1') = v_x'(t_2' - t_1') \quad (3.38)$$

Here v_x' is the velocity of the particle in direction x-x' in S' frame. Using Lorentz transformation relations (3.36):

$$\gamma(x_2 - Vt_2) - \gamma(x_1 - Vt_1)$$

$$= v_x'\left[\gamma\left(t_2 - \frac{Vx_2}{c^2}\right) - \gamma\left(t_1 - \frac{Vx_1}{c^2}\right)\right],$$

or $$(x_2 - x_1)\left(1 + \frac{v_x'V}{c^2}\right) = (t_2 - t_1)(v_x' + V) \quad (3.39)$$

where $$\gamma = 1/\sqrt{1 - V^2/c^2} = 1/\sqrt{1 - \beta^2} \quad (3.39a)$$

Therefore, the velocity of the particle in S comes out to be:

$$v_x = \frac{x_2 - x_1}{t_2 - t_1} = \frac{v_x' + V}{1 + \frac{v_x'V}{c^2}} \quad (3.40)$$

This is the relativistic velocity addition theorem.

Question: Is S' frame a proper frame for the particle under consideration ? What about S ?

We see from Eq. (3.40) that if v_x' and V are small compared to c, this relation reduces to Eq. (3.9), the classical formula. On the other hand if $v_x' = c, v_x = c$ for all values of V. Of course, when $v_x' = c$, our particle is a photon and its velocity in S should also be c as required by the relativity principle. It also shows that any velocity (less than c) added to c gives a resultant c.

Example 3.8 **A radioactive material emits two electrons in opposite directions. Each electron has a speed 0.67 c as measured by a laboratory observer. What is the speed of one electron as measured from the other?**

Here the reference electron is in the S frame and laboratory is the S' frame. Further, $V = 0.67\,c$, and $v' = 0.67\,c$. Using Eq. (3.40):

$$v = \frac{0.67c + 0.67c}{1 + \frac{0.67c \times 0.67c}{c^2}}.$$

$$= \frac{(0.67 + 0.67)c}{1 + (0.67)^2}$$

$$= \frac{1.34c}{1.45}$$

$$= 0.92c$$

Thus the speed of one electron relative to the other is less than c.

Question: What will this speed be according to Galilean transformation?

Let us now consider the case when the velocity of the particle is perpendicular to x-x' direction (say along y-y' direction) in the S' frame. Let the particle be observed at y_1' and y_2' at times t_1' and t_2'. Therefore:

$$v_y' = \frac{y_2' - y_1'}{t_2' - t_1'} \tag{3.41}$$

Now

$$y_2' - y_1' = y_2 - y_1$$

$$t_2' - t_1' = \gamma\left[(t_2 - t_1) - \frac{V}{c^2}(x_2 - x_1)\right]$$

Therefore

$$v_y' = \frac{y_2' - y_1'}{t_2' - t_1'}$$

$$= \frac{y_2 - y_1}{\gamma\left[(t_2 - t_1) - \frac{V}{c^2}(x_2 - x_1)\right]}$$

$$= \frac{\frac{y_2 - y_1}{t_2 - t_1}\sqrt{1 - \frac{V^2}{c^2}}}{1 - \frac{V}{c^2}\left(\frac{x_2 - x_1}{t_2 - t_1}\right)},$$

or $$v_y' = \frac{v_y\sqrt{1 - \frac{V^2}{c^2}}}{1 - \frac{V}{c^2}v_x} \tag{3.42}$$

To get the inverse transformation we merely substitute $-V$ for V and get:

$$v_y = \frac{v_y'\sqrt{1 - \frac{V^2}{c^2}}}{1 + v_x'\left(\frac{V}{c^2}\right)} \tag{3.43}$$

The expressions for velocities in z-z' direction will be exactly similar to Eqs. (3.42) and (3.43). The complete set is:

$$\left.\begin{aligned}
v_x' &= \frac{v_x - V}{1 - v_x\left(\frac{V}{c^2}\right)} & v_x &= \frac{v_x' + V}{1 + v_x'\left(\frac{V}{c^2}\right)}\\
v_y' &= \frac{v_y\sqrt{1 - \frac{V^2}{c^2}}}{1 - v_x\left(\frac{V}{c^2}\right)} & v_y &= \frac{v_y'\sqrt{1 - \frac{V^2}{c^2}}}{1 + v_x\left(\frac{V}{c^2}\right)}\\
v_z' &= \frac{v_z\sqrt{1 - \frac{V^2}{c^2}}}{1 - v_x\left(\frac{V}{c^2}\right)} & v_z &= \frac{v_z'\sqrt{1 - \frac{V^2}{c^2}}}{1 + v_x\left(\frac{V}{c^2}\right)}
\end{aligned}\right\} \tag{3.44}$$

We see from the above that in the transverse velocity transformations, the longitudinal components also get mixed up.

Question: In Eqs. (3.44) if $v_x' = 0$, $v_y = v_y'\sqrt{1 - \frac{V^2}{c^2}}$. But there is no length contraction in this direction. Then how do you explain the factor $\sqrt{1 - \frac{V^2}{c^2}}$?

3.12 General Theory of Relativity

Two definitions of mass based on two fundamentally different principles have been given in Chapter 2. From experience the two masses of a body – the inertial mass m_i and the gravitational mass m_g — seem to be equal. Equivalence of the two masses has been established experimentally upto an accuracy of 1 part in 10^{11}. Experiments further indicate that for all substances the two masses are identically equal. But there is no theoretical basis for this in mechanics.

The above equivalence becomes clear on giving a wider meaning to the principle of relativity. It has been shown in Section 2.6 that

$$m_i \times \text{acceleration} = m_g \times \text{gravitational field intensity}$$

m_i being equal to m_g the gravitational acceleration of a body does not depend on its mass. It has been shown below that an accelerated frame in gravity free space is equivalent to an unaccelerated frame in gravitational field.

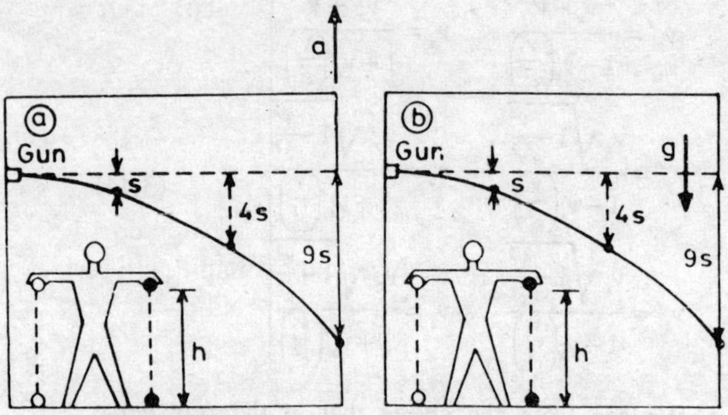

Fig. 3.19 (a) *Accelerated frame* (b) *A frame at rest in gravitational field*

Take a cabin in gravity-free space [Fig. 3.19(a)]. It is closed on all sides and the observer inside it has no contact with the outside world. The cabin is going up with an acceleration **a** = −**g**. It has been seen in Section 3.4 that to the observer a body of mass m_i will experience a force, $\mathbf{F} = -m_i\mathbf{a} = m_i\mathbf{g}$ downwards. If the observer releases two bodies of unequal masses from the same height h, both the bodies will strike the floor simultaneously after time $\sqrt{2h/g}$. If a bullet is fired from a gun horizontally it will follow a parabolic path. After three successive equal time intervals the bullet will fall distances $s = g\Delta t^2/2$, $4s$ and $9s$, respectively.

Let us now consider a similar cabin at rest in a space where the gravitational field is **g** [Fig. 3.19(b)]. For an observer in this cabin force on a body of mass m_g is $\mathbf{F} = m_g\mathbf{g}$. On repeating the experiments done above, this observer will get the same results, i.e. the two bodies will fall on the floor simultaneously after time $\sqrt{2h/g}$, and the bullet will follow a parabolic path. Einstein, in 1911, extended this experimental fact to light also and postulated the general theory of relativity. According to this principle, not only for mechanical experiments but also for optical experiments conducted in the two frames the results will be the same. No experiment is possible which can distinguish situation-1 from situation-2. Equivalence between an accelerated frame and an appropriate gravitational field is called the principle of equivalence. Equivalence of inertial and gravitational masses is inherent in this principle. The beauty of this viewpoint is that one mass can explain inertial, gravitational and their combined effects.

On the basis of the above equivalence principle the gravitational field at a given place can be annulled by the choice of a frame moving with appropriate acceleration. In this manner one can obtain a local inertial frame in a gravitational field or near a gravitational body. As an example, a frame falling freely in gravitational field becomes an inertial frame. Strictly speaking, completely inertial frame can be had at a point only, because gravitational field at two points inside the cabin will not exactly be the same.

Let us now conduct another experiment inside a cabin moving in gravity free space with an acceleration **a** = − **g**, as in Fig. 3.19(a). We now replace the gun by a light-source (Fig. 3.20). Fix three vertical fluorescent screens at equal distances b inside the cabin. During the time $\Delta t (= b/c)$ taken by light in going from one screen to another, the cabin moves up by a distance $s = g\Delta t^2/2$. As a result, a fluorescent spot will appear on the first screen at a distance s below the initial direction of light. On the other two screens fluorescence will appear at distances $4s$ and $9s$, respectively, below the initial direction of light. Hence to this observer, light does not move in a straight line, but in a parabolic path like a bullet.

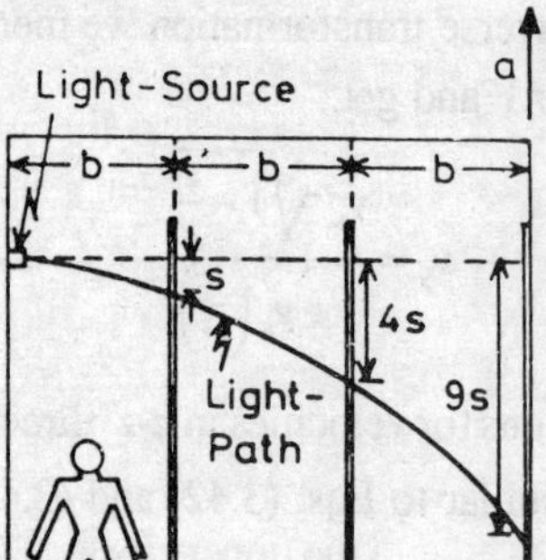

Fig. 3.20 *Deviation of light from a straight line path in an accelerated frame*

The principle of equivalence is applicable to both material particles and light. (The reader should recapitulate how Einstein had extended the principle of relativity of Galileo to light also.) Hence, the path of light in gravitational field will be curved and not a straight line. On the basis of this argument Einstein predicted that the light coming to earth from a far-off star should get deflected through an angle 1.75" while passing near the sun (Fig. 3.21). Such experiments were conducted during total solar eclipse, and the principle of equivalence was verified.

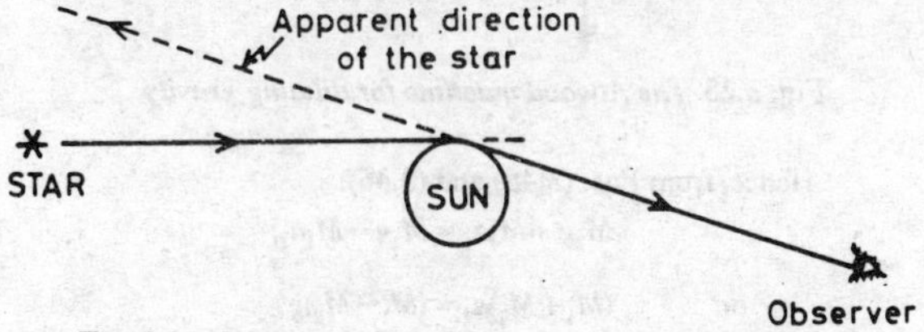

Fig. 3.21 *Bending of light in gravitational field of the sun*

To explain the above bending of light in a more general and clear manner, Einstein further postulated that the four dimensional space-time develops curvature near a gravitational object. As a line joining two points on a curved surface is also curved, light follows a curved path. These curved paths are called geodesics. Precession of orbits of planets revolving around the sun has also been explained using this picture.

Another consequence of general theory of relativity is that wavelength of light gets increased while leaving a gravitational object. This phenomenon is called the *gravitational red-shift*. This shift has been observed in light emitted by sun and other heavy stars. It means that the vibrations of atomic clocks at the surface of sun are slower than the vibrations of the same atomic clocks at the surface of earth. Hence, all physical phenomena slow down in gravitational field. In two clocks separated by 1 m distance vertically at the surface of earth the lower clock moves slower by 1 part in 10^{16} as compared to the upper clock. This fact has been verified in the laboratory using Mossbauer effect.

General theory of relativity predicts that an accelerated mass should emit gravitational waves. Very high acceleration can be had in a collapsing star, or in the collision of two stars. Efforts are going on for the detection of these waves.

According to this theory, after nuclear fuel of a star is finished its collapse under appropriate conditions is possible. A huge collapse will give rise to a black-hole. No signals including light can leave a black-hole. Such objects can be detected indirectly. Thus general theory of relativity has provided a new direction to the researches in astrophysics.

Discussion and Problems

3.1 Does a laboratory fixed on the surface of the earth provide a good inertial frame of reference? If not, why?

3.2 Write the Galilean transformations when OO' in Fig. 3.3 is represented by a vector **R** at the time $t = t' = 0$.

3.3 A satellite is circling the earth at a height of 1500 km. A hydrogen bomb is tied to it with the help of a string. The string is cut when the satellite is above Delhi and is going towards north. The bomb explodes after one hour. Where will it do the maximum harm?

3.4 Two cars A and B are moving with the same speed. The driver of A feels that the motion of B is perpendicular to the motion of his car. How will B look at A? Draw the paths of A and B when they are: (i) receding away from each other, (ii) approaching.

3.5 A plane flying south at 400 km/hr passes past another plane flying at 300 km/hr towards west. What is the velocity of the second plane as seen by the pilot sitting in the first plane? Assume that they are flying at the same height.

3.6 A ship is sailing north at a speed of 20 km/hr. A passenger on deck walks towards the rear of it at a speed of 5 km/hr. What is the velocity of the man relative to the water surface? He then turns east and continues walking at the same speed. What is his velocity now with respect to the water surface?

3.7 Discuss the motion of a projectile as seen by an observer: (i) standing on the ground, (ii) sitting in a car moving with a constant velocity, and (iii) sitting in an elevator moving with an acceleration a.

3.8 Water level in a vessel remains parallel to the base when the vessel is on a horizontal plane [Fig. 3.22(a)]. If the plane is inclined by an angle θ and the vessel is kept stationary, the water level makes an angle θ with the base [Fig. 3.22(b)]. If the inclined plane offers no friction, and the vessel is allowed to slide freely, the water level remains parallel to the base as in Fig. 3.22(c). If the vessel is pushed up this frictionless inclined plane and let go, the level of water will again be parallel to the base [Fig. 3.22(d)]. Explain all these situations.

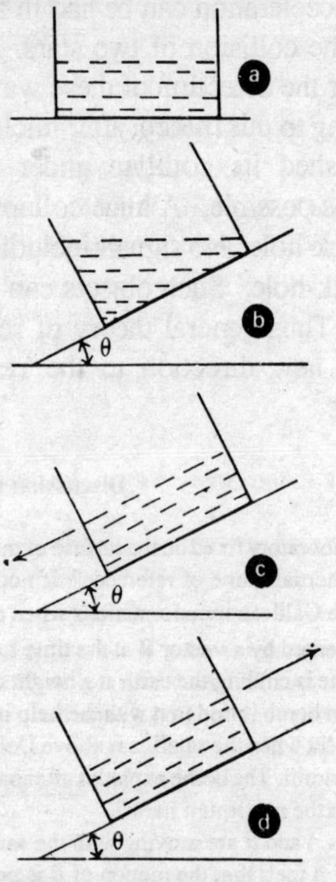

Fig. 3.22 *For Problem 3.8*

3.9 In this chapter you have learnt how the value of acceleration due to gravity changes for a man inside an elevator with the accelerated motion of the frame. Can you suggest some other method to reduce the value of g?

Two bodies of masses M_1 and M_2 are connected by a light in extensible string over a pulley of negligible mass as shown in Fig. 3.23. If the mass of the string is considered to be zero, the net force acting on the string must be zero. Hence, the tension T in the string must be the same throughout and should be directed as shown in Fig. 3.23. The net force on M_1 is:

$$M_1g - T = M_1a_1$$

where a_1 is M_1's downward acceleration,

or $$T = M_1g - M_1a_1 \quad (3.45)$$

The net force on M_2 is:

$$M_2g - T = M_2a_2$$

where a_2 is M_2's downward acceleration.

Since the length of the string is fixed, $a_1 = -a_2$, and we get:

$$m_2g - T = -M_2a_1,$$

or $$T = M_2g + M_2a_1 \quad (3.46)$$

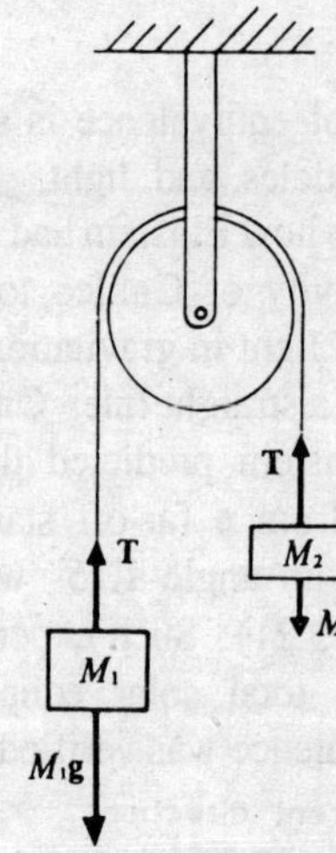

Fig. 3.23 *The Atwood machine for diluting gravity*

Hence, from Eqs. (3.45) and (3.46):

$$M_2g + M_2a_1 = M_1g - M_1a_1,$$

or $$(M_1 + M_2)a_1 = (M_1 - M_2)g,$$

or $$a_1 = \frac{M_1 - M_2}{M_1 + M_2}g \quad (3.47)$$

We see that the acceleration can be made as low as desired.

3.10 A 1 kg stone at the end of a meter long string makes 10 revolutions per second. Calculate the force on the stone as measured in an inertial frame, and in a frame which is rotating with the string.

3.11 A massless rope is hanging over a frictionless pulley. A monkey holds one end of the rope and a mirror having the same weight as that of the monkey is attached to the other end of the rope at the monkey's level. Can the monkey get away from his image seen in the mirror: (i) by climbing up the rope? (ii) by climbing down the rope? (iii) by releasing the rope?

3.12 Two spring balances are hung at the ends of the string of the Atwood machine of Fig. 3.23. Attached to these balances are two weights of masses 3 kg and 1 kg. (i) What is the acceleration of the system when let go? Neglect masses of the balances. (ii) What will be the readings of the two balances when the system is in motion?

3.13 When a car suddenly negotiates a turn, persons sitting in it are pushed outwards. Explain this: (i) in terms of an observer sitting in the car, and (ii) in terms of an observer sitting on the ground.

3.14 What should an observer in an accelerated frame do if he wants to study the motion of the particles in it with the help of Newton's laws?

3.15 Show that the relative motion of two particles of masses m_1 and m_2 under their mutual interaction only is equivalent to the motion of a particle of mass $m_1m_2/(m_1+m_2)$ (called the reduced mass) under a force equivalent to their interaction.

3.16 Derive Eqs. (3.18) and (3.19) for an observer sitting at the apparatus.

3.17 Derive time dilation and length contraction equations starting from Lorentz transformation equations.

3.18 Suppose we perform a Michelson and Morley type experiment with sound. Would an effect be observed when the apparatus is rotated through 90°?

3.19 An airplane takes a round trip to a point 1000 km away due east. Its air speed is 300 km/hr. There is an east-west wind of 50 km/hr. Calculate the time taken. What will this time be if the wind is north-south?

3.20 Is freely falling lift an inertial frame?

3.21 Average life time (proper) of a muon is about 2.2×10^{-6} s. But it can travel about 10^4 m. Has it travelled faster than light? Explain from the point of view of observer sitting on the ground.

3.22 The mean radius of our galaxy is about 4×10^{20} m. With what velocity will a man in spaceship move to make the trip from the center to the edge in 30 years in his frame of reference? How much will this time be for a man at rest with respect to the galaxy?

3.23 A spaceship 100 m long is moving at $0.8c$ with respect to another spaceship. What is the difference in clock readings at the two ends of the first ship (and synchronized in this frame) to an observer in the other ship?

3.24 Show that two events simultaneous in one inertial frame will be simultaneous in another inertial frame moving with respect to it only if they occur at the same place.

3.25 In an inertial reference frame an event occurs at a point (say A on the x-axis) and then 10^{-6} s later another event occurs at some other point (say B, farther out on the x-axis). The distance between the two points is 600 m. Find the velocity of inertial frame S' (moving parallel to the x-axis) so that the above two events are simultaneous in it. What is the separation between the events?

3.26 (a) How much is an actual train 100 m long moving at 100 km/hr shortened? (b) With what velocity has a train to move in order to get shortened in length by 1%?

3.27 A train moves past a pole on the ground with velocity V. Taking pole as a reference, the length of the train is measured by the observers in the train and on the ground separately. How do the two measurements compare?

3.28 Suppose that an event occurs in S at $x = 100$ m, $y = 10$ m, $z = 1000$ m at $t = 1\ \mu s$. Let S' move relative to S at speed $0.5c$ along the common x-x' axis, the origins coinciding at $t = t' = 0$. What are the coordinates of this event in S'?

3.29 An object in a spaceship is moving at $0.5c$ and the spaceship is moving at $0.5c$ in the same direction with respect to an observer. What is the speed of the object for the observer?

3.30 A bullet is fired from a moving train in its direction of motion. An observer A is in the train and an observer B is on the ground. How do the velocities of the bullet and the sound depend upon the velocity of the train for observers A and B?

3.31 To an observer sitting on a particle, it appears to decay and emit another particle to the right with a velocity $0.5c$. Suppose we observe this event when the parent particle is moving to the right with velocity $0.4c$. What will be the velocity of the decay product?

3.32 The concept of an ideal rigid body is untenable in relativity. Discuss.

3.33 Show that the transformation equation for x-x' components of acceleration is

$$a_x' = a_x \frac{\left(1-\frac{v^2}{c^2}\right)^{3/2}}{\left(1-v_x\frac{v}{c^2}\right)^3}.$$

Chapter 4

Linear Momentum

4.1 Linear Momentum

When a force **F** acts on a body of mass *m*,

$$\mathbf{F} = m \times \frac{\text{change in velocity}}{\text{time}}$$

If the force acts on the body for a duration of time Δt and, thus, changes the velocity by $\Delta \mathbf{v}$, then:

$$\mathbf{F} = m\frac{\Delta \mathbf{v}}{\Delta t},$$

or

$$\mathbf{F}\Delta t = m\Delta \mathbf{v} \qquad (4.1)$$

Eq. (4.1) shows clearly that the change in the velocity of a body depends not only on the force, but also on the duration for which the force acts on the body. Actually, it is the product $\mathbf{F}\Delta t$ which is important. Effect of a force **F** for a time Δt is the same as the effect of a force $\mathbf{F}/2$ for a time $2\Delta t$. The quantity $\mathbf{F}\Delta t$ is known as the impulse of the force acting on the body. We can exert a given impulse in many different ways: by applying a big force for a short time, or by applying a smaller force for a longer time, or even by a force that changes while it acts. When we hit a ball by a hockey stick we apply the impulse by a big force for a short time. Whenever the state of motion of a body changes instantaneously, the impulse is applied in a short time and the force involved is a big force. A cyclist is in motion. He can stop his cycle in two ways: (i) by applying brakes suddenly and bringing the cycle to rest immediately, and (ii) by applying the brakes slowly, thus progressively decreasing the velocity of the vehicle so that it stops finally. In the first case a large force acts for a short time, while in the second case a small force acts for a long time. In both the cases the cycle finally comes to rest. Total impulse is the same in both the cases, the only difference being in the magnitude of the force and its duration.

Example 4.1 A 20 kg wagon is pushed with a force of 7 N for 1.5 s, then with a force 5 N for 1.7 s, then with a force of 10 N for 3 s. (i) What is the total impulse applied to the wagon? (ii) What is the change in its velocity?

Total impulse $= 7 \times 1.5 + 5 \times 1.7 + 10 \times 3$

$= 10.5 + 8.5 + 30 = 49$ Ns

$\therefore \quad m\Delta v = 49$ Ns, or $\Delta v = 49$ Ns /20 kg

$= 2.45$ m/s

For a constant force acting for a given time interval, we can always get the magnitude of the impulse as the area under the force–time curve [Fig. 4.1(a)]. Now, suppose that the force changes from a constant value $\mathbf{F}_1$ to a constant value $\mathbf{F}_2$. For an interval Δt_1 the constant force is $\mathbf{F}_1$ giving the impulse $\mathbf{F}_1\Delta t_1$. This impulse makes a change $m(\Delta \mathbf{v})_1$ in the motion. In the next time interval Δt_2 the force is again constant, this time $\mathbf{F}_2$, and there is an impulse $\mathbf{F}_2\Delta t_2$ which results in a change $m(\Delta \mathbf{v})_2$ in the motion. That is:

$$\mathbf{F}_1\Delta t_1 = m(\Delta \mathbf{v})_1,$$

and

$$\mathbf{F}_2\Delta t_2 = M(\Delta \mathbf{v})_2$$

If we add these two statements vectorially:

$$\mathbf{F}_1\Delta t_1 + \mathbf{F}_2\Delta t_2 = m[(\Delta \mathbf{v})_1 + (\Delta \mathbf{v})_2]$$

$$= m\Delta \mathbf{v} \qquad (4.2)$$

where $\Delta \mathbf{v} = (\Delta \mathbf{v})_1 + (\Delta \mathbf{v})_2$

= total change in velocity in the duration $(\Delta t_1 + \Delta t_2)$

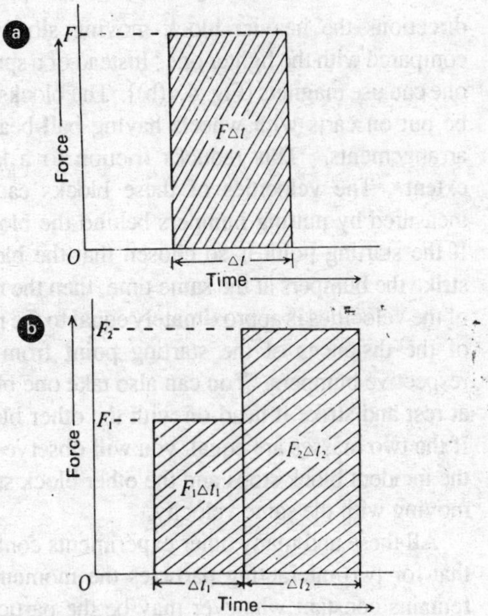

Fig. 4.1 (a) *For a constant force along a straight line the area $F\Delta t$ under the curve gives the impulse of this force in time Δt.* (b) *Total impulse is equal to the area of the two rectangles.*

Thus, the total impulse in the duration $\Delta t (= \Delta t_1 + \Delta t_2)$ is mass times the total change in velocity. This is given by the shaded area in Fig. 4.1 (b). If the force changes continuously during the time it is acting on the body, we can again draw a force-time curve*. The total impulse can be found by dividing the duration of time into small intervals, and then summing up all the small areas $F\Delta t$. The sum is equal to the area contained by the curve and the time axis between t_1 and t_2 (Fig. 4.2). If before the application of the impulse the velocity is **v**, and after the impulse the velocity is **v**′, we have:

$$\text{Total impulse} = m(\mathbf{v}' - \mathbf{v}) = m\mathbf{v}' - m\mathbf{v} \quad (4.3)$$

The quantity $m\mathbf{v}$ is quite a different measure of motion than velocity. Velocity tells us how fast a body moves and in which direction. It does not tell us anything about the effort required to get the body in motion, or to stop it. But $m\mathbf{v}$ is the quantity which tells us about this effort. The product $m\mathbf{v}$ does not depend upon the manner in which impulse was applied. Therefore, this important parameter is given a special name *momentum*, and a symbol **p**.

*This graphical analysis can be done only for motion along a straight line.

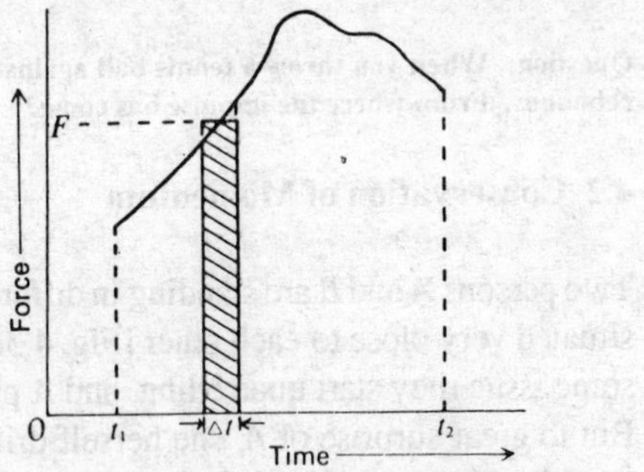

Fig. 4.2 *For a continuously varying force along a straight line, the total impulse* is equal to the area under the *F-t* curve.

Example 4.2 A ball weighing 160 g and travelling horizontally with a velocity of 25 m/s is struck by a stick. The ball returns with a velocity of 30 m/s. What is the impulse of the blow?

Impulse = Change in momentum

$$= m\mathbf{v}' - m\mathbf{v} = m(\mathbf{v}' - \mathbf{v})$$

$$= 160(-30 - 25)/1000 \text{ kg m/s}$$

$$= -8.8 \text{ Ns}$$

Fig 4.3 *Two persons are in carts which can move frictionlessly.* (a) *A pushes B.* (b) *Both start moving in opposite directions.*

The negative sign shows that the impulse is in a direction opposite to the direction of original velocity.

Eq. (4.3) can be written as:

$$\mathbf{F}\Delta t = \mathbf{p}' - \mathbf{p} = \Delta\mathbf{p} \quad (4.4a)$$

where $\mathbf{p} = m\mathbf{v}$ and $\mathbf{p}' = m\mathbf{v}'$,

or $$\mathbf{F} = \Delta\mathbf{p}/\Delta t \quad (4.4b)$$

is the average force causing a change in momentum. The instantaneous force is given by:

$$\lim_{\Delta t \to 0} \frac{\Delta \mathbf{p}}{\Delta t} \quad \text{(see Example 2.2)} \quad (4.4c)$$

Question: When you throw a tennis ball against a wall, it rebounds. From where the impulse has come?

4.2 Conservation of Momentum

Two persons *A* and *B* are standing in different carts situated very close to each other [Fig. 4.3(a)]. On some issue they start quarrelling, and *A* pushes *B*. But to great surprise of *A*, she herself drifts away from *B* and is pushed back in the opposite direction without any effort on the part of *B*. Both start moving in opposite directions [Fig. 4.3 (b)]. If both the cart-systems have the same mass, they seem to move with equal and opposite velocities. If the cart system *A* is heavier, it moves with a velocity smaller than the velocity of the cart-system *B*. If the cart-system *A* is lighter, it moves faster. Actually, it is the momenta of the cart-systems which are exactly equal and opposite, i.e.:

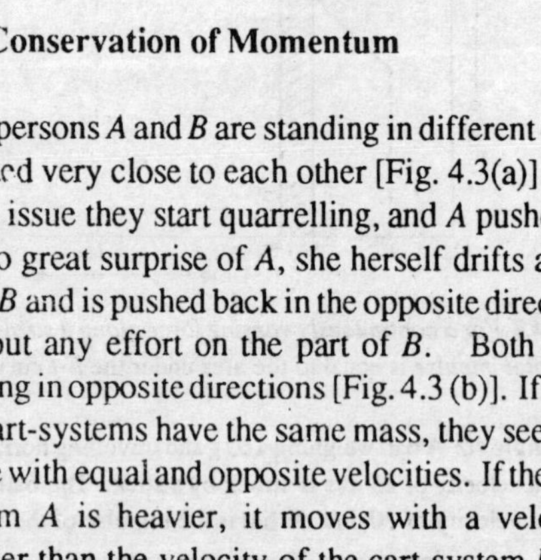

Fig. 4.4 *Simple kinds of explosion* : **(a)** *with the help of a spring*, **(b)** *with the help of magnets*

$$\mathbf{p}_A = -\mathbf{p}_B, \quad \text{or} \quad \mathbf{p}_A + \mathbf{p}_B = 0$$

Before *A* pushes *B*, the sum of the momenta was zero, and it remains zero even after they come in motion.

The above conclusions can be checked in the laboratory. Take two blocks of masses M_1 and M_2 on a horizontal surface, attach them with a spring, compress the spring, and tie the two blocks with a thread [Fig. 4.4(a)]. Now burn the thread and watch the motion of M_1 and M_2. They go in opposite directions–the heavier block moving slowly as compared with the lighter one. Instead of a spring, one can use magnets [Fig. 4.4(b)]. The blocks can be put on carts with wheels having ball-bearing arrangements. This reduces friction to a large extent. The velocities of these blocks can be measured by putting bumpers behind the blocks. If the starting point is so chosen that the blocks strike the bumpers at the same time, then the ratio of the velocities is approximately equal to the ratio of the distances of the starting point from the respective bumpers. You can also take one block at rest and strike it head-on with the other block. If the two masses are equal, you will observe that the incident block stops and the other block starts moving with the same velocity.

All these and many other experiments confirm that for two interacting particles the momentum remains constant whatever may be the particles' initial state of motion. If both of them are at rest and the interaction starts, the momentum gained by each is equal and opposite. If they are in the state of motion, then also the changes in momenta are equal and opposite. We can now say that the total momentum of two interacting bodies.

$$\mathbf{p} = \mathbf{p}_1 + \mathbf{p}_2 = \text{constant} \quad (4.5)$$

where $\mathbf{p}_1$ and $\mathbf{p}_2$ are the momenta of first and second body respectively. Alternatively:

$$\Delta\mathbf{p} = \Delta\mathbf{p}_1 + \Delta\mathbf{p}_2 = 0 \quad (4.6)$$

Thus, the change in total momentum is zero. Or:

$$\Delta\mathbf{p}_1 = -\Delta\mathbf{p}_2 \quad (4.7)$$

Let us consider the interaction of two bodies which come into contact during collision. Now, the change in momentum of one is equal and opposite to the change in momentum of the second. This change in momentum must have taken place during the interval Δt when the two bodies were in contact. Hence:

$$\Delta\mathbf{p}_1 = \mathbf{F}_1 \Delta t$$

where $\mathbf{F}_1$ is the force acting on the first body; and

$$\Delta\mathbf{p}_2 = \mathbf{F}_2 \Delta t$$

where $\mathbf{F}_2$ is the force acting on the second body. From Eq. (4.7):

$$\mathbf{F}_1 = -\mathbf{F}_2 \tag{4.8}$$

which is just the third law of Newton. When we have formulated a more fundamental law of physics, we have no necessity of keeping Newton's third law*. The *law of conservation of momentum* is valid even in atomic and nuclear domains where Newtonian mechanics fails.

The law of momentum can be extended further to a system containing more than two bodies. We state this law of nature as : *In an isolated system of mutually interacting particles, the total linear momentum remains constant. This is known as the law of conservation of momentum.*

It should be made clear that the law of conservation of momentum holds for isolated systems only. It does not hold for the systems which are under the influence of the force fields of other systems. Take a stone and throw it up. It stops at a certain height, and then starts coming back. Thus, the momentum of the stone is changing continuously. The reason is that it is interacting with the earth through the gravitational field. If the stone and the earth are treated as one system, then the momentum will be conserved. When the stone goes up, the earth goes down, and when the stone comes down, the earth comes up. It is for this reason that all experiments suggested above are to be performed on horizontal surfaces. Motions on these planes being perpendicular to the gravitational force are not influenced by it. Friction between two systems also amounts to an interaction between them. If a system is under the influence of an external force, the change in its momentum is given by Eq. (4.4).

Example 4.3 A ball of 100 g, moving horizontally with a velocity of 5 m/s, strikes another ball of the same mass and gets deflected by an angle of 30°. Its speed decreases to 3 m/s. Find the speed and the direction of motion of the second ball.

Let the collision be represented by Fig. 4.5. As momentum is a vector, both magnitude and direction should be conserved.

Momentum in the direction xox' before collision

$$= \frac{100}{1000} \text{ kg} \times 5 \text{ m/s} = 0.5 \text{ N s}$$

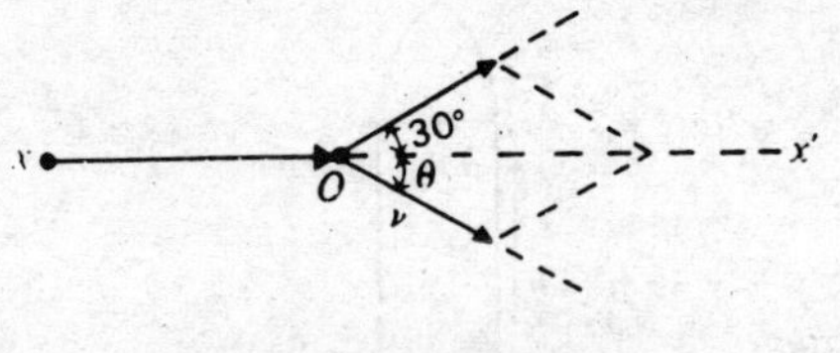

Fig. 4.5 *For Example 4.3*

Momentum in the direction xox' after collision

$= 0.3 \cos 30° + 0.1\, v \cos\theta$

$= 0.3 \times 0.866 + 0.1\, v \cos\theta$, therefore $0.260 + 0.1\, v\cos\theta$

$= 0.5$, or $v \cos\theta = 2.4$

As the initial momentum in the direction perpendicular to xox' is zero, the momentum in this direction after collision should also be zero.

Therefore $\quad 0.3 \sin 30° = v \sin\theta \times 0.1,$

or $\quad v\sin\theta = 0.3 \times 0.5 \times 10 = 1.5$

Hence $\quad \tan\theta = 1.5/2.4 = 0.625,$

or $\quad \theta = 32°$(from tables)

And $\quad v = 1.5/\sin 32° = 1.5/0.53 = 2.8$ m/s

4.3 A Rocket

A rocket is shown in Fig. 4.6. Let its mass at time t be m, and its velocity for an observer on earth be $\mathbf{v}$. If the exhaust velocity (velocity relative to the rocket) of the fuel is $\mathbf{u}$, then this velocity for the observer will be $\mathbf{u} + \mathbf{v}$. As the rocket continuously ejects fuel gases, its mass will change continuously. Let at time $t + \Delta t$ the mass of the rocket become $m + \Delta m$ (Δm will be negative), and its velocity become $\mathbf{v} + \Delta\mathbf{v}$. For the rocket-fuel system,

momentum at time $t = m\mathbf{v}$

momentum at time $t + \Delta t = (m + \Delta m)(\mathbf{v} + \Delta\mathbf{v})$

$$+(-\Delta m)(\mathbf{v} + \Delta\mathbf{v} + \mathbf{u}) = m\mathbf{v} + m\Delta\mathbf{v} - \Delta m\mathbf{u}$$

$\therefore$ Change in momentum $= m\Delta\mathbf{v} - \Delta m\mathbf{u}$

*Electromagnetic forces between moving charges do not obey Newton's third law. Take, for example, the case of a charge moving away from a current-carrying wire.

Hence

$$\mathbf{F}_{ext} = m\frac{\Delta \mathbf{v}}{\Delta t} - \mathbf{u}\frac{\Delta m}{\Delta t}$$

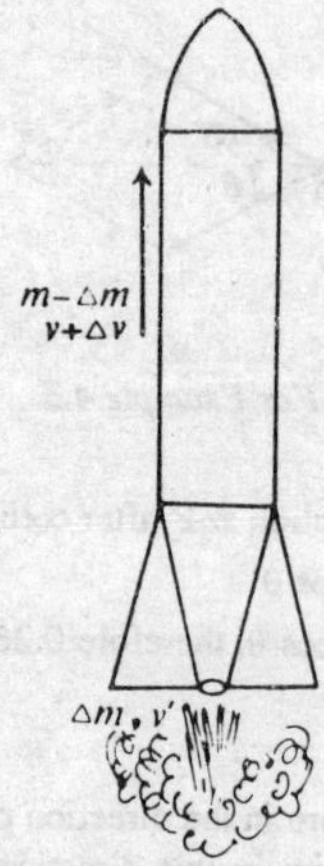

Fig. 4.6 *A rocket*

If the observer concentrates on the rocket alone,

$$m\frac{\Delta \mathbf{v}}{\Delta t} = \mathbf{F}_{ext} + \mathbf{u}\frac{\Delta m}{\Delta t}$$

The second term on the right-hand side is called the thrust of the rocket-motor. As $\Delta m/\Delta t$ is negative, the thrust will be in a direction opposite to the exhaust velocity. $\mathbf{F}_{ext}$ may represent air resistance, gravitational force, or any other force acting on the rocket.

When there is no external force,

$$m\frac{\Delta \mathbf{v}}{\Delta t} = \mathbf{u}\frac{\Delta m}{\Delta t}$$

Let us now take a simple case of a rocket moving in a straight line. As **u** and **v** have opposite directions,

$$m\frac{\Delta v}{\Delta t} = -u\frac{\Delta m}{\Delta t}$$

Multiplying both sides by $\Delta t/m$, and assuming u to be constant, the integration gives

$$\int_{v_o}^{v} dv = -u\int_{m_o}^{m}\frac{dm}{m},$$

or $$v - v_o = u \ln (m_o/m)$$

Here v_o and m_o are initial velocity and initial mass of the rocket, respectively.

For a rocket moving in a gravitational field **g**,

$$m\frac{\Delta \mathbf{v}}{\Delta t} = \mathbf{u}\frac{\Delta m}{\Delta t} + m\mathbf{g}$$

When the motion is vertically upwards,

$$m\frac{\Delta v}{\Delta t} = -u\frac{\Delta m}{\Delta t} - mg$$

For a rocket starting from rest the integration gives

$$v = u\ln(m_o/m) - gt$$

Question: Why do we use multi-stage rockets?

4.4 Centre of Mass

Let M_1 and M_2 (Fig. 4.7) be masses of two bodies, and O be a reference point. If their positions at any moment are given by x_1 and x_2, respectively, then the position of the centre of mass is by definition:

$$x \equiv \frac{M_1x_1 + M_2x_2}{M_1 + M_2} \tag{4.9}$$

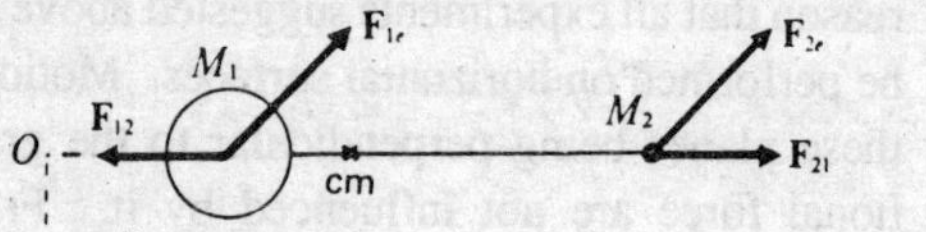

Fig. 4.7 *Definition of the centre of mass*

In three dimensions, if (x_1, y_1, z_1) are the coordinates of M_1 and (x_2, y_2, z_2) are the coordinates of M_2, the coordinates (x, y, z) of the centre of mass are given by Eq. (4.9) and,

$$\left.\begin{aligned} y &= \frac{M_1y_1 + M_2y_2}{M_1 + M_2} \\ z &= \frac{M_1z_1 + M_2z_2}{M_1 + M_2} \end{aligned}\right\} \tag{4.9a}$$

From Eqs. (4.9) and (4.9a), the position vector **r** of the centre of mass becomes:

$$\mathbf{r} = x\mathbf{i} + y\mathbf{j} + z\mathbf{k}$$

$$= \frac{M_1\mathbf{r}_1 + M_2\mathbf{r}_2}{M_1 + M_2} \qquad (4.10)$$

where $\mathbf{r}_1$ $(= x_1\mathbf{i} + y_1\mathbf{j} + z_1\mathbf{k})$ is the position vector of M_1, and $\mathbf{r}_2 (= x_2\mathbf{i} + y_2\mathbf{j} + z_2\mathbf{k})$ is the position vector of M_2.

From Eq. (4.10) we can easily see that a small change in **r** is:

$$\Delta\mathbf{r} = \frac{M_1\Delta\mathbf{r}_1 + M_2\Delta\mathbf{r}_2}{M_1 + M_2},$$

or $$\frac{\Delta\mathbf{r}}{\Delta t} = \frac{M_1\frac{\Delta r_1}{\Delta t} + M_2\frac{\Delta r_2}{\Delta t}}{M_1 + M_2},$$

or $$\mathbf{v} = \frac{M_1\mathbf{v}_1 + M_2\mathbf{v}_2}{M_1 + M_2} \qquad (4.11)$$

Eq. (4.11) gives the velocity of the centre of mass in terms of the velocities of the particles. Similarly, we can show that the acceleration of the centre of mass is given by:

$$\mathbf{a} = \frac{M_1\mathbf{a}_1 + M_2\mathbf{a}_2}{M_1 + M_2},$$

or $$(M_1 + M_2)\mathbf{a} = M_1\mathbf{a}_1 + M_2\mathbf{a}_2 = \mathbf{F}_1 + \mathbf{F}_2$$

$$= (\mathbf{F}_{12} + \mathbf{F}_{1e}) + (\mathbf{F}_{21} + \mathbf{F}_{2e})$$

$$= (\mathbf{F}_{1e} + \mathbf{F}_{2e}) + (\mathbf{F}_{12} + \mathbf{F}_{21})$$

$$= (\mathbf{F}_{1e} + \mathbf{F}_{2e}) = \mathbf{F}_e \qquad (4.12)$$

Here, $\mathbf{F}_1$ and $\mathbf{F}_2$ are the total forces acting on M_1 and M_2 respectively. $\mathbf{F}_{12}$ and $\mathbf{F}_{21}$ are the forces of interaction between M_1 and M_2 and we know that if Newton's third law of motion holds good $\mathbf{F}_{12} = -\mathbf{F}_{21}$, i.e. $\mathbf{F}_{12} + \mathbf{F}_{21} = \mathbf{0}$. $\mathbf{F}_{1e}$ and $\mathbf{F}_{2e}$ are the external forces acting on M_1 and M_2, and $\mathbf{F}_e = \mathbf{F}_{1e} + \mathbf{F}_{2e}$ is the total external force on the system. Thus we have:

Total mass of the system × acceleration of the centre of mass = total external force

We see that the acceleration of the centre of mass depends only on the external force, and not on the forces of mutual interaction of the particles. When no external force is acting, the centre of mass remains at rest, or in a state of motion with constant velocity.

We can easily extend the definition of the centre of mass to a system of N particles. If the positions of masses $M_1, M_2....M_N$ are $\mathbf{r}_1, \mathbf{r}_2....\mathbf{r}_N$ and velocity vectors are $\mathbf{v}_1, \mathbf{v}_2....\mathbf{v}_N$, respectively, the position vector ($\mathbf{r}_{cm}$) of the centre of mass is:

$$\mathbf{r}_{cm} = \frac{M_1\mathbf{r}_1 + M_2\mathbf{r}_2 + ... + M_N\mathbf{r}_N}{M_1 + M_2 + ... + M_N}$$

$$= \frac{\sum_{n=1}^{N} M_n\mathbf{r}_n}{\sum_{n=1}^{N} M_n}$$

Similarly, for the velocity vector of the centre of mass we have:

$$\mathbf{v}_{cm} = \frac{\sum_{n=1}^{N} M_n\mathbf{v}_n}{\sum_{n=1}^{N} M_n} = \frac{\sum_{n=1}^{N} \mathbf{p}_n}{\sum_{n=1}^{N} M_n},$$

or $$M\mathbf{v}_{cm} = \sum_{n=1}^{N} \mathbf{p}_n$$

where $\sum_{n=1}^{N} M_n = M$ is the total mass of the system.

Thus, to find out linear momentum of a system of particles we can treat the masses of the constituent particles to be kept at the centre of mass, and moving along with it.

Example 4.4 Three particles of masses 2.0 g, 3.0 g, and 5.0 g are situated at the points (2i, – 3j, 4k), (i, 2j, 3k) and (3i, j, –k) respectively. Find the position of the centre of mass.

$$M_1\mathbf{r}_1 = 4\mathbf{i} - 6\mathbf{j} + 8\mathbf{k}$$

$$M_2\mathbf{r}_2 = 3\mathbf{i} + 6\mathbf{j} + 9\mathbf{k}$$

$$M_3\mathbf{r}_3 = 15\mathbf{i} + 5\mathbf{j} - 5\mathbf{k}$$

$$\sum_{n=1}^{3} M_n\mathbf{r}_n = 22\mathbf{i} + 5\mathbf{j} + 12\mathbf{k}$$

Hence, $$\mathbf{r}_{cm} = \frac{22\mathbf{i} + 5\mathbf{j} + 12\mathbf{k}}{2 + 3 + 5} = 2.2\mathbf{i} + 0.5\mathbf{j} + 1.2\mathbf{k}$$

For an isolated system $\Sigma \mathbf{p}_n$ is a constant. Place a number of glass balls on the floor and put a cracker at their centre (Fig. 4.8). Now explode the cracker. The balls will diverge from the point of explosion in all directions, and the vector sum of their (horizontal) momenta will be zero. Hence, in the absence of an external force the velocity of the centre of mass remains constant. If it is zero in the beginning it will continue to be zero thereafter.

An external force changes the momentum of the system and, hence, the velocity of the centre of mass changes. We have, as in Eq. (4.12):

$\mathbf{F}_e = M \times$ acceleration of the centre of mass

Thus, the centre of mass of a system of particles moves as though the masses of the particles were concentrated at this point, and all the external forces were applied here.

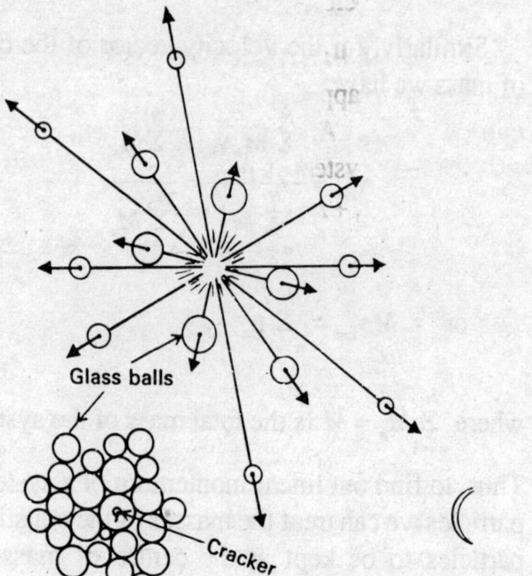

Fig. 4.8 *When a bomb explodes, the fragments go in all directions such that the total momentum after the explosion is the same as before, i.e. zero.*

Example 4.5 A man weighing 60 kg is standing at the centre of a flat boat of mass 300 kg. The man, who is at a distance of 10 m from the shore, walks 2 m towards it and stops. How far will he be from the shore? Assume the boat to be of uniform thickness, and neglect friction between it and the water.

Here, our system consists of the boat and the man. The centre of mass of this system, which is 10 m away from the shore, is at rest to start with. As no external forces are acting on this system, the centre of mass will remain stationary at this position. Let us take this point as origin, and the direction towards the shore as x-axis. If $\mathbf{x}_1 = x_1\mathbf{i}$ and $\mathbf{x}_2 = x_2\mathbf{i}$ are the positions of the man and the centre of the boat, respectively, we get:

$$0 = \frac{60\mathbf{x}_1 + 300\mathbf{x}_2}{60 + 300},$$

or $$\mathbf{x}_1 + 5\mathbf{x}_2 = 0, \quad \text{or } x_1 + 5x_2 = 0,$$

also $$\mathbf{x}_1 - \mathbf{x}_2 = 2\mathbf{i}, \quad \text{or } x_1 - x_2 = 2$$

Therefore,

$$x_1 = 5/3 \text{ m}$$

Hence, the distance of the man from the shore = 10 − 5/3 = 8.3 m

Example 4.6 If an observer sits in a frame of reference in which the centre of mass is at rest (C-frame), discuss the problem given in Fig. 4.9.

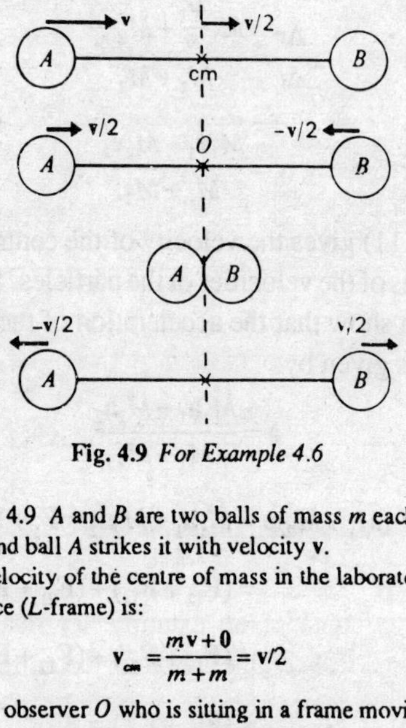

Fig. 4.9 *For Example 4.6*

In Fig. 4.9 A and B are two balls of mass m each. Ball B is at rest and ball A strikes it with velocity $\mathbf{v}$.

The velocity of the centre of mass in the laboratory frame of reference (L-frame) is:

$$\mathbf{v}_{cm} = \frac{m\mathbf{v} + 0}{m + m} = \mathbf{v}/2$$

For an observer O who is sitting in a frame moving with a velocity $\mathbf{v}/2$ with respect to the laboratory frame the velocity of $A = \mathbf{v} - \mathbf{v}/2 = \mathbf{v}/2$ and the velocity of $B = (0 - \mathbf{v}/2) = -\mathbf{v}/2$. Thus both A and B approach the observer with a speed $|\mathbf{v}|/2$ from opposite directions. The collision takes place just by the center of mass and, after the collision, A and B recede from the centre of mass with equal speeds $|\mathbf{v}|/2$.

We see from Eq. (4.12) that in the absence of external forces the centre of mass of the system moves with uniform velocity (which could also be zero). Therefore to study the motion of the constituent particles of an isolated system, its centre of mass can be used as an inertial frame.

When we make observations from the centre of mass what we mean is that our observations are made in an inertial frame in which the velocity of the centre of mass $\mathbf{v}_{cm} = 0$. Or:

$$\sum_{n=1}^{N} \mathbf{p}_n = \sum_{n=1}^{N} M_n \mathbf{v}_n = M \mathbf{v}_{cm} = 0$$

Thus in the reference frame with the centre of mass as the origin (centre of mass coordinate system), the total momentum remains zero, whatever may be the velocity of the constituent particles.

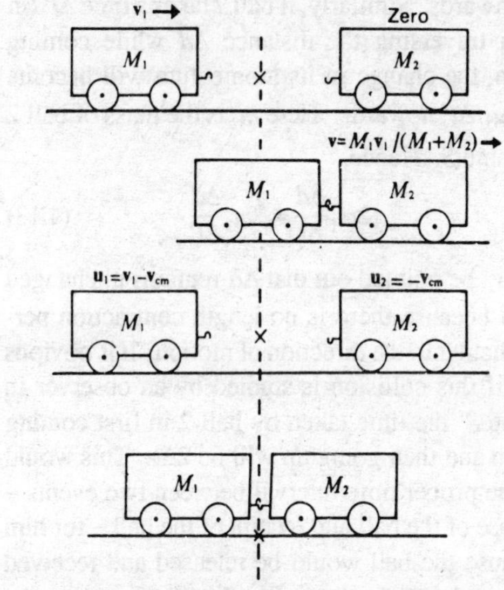

Fig. 4.10 *Relationship between the observations made from the L- and the C-frames of reference*

Let us consider an example to illustrate the difference between the laboratory coordinate system and the centre of mass system. Let two bodies of masses M_1 and M_2 collide in such a way that after the collision they are attached together (Fig. 4.10). You might have observed such processes in railway yards when two coaches are connected together.

In the laboratory system, if the velocity of M_1 is $\mathbf{v}_1$ before collision and M_2 is at rest; and after the two masses are attached together they move with a velocity $\mathbf{v}$, then:

$$\text{Momentum before collision} = M_1\mathbf{v}_1 + M_2 \times 0 = M_1\mathbf{v}_1$$

$$\text{Momentum after collision} = M_1\mathbf{v} + M_2\mathbf{v} = (M_1 + M_2)\mathbf{v}$$

From conservation of momentum:

$$(M_1 + M_2)\mathbf{v} = M_1\mathbf{v}_1,$$

or

$$\mathbf{v} = \frac{M_1}{M_1 + M_2}\mathbf{v}_1$$

Both bodies move with the same velocity $\mathbf{v}$.

Velocity of the centre of mass is:

$$\mathbf{v}_{cm} = \frac{M_1\mathbf{v}_1}{M_1 + M_2}$$

In the centre of mass system the velocity of the centre of mass is zero. Therefore, in this frame M_1 moves with a velocity $\mathbf{u}_1 = \mathbf{v}_1 - \mathbf{v}_{cm}$ and M_2 moves with a velocity $\mathbf{u}_2 = -\mathbf{v}_{cm}$, before collision. Both bodies seem to approach the observer sitting at the centre of mass. After collision the velocity $\mathbf{u}$ of the combined system is given by:

$$(M_1 + M_2)\mathbf{u} = M_1\mathbf{u}_1 + M_2\mathbf{u}_2$$

$$= M_1(\mathbf{v}_1 - \mathbf{v}_{cm}) + M_2(-\mathbf{v}_{cm})$$

$$= M_1\mathbf{v}_1 - (M_1 + M_2)\mathbf{v}_{cm}$$

$$= M_1\mathbf{v}_1 - (M_1 + M_2)\frac{M_1\mathbf{v}_1}{M_1 + M_2}$$

$$= 0,$$

or

$$\mathbf{u} = 0$$

This is obvious because both the masses are now united with each other. Whatever is the velocity of the combination is the velocity of the centre of mass. Hence, in the centre of mass system the combination appears to be at rest.

4.5 Relativistic Variation of Mass

Suppose we have two perfectly elastic and smooth spherical balls of equal mass. Let they be released towards each other (in the laboratory frame) with equal and opposite velocities [Fig.

4.11(a)]. Their total momentum is zero. After collision their directions of motion must again be opposite to each other to keep the momentum zero. Also they must have equal speeds since the balls are identical. Further, this speed will be the same they started with because of conservation of energy. Now fix the x-axis in the plane of the collision in such a way that the collision looks symmetrical about it. Let the initial perpendicular distance of each ball from this axis be Δd. Further, let the plane of the collision be taken as the xy-plane.

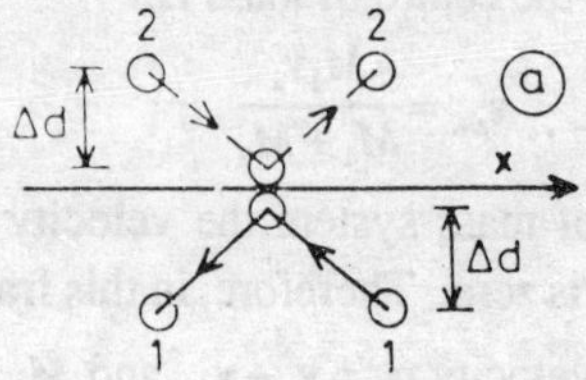

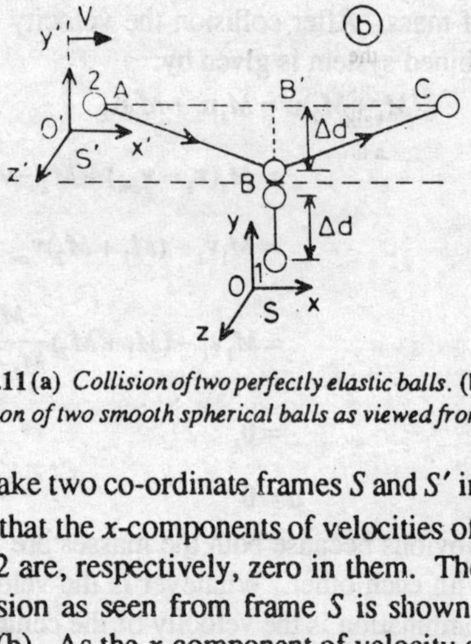

Fig. 4.11 (a) *Collision of two perfectly elastic balls.* **(b)** *Elastic collision of two smooth spherical balls as viewed from frame S*

Take two co-ordinate frames S and S' in such a way that the x-components of velocities of balls 1 and 2 are, respectively, zero in them. The above collision as seen from frame S is shown in Fig. 4.11(b). As the x-component of velocity of ball 1 is zero in this frame it will go straight up and come back. Ball 2 will appear to fly by at a much higher speed and at a smaller angle. But the angles it makes with x-axis before and after the collision will be equal.

Exercise: Draw a diagram showing this collision as seen from frame S'.

If the law of conservation of momentum is to hold good in frame S also, then the change in momentum of ball 1 must be equal and opposite to the change in momentum of ball 2. Clearly the x-component of momentum is conserved – it is always zero for ball 1 and remains unchanged for ball 2. Let us now look into the y-component. If ball 1 takes a time Δt in traversing the distance Δd while going up, its momentum will be $m_1 \Delta d/\Delta t$. Here m_1 is the mass of ball 1 in frame S. Therefore, the change in momentum will become $2m_1 \Delta d/\Delta t$ downwards. Similarly, if ball 2 takes a time $\Delta t'$ (in S) in traversing the distance Δd while coming down, the change in its momentum will become $2m_2 \Delta d/\Delta t'$ upwards. Here m_2 is the mass of ball 2 in frame S. Hence,

$$2m_1 \frac{\Delta d}{\Delta t} = 2m_2 \frac{\Delta d}{\Delta t'} \tag{4.13}$$

It may be pointed out that Δd remains unchanged for S because there is no length contraction perpendicular to the direction of motion. It is obvious that if this collision is studied by an observer in frame S' the time taken by ball 2 in first coming down and then going up will be $2\Delta t$. This would be the proper time interval between two events – release of the ball and receipt of the ball – for him because the ball would be released and received back at the same place. But these two events take place at different places for an observer in S, and the time interval between the two is $2\Delta t'$. Hence $2\Delta t'$ is the improper time interval corresponding to the proper time interval $2\Delta t$. Eq. (4.13) then becomes

$$m_2 = m_1 \frac{2\Delta t'}{2\Delta t},$$

or

$$m_2 = \frac{m_1}{\sqrt{1 - V^2/c^2}} \tag{4.14}$$

Use of Eq. (3.23) has been made here. Here V is the relative speed between the two frames S and S'. This is also the x-component of velocity of ball 2 (to S).

The collision can be planned in such a way that the y- (or y'-) components of velocities become infinitesimally small. Under this limit the speed of ball 1 (to S) will approach zero and the speed of ball 2 (to S) will approach V. If the masses of balls 1 and 2 become m_0 and m_V (to S), respectively, Eq. (4.14) reduces to

$$m_V = \frac{m_0}{\sqrt{1 - V^2/c^2}} \qquad (4.15)$$

The mass m_0 of an object measured by an observer with respect to whom it is at rest is called the rest mass or the proper mass of the object. The mass m_V of this object when moving with speed V with respect to the observer is called the improper mass. The improper mass is always greater than the proper mass. The subscript V is usually dropped from m.

The result of Eq. (4.14) can also be achieved from Eq. (4.13) by finding an appropriate relation between the upward (or downward) velocities $\Delta d/\Delta t$ and $\Delta d/\Delta t'$ of the two balls. The velocity transformation equations of Section 3.11 can be employed for this purpose. This has been left as an exercise to the reader.

Example 4.7 A rocket goes up with escape velocity. What is the percentage change in the mass of the rocket?

Escape velocity,

$$v_e = 11.2 \times 10^3 \text{ m/s}$$

$$\therefore \quad \frac{v_e}{c} = \frac{11.2 \times 10^3}{3.0 \times 10^8} = 3.7 \times 10^{-5},$$

and

$$\left(\frac{v_e}{c}\right)^2 \cong 10^{-9},$$

or

$$1 - \left(\frac{v_e}{c}\right)^2 \cong 1$$

Therefore, there will be no change in mass for all practical purposes.

We see from the above example that variation in mass cannot be detected even at as high velocities as the escape velocity. But with the help of modern accelerators extremely high speed particles can be produced. Experiments with these particles confirm Eq. (4.15). Violation of the principle of conservation of momentum has not been observed so far.

According to Eq. (4.15) when V approaches c, the mass tends to infinity. Thus it is impossible to make a material particle move with the speed of light. Then, what about light particles, photons? They have no rest mass. They have to keep running at the speed of light if they want to continue their existence. Neutrinos are another example.

Newton's law in the form $\mathbf{F} = m(\Delta \mathbf{v}/\Delta t)$ may not hold good, because during the accelerated motion of the particle, the mass may keep on changing. Further, it can be shown that the force vector and the acceleration vector may have different directions. In relativity the proper form of this law is:

$$\mathbf{F} = \frac{\Delta \mathbf{p}}{\Delta t} \qquad (4.16)$$

Discussion and Problems

4.1 Show that the law of conservation of momentum holds good in all inertial frames.

4.2 A man is walking in a gravity-free space. What will he do to take a turn ?

4.3 Vehicles stop on applying brakes. Does this phenomenon violate the principle of conservation of momentum?

4.4 A trolley of 50 kg is running on rails at a speed of 10 m/s. A boy of 40 kg jumps vertically into the trolley. What is the change in the speed of the trolley?

4.5 A tank of water is running on frictionless rails and leaks at the bottom. Discuss its motion.

4.6 In the first second of its flight, a rocket ejects 1/50 of its mass with a relative velocity of 1 km/s. What is the acceleration of the rocket?

4.7 A radioactive nucleus initially at rest decays by emitting an electron and an antineutrino at right angles to each other. The momentum of the electron is 1.5×10^{-23} kg-m/s, and that of antineutrino is 6.25×10^{-23} kg-m/s. Find the magnitude and direction of the momentum of the recoiling nucleus.

4.8 A projectile is fired from a gun placed at an angle of 60° with horizontal and with a muzzle speed of 500 m/s. At the highest point of its flight the projectile explodes into two fragments of equal mass. If one fragment comes back to the gun, how far away from the gun does the other fragment land?
[Hint: Motion of the centre of mass of the projectile is not affected by the explosion.]

4.9 A machine-gun fires 50 g bullets at a speed of 1200 m/s. The gunner holding the machine-gun in his hands, can exert an average force of 200 N against the gun. Determine the maximum number of bullets he can fire per minute.

4.10 Two particles A and B of mass 1.0 kg each have velocities $\mathbf{v}_A = 10\mathbf{i} + 20\mathbf{j}\ \text{ms}^{-1}$ and $\mathbf{v}_B = -5\mathbf{i} + 5\mathbf{j}\ \text{ms}^{-1}$ respectively before collision. The velocity of A after collision is : $\mathbf{v}_A' = -5\mathbf{i} + 5\mathbf{j}\ \text{ms}^{-1}$. What is the velocity of B after collision ?

4.11 A 10 kg body has been accelerated from 10 m/s to 20 m/s by a constant force of 10 N. Calculate the impulse given to the body.

4.12 Eighty per cent of the mass of a 16 ton rocket is fuel. If the fuel is ejected as exhaust gases with an average velocity of 1.5 km/s with respect to the earth's surface, calculate the final velocity of the rocket. Assume that the rocket is located in gravity free space.

4.13 For how long a force of 15,000 N will have to be applied to a 50,000 kg rocket to increase its speed from 20,000 m/s to 30,000 m/s? If the exhaust velocity is 3,000 m/s, at what rate the fuel of the rocket should burn to develop the above thrust?

4.14 A 20 kg rocket mounted on a launching pad can exhaust gases at the speed of 2 km/s. At what rate must it burn the fuel to get an acceleration of 20 m/s^2?

4.15 From the law of conservation of momentum, prove that both the earth and the moon rotate with the same angular velocity about their centre of mass. Consider earth-moon as an isolated system.

4.16 One steel ball of 10 g is moving towards east with a speed of 10 m/s, and another steel ball of 20 g is moving towards north with a speed of 5 m/s. What is the velocity of the centre of mass ? What will be the velocities of these balls for an observer sitting at the centre of mass?

4.17 Does the centre of mass of a solid always lie within the body?

4.18 Two blocks of 10 kg and 5 kg have been placed at the middle of a smooth platform of length 100 m and mass 100 kg. This platform can move frictionlessly. An explosion pushes the two blocks apart and the 5 kg block goes off with a speed of 8 m/s along the length of the platform. These blocks after reaching the ends of the platform get stuck up there. Assume the platform to have uniform width and thickness. (i) Where is the centre of mass of the whole system after 5 sec ? After 10 sec ? After 15 sec ? (ii) How far has the platform moved ?

4.19 Show that Newton's second law in the form–*the rate of change of momentum of a body is equal to the impressed force*–is invariant under Galilean transformation.

The above form of Newton's second law remains unchanged even under Lorentz transformation. But the law in the form $\mathbf{F} = m\mathbf{a}$ though invariant under Galilean transformation, does not hold good under Lorentz transformation. Thus $\mathbf{F} = \Delta\mathbf{p}/\Delta t$ is the proper form of one of the basic laws of physics.

4.20 A grenade when falling vertically down with speed 40 ms^{-1} explodes into two equal fragments. Immediately after explosion one fragment goes up with speed 40 ms^{-1}. (i) What is the distance travelled by the centre of mass in next two seconds? (ii) What are the distances travelled by the fragments in these two seconds?

4.21 The linear momenta of a system of particles in inertial frames S and S' are $\mathbf{P}$ and $\mathbf{P}'$, respectively. Show that:

$$\mathbf{P} = \mathbf{P}' + M\mathbf{V},$$

where M is the total mass of the system and $\mathbf{V}$ is the velocity of S' with respect to S. Use Galilean transformations.

4.22 For a system of particles the motion about the centre of mass is known as the internal motion. How much is the internal linear momentum in C-frame?

4.23 Show that for calculating the centre of mass of two bodies, their masses may be treated as concentrated at their respective centres of mass.

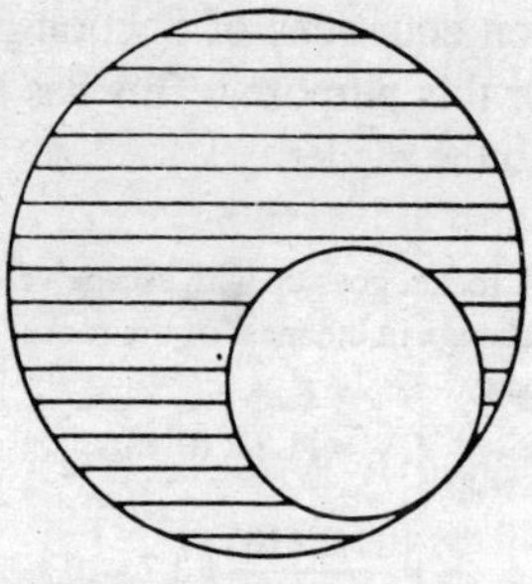

Fig. 4.12 *For Problem 4.24*

4.24 A circular sheet of diameter 10 cm has been cut out of a circular sheet of diameter 20 cm as shown in Fig. 4.12. Find the centre of mass of the sheet left behind.
[Hint: Use the results of Problem 4.23 and the arguments of symmetry.]

Chapter 5

Work, Power and Energy

Why do we take food? The obvious answer is that we supplement the energy used by us. We use energy in moving our limbs, in using our muscles for doing various jobs, in the movement of blood, etc. Body also radiates energy. The food is a fuel for the machinery of our body, and the jobs we do are the results of the transfer of energy from the form of food to the muscular action. In running the trains, the machinery, and the factories we use coal, oil and petroleum etc. as fuels. In performing a job there is always a transfer of energy. Though we seem to be quite familiar with energy it is very hard to define it. The only thing we can say is that it is *something* which is conserved.*

5.1 Work

The job performed is given by the term *work*, which is used in a very restricted sense in physics. If a force F is applied on a body and the body moves a distance s in the direction of the force [Fig. 5.1(a)], the work done on the body is, by definition:

$$W \equiv Fs \qquad (5.1)$$

If this force makes an angle θ with its direction of motion [Fig. 5.1(b)], the work will be:

$$W = (F\cos\theta)s \qquad (5.2)$$

In general, the work of a force is equal to the product of the displacement of its point of application and the component of the force in that direction.

*For an interesting discussion on this topic refer to R.P. Feynman, R.B. Leighton & M. Sands, *The Feynman Lectures on Physics*, Vol. I (Addison-Wesley, 1963).

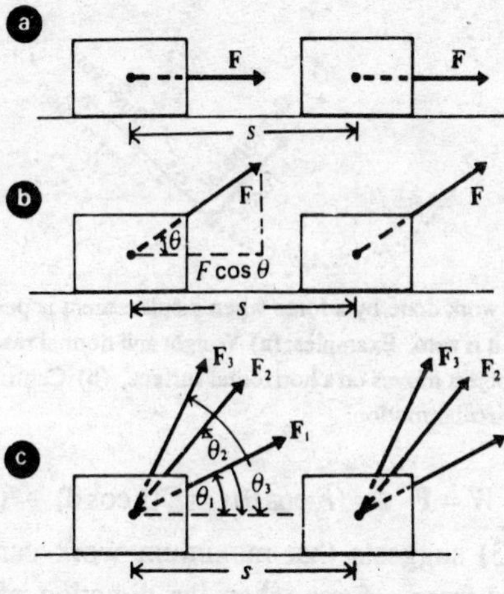

Fig. 5.1 *Work done by a force.* **(a)** *Force and displacement are in the same direction.* **(b)** *Force and displacement make an angle θ with each other.* **(c)** *Work done by more than one forces is equal to the sum of the work done by individual forces.*

Instead of one single force, there may be a number of forces $F_1, F_2, F_3, \ldots$, inclined respectively at angles $\theta_1, \theta_2, \theta_3, \ldots$, to the direction of motion [Fig. 5.1(c)]. The work of each force on the body will be different, viz:

$$W_1 = (F_1\cos\theta_1)s, \quad W_2 = (F_2\cos\theta_2)s,$$
$$W_3 = (F_3\cos\theta_3)s, \ldots$$

and the total work on the body is just the algebraic sum of $W_1, W_2, W_3, \ldots$, i.e.

$$W = (F_1\cos\theta_1 + F_2\cos\theta_2 + F_3\cos\theta_3 + \ldots)s$$

Thus the works of different forces on the body add up like scalar quantities. If we represent the distance moved by the body by a vector **s**, the work done on the body is represented as:

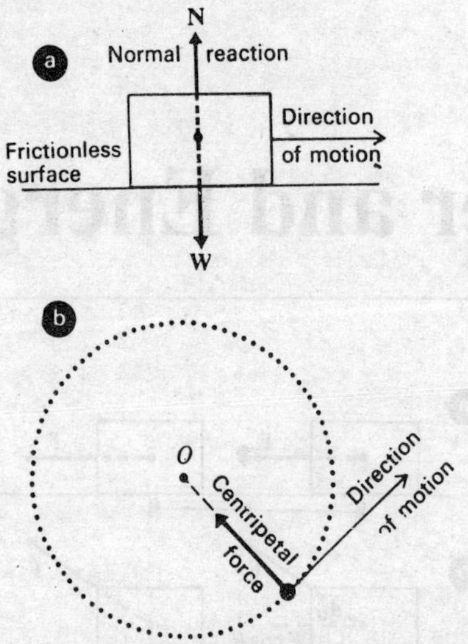

Fig. 5.2 work done by a force when displacement is perpendicular to it is zero. Examples: **(a)** Weight and normal reaction when an object moves on a horizontal surface, **(b)** Centripetal force in circular motion.

$$W = \mathbf{F} \cdot \mathbf{s} = (F \cos\theta)s = F(s\cos\theta) \qquad (5.3)$$

Eq. (5.3) suggests that maximum work can be obtained from a force when the direction of the force is the same as that of the displacement. When the force acting on a body is perpendicular to the direction of motion, the work done is zero. When a body slides on a frictionless horizontal surface no work is done by either the weight **W** of the body or the normal force **N** [Fig. 5.2(a)], because both the forces are perpendicular to the direction of motion. In a circular motion the body moves along the circumference and the only force is the centripetal force acting along the radius [Fig. 5.2(b)]. Here again the force is acting perpendicular to the direction of motion.

Question: Can you estimate the amount of work done when a satellite completes its orbit around the earth (neglect the air resistance)?

Will there be any work done by the earth in revolving around the sun?

Eq. (5.3) also suggests that the work is positive whenever the force makes an acute angle with the

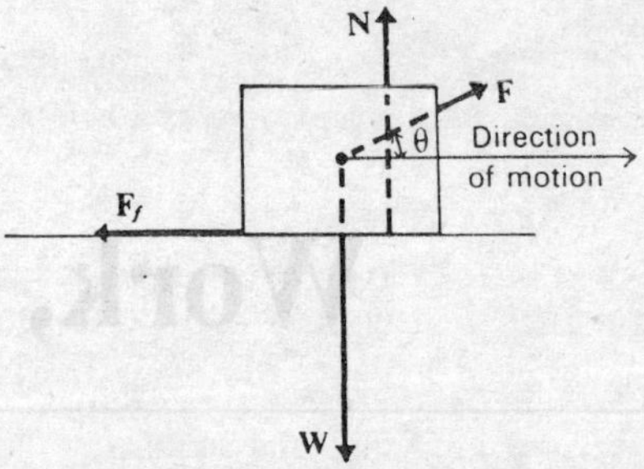

Fig. 5.3 *Work done by a frictional force is negative*

direction of motion. When the force makes an obtuse angle the work is negative. All those forces which try to oppose the motion of a body do a negative work on the body. Let us find the work on a body moving on a rough horizontal surface (Fig. 5.3). The forces **W** and **N** are perpendicular to the direction of motion. The force **F** acts at an angle θ and the frictional force $\mathbf{F}_f$ is opposite (at 180°) to the direction of motion. If the displacement vector is given by **s**, the work done on the body is:

$$W = \mathbf{F}\cdot\mathbf{s} + \mathbf{W}\cdot\mathbf{s} + \mathbf{N}\cdot\mathbf{s} + \mathbf{F}_f\cdot\mathbf{s} \qquad (5.4)$$

$$= Fs\cos\theta + 0 + 0 + \mathbf{F}_f s\cos 180°$$

$$= (F\cos\theta - F_f)s,$$

$$\text{or} \quad W = (F\cos\theta - \mu_k N)s$$

where μ_k is the coefficient of kinetic friction.

The work done by the frictional force on a body is always negative because the direction of the frictional force is always opposite to the direction of motion.

The unit of work is joule and it is defined as the amount of work done when a constant force of 1 N exerted on a body moves it by a distance of 1 metre in the direction of the force. Therefore:

$$1 \text{ joule} = 1 \text{ newton} \times 1 \text{ metre}$$

$$= 10^5 \text{ dyn} \times 10^2 \text{ cm} = 10^7 \text{ dyn cm}$$

$$= 10^7 \text{ erg}$$

Erg is the unit of work in the cgs system. In the British Engineering system the unit of work is foot-pound.

Question: A coolie carries a load from one place to another on a horizontal surface. Has he done any work on the load?

5.2 Graphical Representation

If a constant force F acts on a particle in the direction x and moves it from x_1 to x_2, the work is given by $F \times (x_2 - x_1)$, which is the area under the F-x curve [Fig. 5.4(a)]. If the force is not constant, the F-x curve may be of the form shown in Fig. 5.4(b). To calculate the total work done we divide the total displacement into a large number of small equal segments, Δx. Consider the small displacement Δx from x_1 to $x_1 + \Delta x$. We can consider the force to be constant, F_1, during this interval. The work done is:

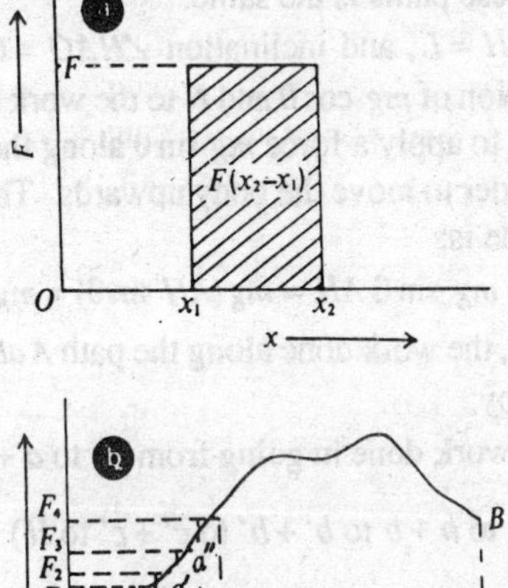

Fig. 5.4 When force and displacement are in the same direction, work done is equal to the area under the force-displacement curve. **(a)** *Constant force,* **(b)** *Continuously varying force.*

$$\Delta W_1 = F_1 \Delta x = \text{area of the small rectangular strip}$$
$$AacC$$

Similarly, the work done for displacement from $x_1 + \Delta x$ to $x_1 + 2\Delta x$ is:

$$\Delta W_2 = F_2 \Delta x = \text{area of the small rectangular strip}$$
$$aa'c'c$$

Continuing in this way we can find the work done for all the intervals. The total work done in moving from x_1 to x_2 is then given by:

$$W = F_1 \Delta x + F_2 \Delta x + F_3 \Delta x + \ldots..$$

= area of the rectangular strips:

$$AacC + aa'c'c + a'a''c''c' + \ldots..$$

= area under the F-x curve from x_1 to x_2 (5.5)

Hence, even when the force is varying along the direction of motion of a body, the work done can be conveniently given by the area under the F-x curve.

If a spring is stretched so that its end point moves to a position x [Fig. 5.5(a)], the force exerted by the spring on the agent stretching it is

$$\mathbf{F} = -k\mathbf{x} \quad (5.6)$$

where k is a constant called the force constant of the spring*.

The direction of the force is always opposite to the displacement of the end point. To stretch a spring we must exert a force kx on the spring. As we continue to stretch the spring we will have to increase the force in proportion to the stretching. If we plot this force as a function of x, we get a straight line [Fig. 5.5(b)]. From this plot it is clear that the work done in stretching the spring from $x = 0$ to $x = x$ is given by:

$$W = \text{ area of the triangle } OxA$$

$$= \frac{1}{2} x \times kx = \frac{1}{2} kx^2 \quad (5.7)$$

and the work done in stretching the spring from a point $x = x_1$ to a point $x = x_2$ is given by:

***It will hold good only for a limited magnitude of the force. Beyond a certain limit, the extension will be much more than given by Eq. (5.6). Verify this relation in the laboratory.**

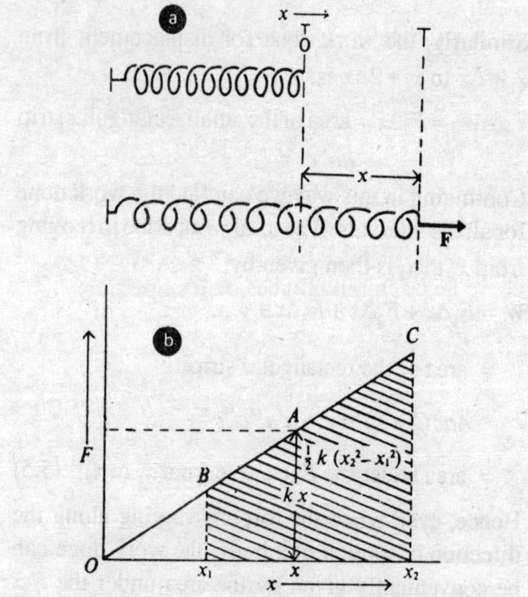

Fig. 5.5 (a) *Spring being stretched outwards* (b) *Work done in stretching the spring*

$$W_{12} = \text{ area of the trapezium } Bx_1x_2C$$

$$= \frac{1}{2}(x_2+x_1)k \times (x_2-x_1) = \frac{1}{2}k(x_2^2-x_1^2)$$

$$= \frac{1}{2}kx_2^2 - \frac{1}{2}kx_1^2, \quad (5.8)$$

or $\quad W_{12} = f(x_2) - f(x_1), \quad (5.9)$

where $\quad f(x) = \frac{1}{2}kx^2$

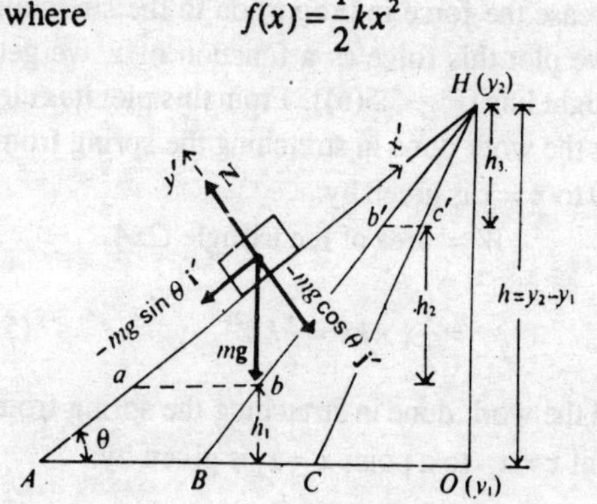

Fig. 5.6 *Work done in taking a body from one point to another in the gravitational field is independent of the path taken.*

5.3 Work in Gravitational Field

Suppose we want to lift a body vertically upward by a distance h (OH in Fig. 5.6). We will have to apply a force against the gravitational pull $m\mathbf{g} = \mathbf{F}_g$, which is acting vertically downward. The work necessary for lifting the body is:

$$\Delta W = \mathbf{F} \cdot \Delta \mathbf{s}$$

$$= -\mathbf{F}_g \cdot \Delta \mathbf{s}$$

$$= F_g \Delta s$$

$$= mgh \quad (5.10)$$

Question: What is the work done by the gravitational force?

We can reach from the ground to the point H by sliding the body on various frictionless inclined planes (AH, BH or CH) also. We may slide the body along the frictionless path $Aabb'c'H$. We find that the work done in carrying the body along any of these paths is the same.

Let $AH = L$, and inclination $\angle HAO = \theta$. The contribution of $mg\cos\theta$ and N to the work is zero. We have to apply a force $mg\sin\theta$ along the plane AH in order to move the body upwards. Thus, the work done is:

$$W = mg\sin\theta\, AH = mg(AH\sin\theta) = mgh$$

similarly, the work done along the path $Aabb'c'H$ is given by:

$W =$ work done in going from (A to $a + a$ to $b + b$ to $b' + b'$ to $c' + c'$ to.H)

$$= mgh_1 + 0 + mgh_2 + 0 + mgh_3$$

$$= mg(h_1 + h_2 + h_3) = mgh$$

If the height of O from some reference plane is y_1 and that of H is y_2, then $h = y_2 - y_1$. We can therefore write

$$W = mg(y_2 - y_1) = mg\,y_2 - mg\,y_1$$

$$= f(y_2) - f(y_1) \quad (5.11)$$

Here $f(y) = mgy$ provided we consider g to be constant. This holds true if h is small as compared with the radius of the earth (Example 5.3.)

Question: Does the work done in raising a box to a higher storey in a building depend on how fast it is raised?

From Eq. (5.11) we see that the work done in going from point O to point H depends only on the position coordinates of the two points, and has nothing to do with the manner in which the body has been lifted from O to H. Similarly, in stretching a spring from one point to another, as we saw in Eq. (5.9), the work done depends only on the coordinates of the two points. We shall come back to this point in Section 5.7 when we discuss conservative and non-conservative forces.

5.4 Kinetic Energy

What change takes place in a particle when work is done on it? When a positive work is done some quantity must increase, while a negative work on the body should reduce something. If a body is slid on a rough surface with a certain velocity, it stops after covering some distance. In this case the frictional force does a negative work on the body and the velocity becomes zero. The negative work has reduced a quantity which depends upon the velocity of the body. As soon as this quantity is zero (velocity zero), no more negative work can be done on the body. Let us find this function of velocity which is always positive.

Let a sliding body of mass m moving with velocity v come to rest after travelling a distance s under the influence of force F_f. Acceleration $a = F_f/m$ is given by:

$$0 - v^2 = 2as,$$

or

$$a = -v^2/2s,$$

therefore

$$F_f = -mv^2/2s$$

and work done by this force on the body is:

$$F_f s = -\frac{1}{2}mv^2$$

The negative sign means that F_f and s are in opposite directions. This force could also be due to agents other than friction, e.g. gravitational force on bodies going away from the earth. In all such cases the work done on the agent which stops the body will be given by $mv^2/2$. In other words, when the velocity of the body was v, a total work of $mv^2/2$ could be taken out of it before it stops. Or

$$T = \frac{1}{2}mv^2 \tag{5.12}$$

represents the capacity of a body to do work. This quantity T is called the *kinetic energy* of the body, and is always positive.

Suppose a body of mass m, moving with a velocity $\mathbf{v}_1$, is acted upon by a force $\mathbf{F}$ for a time Δt. Suppose this force changes the velocity to $\mathbf{v}_2$. We then have the impulse:

$$\mathbf{F}\Delta t = m(\mathbf{v}_2 - \mathbf{v}_1) \tag{5.13}$$

During the time Δt the displacement of the body is:

$$\mathbf{s} = \frac{\mathbf{v}_2 + \mathbf{v}_1}{2}\Delta t \tag{5.14}$$

where $(\mathbf{v}_1 + \mathbf{v}_2)/2$ is the average velocity during the interval Δt. The work done on the body is:

$$W = \mathbf{F} \cdot \mathbf{s} = m\frac{(\mathbf{v}_2 - \mathbf{v}_1)}{\Delta t} \cdot \frac{1}{2}(\mathbf{v}_2 + \mathbf{v}_1)\Delta t$$

$$= \frac{1}{2}m(\mathbf{v}_2 - \mathbf{v}_1) \cdot (\mathbf{v}_2 + \mathbf{v}_1)$$

$$= \frac{1}{2}m(|\mathbf{v}_2|^2 - |\mathbf{v}_1|^2)$$

$$= T_2 - T_1 \tag{5.15}$$

Thus *the work done by a resultant external force on a body is equal to the increase in the kinetic energy of the body.* When the work is done on a body its kinetic energy increases, and when the work is done by the body (work done on the body is negative) its kinetic energy decreases. In moving on a frictional surface (or moving against gravity) a body has to do work against friction (or gravity) and it continues to move till its whole kinetic energy is consumed. (Friction and gravity are two different types of forces. See Section 5.7.)

Question: Is any work being done on a car moving with a constant speed along: (i) a straight level road, and (ii) a circular path on a horizontal plane?

Example 5.1 A force of 50 N acts on a 10 kg roller initially at rest on a frictionless surface. The roller travels 10 m while the force acts. Find the work done on the roller. Also calculate the kinetic energy of the roller.

$$\text{Work done on the roller} = 50\ \text{N} \times 10\ \text{m} = 500\ \text{J}$$

$$\text{Acceleration of the roller} = \frac{50\ \text{N}}{10\ \text{kg}} = 5\ \text{ms}^{-2}$$

If v is the final velocity of the roller :

$$v^2 - 0 = 2 \times 5 \times 10,$$

or $$v = 10\ \text{ms}^{-1}$$

Therefore the kinetic energy of the roller:

$$T = \frac{1}{2} \times 10\ \text{kg} \times (10)^2\ \text{m}^2\ \text{s}^{-2} = 500\ \text{J}$$

Hence, the gain in kinetic energy of the roller equals the work done on it, *or work is a measure of transfer of energy.*

Example 5.2 Show that the motional energy of a system of particles is the kinetic energy if the whole mass were concentrated at the centre of mass plus the kinetic energy of motion of the particles with respect to the centre of mass. Energy measured in the *C*-frame is called the internal energy.

Let us take a collection of particles as shown in Fig. 5.7. Here O is the origin of the *L*-frame and C is the origin of the *C*-frame. If $\mathbf{r}_i$ and $\mathbf{r}_i'$ are the position vectors of the *i*th particle of mass M_i with respect to O and C, respectively, and $\mathbf{R}$ is the position vector of the centre of mass with respect to O, then:

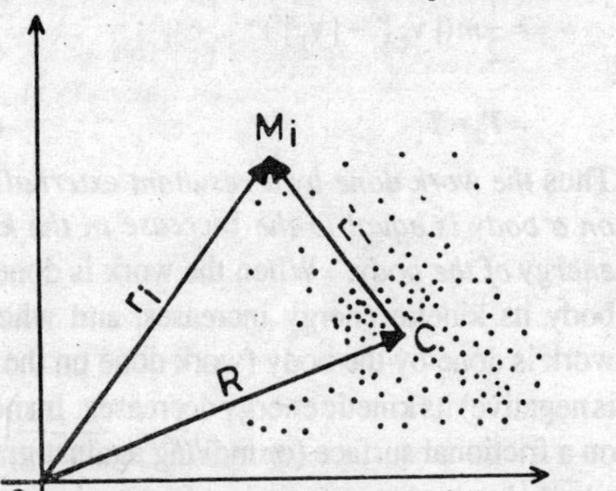

Fig. 5.7 *For Example 5.2*

$$\mathbf{r}_i = \mathbf{r}_i' + \mathbf{R},$$

or $$\mathbf{v}_i = \mathbf{v}_i' + \mathbf{v}$$

Here $\mathbf{v}_i$ and $\mathbf{v}_i'$ are the velocities of M_i with respect to O and C, respectively, and $\mathbf{v}$ is the velocity of the centre of mass with respect to O.

If T_L is the kinetic energy of the system in the *L*-frame,

$$T_L = \sum_i \frac{1}{2} M_i v_i^2$$
$$= \sum \frac{1}{2} M_i (\mathbf{v}_i \cdot \mathbf{v}_i)$$
$$= \sum \frac{1}{2} M_i (\mathbf{v} + \mathbf{v}_i') \cdot (\mathbf{v} + \mathbf{v}_i')$$
$$= \sum_i \frac{1}{2} M_i \left(v^2 + v_i'^2 + 2\mathbf{v} \cdot \mathbf{v}_i'\right)$$
$$= \sum_i \frac{1}{2} M_i v^2 + \sum_i \frac{1}{2} M_i v_i'^2 + \mathbf{v}.\sum_i M_i \mathbf{v}_i'$$

The last term on the right hand side is zero because $\sum_i M_i \mathbf{v}_i' / \sum_i M_i$ gives the velocity of the centre of mass with respect to the centre of mass, and is obviously zero. Therefore,

$$T_L = \frac{1}{2} M v^2 + T_C$$

where M is the total mass of the system, and T_C is the kinetic energy of the system due to the motion around the centre of mass. The term $\frac{1}{2} M v^2$ is clearly the kinetic energy of the system if all the mass were kept at the centre of mass and were moving along with it.

5.5 Potential Energy

If we throw a body upwards, its velocity, and hence its kinetic energy, decreases continuously. The kinetic energy first becomes zero, but then reappears as the body comes back. It means that its kinetic energy was stored somewhere temporarily. If this body has a mass m and is thrown up with a velocity v_1 from a height h_1, and if it has a velocity v_2 at a height h_2, we have

$$v_2^2 - v_1^2 = -2g(h_2 - h_1),$$

or $$\frac{1}{2} m v_1^2 - \frac{1}{2} m v_2^2 = mgh_2 - mgh_1,$$

or the kinetic energy lost $= f(h_2) - f(h_1)$

We know from Section 5.3 that this loss in KE is independent of the path followed. This change in kinetic energy has changed the configuration, or the position, of the particle. Thus, this temporary decrease in kinetic energy has been stored in the new configuration of the system. This stored energy is called the configurational or the *potential energy* and is a function of space coordinates only. In this example we can write the potential energy of a particle at height h as mgh. Then: $U_1 = mgh_1$, and $U_2 = mgh_2$ give us:

$$T_1 - T_2 = U_2 - U_1,$$

or $$T_1 + U_1 = T_2 + U_2 = \text{constant} \qquad (5.16)$$

Thus *the sum of the kinetic energy and the potential energy is constant.* The term $E = (T + U)$ is called the total mechanical energy, and Eq. (5.16) is called the law of conservation of mechanical energy. We can write Eq. (5.16) as:

$$\Delta T + \Delta U = 0,$$

or $$\Delta T = -\Delta U \qquad (5.17)$$

Therefore, the change in kinetic energy is equal and opposite to the change in potential energy. But we know that increase in kinetic energy is given by:

$$\Delta T = F\,\Delta x$$

where F is the force in the direction of Δx. This gives:

$$F\,\Delta x = -\Delta U, \qquad (5.18)$$

or $$F = -\Delta U / \Delta x \qquad (5.19)$$

Thus *the rate of decrease of potential energy with respect to the space in a certain direction gives the force at a point in that direction.* (See also Section 19.2.)

If the potential energy at different points in space is different, the displacement from one point to another involves work. If we move from one point to another without involvement of work, the potential energy at the two points must be the same. We can define ΔU (the difference between potential energy of a particle at two different points) as the amount of work done on the particle in moving from one point to another against the gravitational, elastic or electric force, as the case may be. Consider a system of two mutually interacting particles. The zero of the potential energy of the system can be safely taken as the energy when there is no interaction between the two bodies. If we take the earth and a body then the potential energy will be taken as zero when the earth and the body are infinite distance apart.

Example 5.3 A point mass m is kept at a distance R from another point mass M. Calculate the work done in carrying m to infinity.

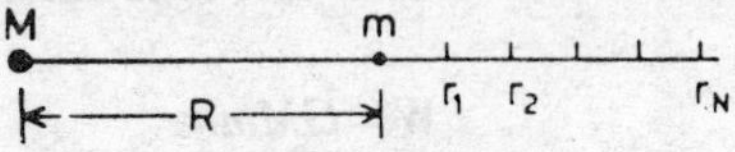

Fig. 5.8 *For Example 5.3*

Gravitational force between the two masses is:

$$\mathbf{F}_g = -\frac{GMm}{r^2}\hat{r}$$

When we want to take m away from M we will have to apply a force against the gravitational force of attraction. The force to be applied is not constant but decreases as the distance from M increases. Let us find the work in moving the body from r_1 to r_N (Fig. 5.8). Divide the distance $r_1 r_N$ into a number of small intervals $r_1 r_2, r_2 r_3, \ldots$. The work W_{12} in moving the body from r_1 to r_2 can be given by:

$$\frac{GMm}{r_1^2}(r_2 - r_1) > W_{12} > \frac{GMm}{r_2^2}(r_2 - r_1),$$

because $\mathbf{F} = -\mathbf{F}_g$, and F_g at $r_1 > F_g$ at r_2

If r_1 and r_2 are close to each other, we can write:

$$W_{12} \cong \frac{GMm}{r_1 r_2}(r_2 - r_1)$$

$$= GMm\left(\frac{1}{r_1} - \frac{1}{r_2}\right)$$

Similarly, the work done in moving from r_2 to r_3 is:

$$W_{23} = GMm\left(\frac{1}{r_2} - \frac{1}{r_3}\right)$$

and from r_3 to r_4

$$W_{34} = GMm\left(\frac{1}{r_3} - \frac{1}{r_4}\right)$$

....

....

and so on

$$W_{N-1,N} = GMm\left(\frac{1}{r_{N-1}} - \frac{1}{r_N}\right)$$

Hence, total work done in moving m from r_1 to r_N is:

$$W = W_{12} + W_{23} + ... + W_{N-1,N}$$
$$= GMm\left(\frac{1}{r_1} - \frac{1}{r_N}\right)$$

If we take $r_1 = R$ and $r_N = \infty$, we get*:

$$W = GMm\left(\frac{1}{R} - \frac{1}{\infty}\right)$$

or $$W = GMm/R$$

We have seen in Example 5.3 that the work done on a mass m in taking it from a point at a distance r from another mass M to infinity is given by:

$$W = GMm/r \tag{5.20}$$

At infinity we have taken the potential energy to be zero. What then is the potential energy of m at r? We know that:

Energy at starting point + work done on the body = energy at the final point

Hence, when a body is taken to infinity (with zero velocity):

$$\text{Potential energy at } r + \frac{GMm}{r} = 0,$$

or $$U(r) = -\frac{GMm}{r} \tag{5.21}$$

If $m = 1$, then $U(r) = -(GM/r)$, and is called the gravitational potential due to M at a distance r. Thus the gravitational potential is nothing but the potential energy per unit mass.

Example 5.4 Calculate the gravitational potential due to a uniform spherical shell of mass M and radius R at a point outside it, and at a point inside it.

Let the geometry be as shown in Fig. 5.9. Let us first calculate the gravitational potential at a point P outside the shell. Take a very thin ring on the shell, with centre N on line OP, as shown. Every element of mass on this ring is at the same distance, say x, from P. As the radius of the ring is $R \sin\theta$, its circumference is $2\pi R \sin\theta$, and surface area is $(2\pi R \sin\theta) \times R\Delta\theta$. If σ is the mass per unit surface area of the shell, then mass of the ring will be $2\pi R^2 \sin\theta\Delta\theta \times \sigma$. Therefore, the gravitational potential at P due to this ring is:

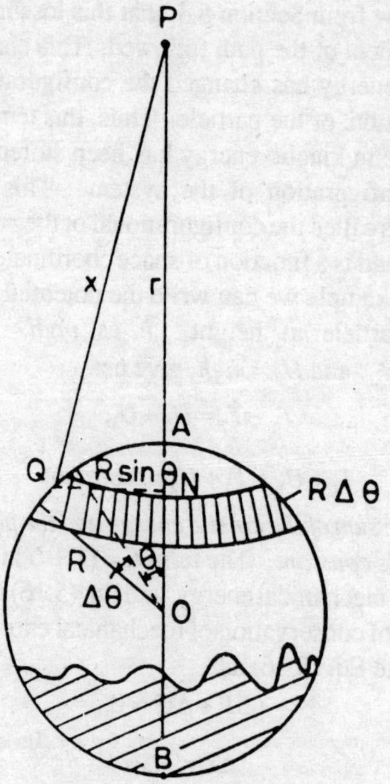

Fig. 5.9 *For Example 5.4*

$$\Delta U = \frac{-G \times 2\pi\sigma R^2 \sin\theta\Delta\theta}{x} \tag{5.22}$$

When this expression is summed up over the whole shell, the result is*:

$$U = \frac{-GM}{r} \tag{5.23}$$

---

*Using calculus : $\Delta W = \mathbf{F} \cdot d\mathbf{s} = -\mathbf{F}_g \cdot d\mathbf{r} = F_g dr$

$$\therefore\ W = \int_R^{\infty} \frac{GMm}{r^2} dr = GMm\left[\frac{-1}{r}\right]_R^{\infty} = \frac{GMm}{R}$$

*In triangle POQ.

$$x^2 = r^2 + R^2 - 2rR\cos\theta$$
$$\therefore\quad 2 \times x dx = 2rR \sin\theta d\theta$$

Hence, Eq. (5.22) becomes :

$$dU = -G\frac{2\pi\sigma R}{r} dx$$

For covering the whole shell one notes that x varies from $(r-R)$ at A to $(r+R)$ at B. Therefore, for the complete shell :

$$U = \int_{r-R}^{r+R} \frac{-G2\pi\sigma R}{r} dx = -\frac{G2\pi\sigma R}{r} |x|_{r-R}^{r+R}$$
$$= \frac{-G2\pi\sigma R \times 2R}{r}$$
$$= \frac{-GM}{r}$$

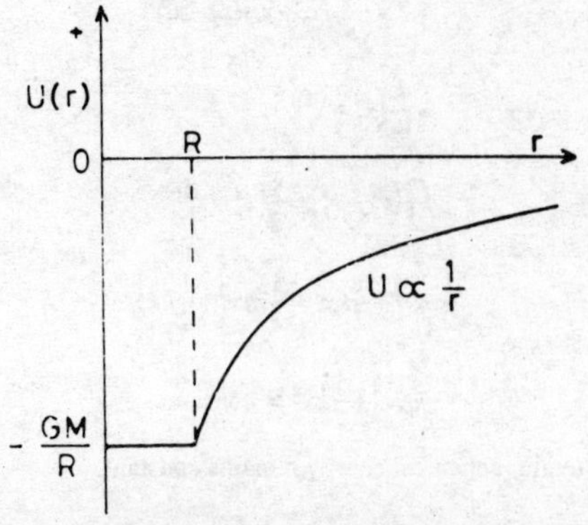

Fig. 5.10 *Variation of gravitational potential due to a uniform spherical shell as a function of distance from its centre*

This result is quite significant in that for calculating the gravitational potential due to a uniform shell at a point outside it, the mass of the shell can be treated to be kept at its centre.

The gravitational potential at a point inside the shell is given as*:

$$U = \frac{-GM}{R} \tag{5.24}$$

Therefore, the gravitational potential inside a uniform shell is constant and is equal to its value at the surface of the shell. The variation of U as a function of distance from the centre of the shell is shown in Fig. 5.10.

Example 5.5 Find the escape velocity of a particle at the surface of the earth.

The escape velocity is that minimum velocity with which the particle should be thrown away from the earth so that it does not return. In other words its upward velocity is such that the kinetic energy of the particle is just sufficient to overcome the gravitational pull.

Divide the earth into concentric shells and assume that in a particular shell the material density is uniform. As a point at the surface of the earth is outside all the shells, the mass of every shell can be treated to be kept at the centre of the earth for calculating potential at its surface.

Let v be the escape velocity of the object of which the mass is m. Therefore, using Eq. (5.16):

$$T + U = 0 + 0,$$

or

$$\frac{1}{2}mv^2 = GMm/R,$$

or

$$v^2 = 2GM/R$$

Hence, escape velocity $v = \sqrt{\dfrac{2GM}{R}}$

Here M is the mass of the earth.

*In this case x varies from $(R - r)$ at A to $(R + r)$ at B.

After substituting the values of various constants v comes out to be about 11.2 km s^{-1}. If the upward velocity of the particle is greater than this value, it will have some velocity at infinity.

Let us consider a massless spring of force constant k with a block of mass m attached to the free end of the spring. What happens when the block is moved from $x = 0$ to $x = x_0$ (Fig. 5.11)? Assume that the block slides on a frictionless horizontal surface. In stretching the spring from $x = 0$ to $x = x_0$ a work $\frac{1}{2}kx_0^2$ is done on the spring, thereby increasing the total energy of the system to $\frac{1}{2}kx_0^2$. When the mass is released it is pulled towards the point $x = 0$ and again the free end of the spring comes to $x = 0$. But this time the block has a velocity v_0 at $x = 0$. The energy $\frac{1}{2}kx_0^2$ is now

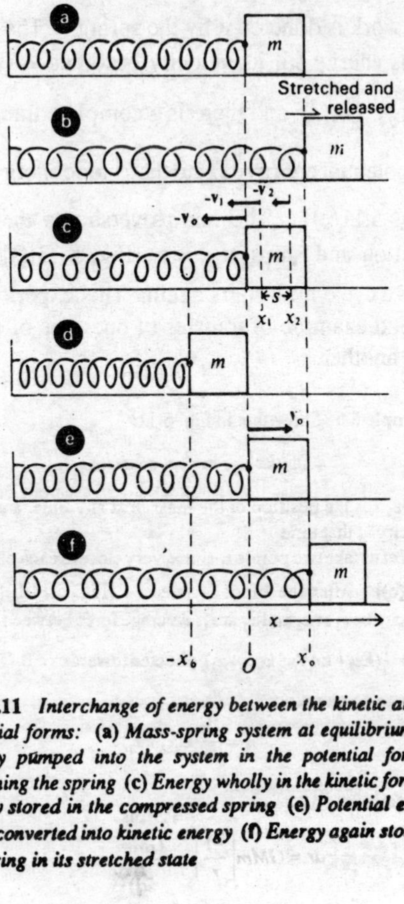

Fig. 5.11 *Interchange of energy between the kinetic and the potential forms:* **(a)** *Mass-spring system at equilibrium* **(b)** *Energy pumped into the system in the potential form by stretching the spring* **(c)** *Energy wholly in the kinetic form* **(d)** *Energy stored in the compressed spring* **(e)** *Potential energy again converted into kinetic energy* **(f)** *Energy again stored in the spring in its stretched state*

in the form of $\frac{1}{2}mv_0^2$. The energy $\frac{1}{2}kx_0^2$ is due to the change in the configuration of the system, while the energy $\frac{1}{2}mv_0^2$ is due to the motion of the mass m. In Fig. 5.11(b) the whole energy is due to change in the configuration, i.e. the potential energy. In Fig. 5.11(c) as the configuration comes back to the original state [Fig. 5.11(a)], the whole energy is kinetic. The block passes through the point $x = 0$ and compresses the spring. In compressing the spring, work is done on the spring and the block continues to lose its kinetic energy till it comes to rest [Fig. 5.11(d)]. Meanwhile, the configuration of the system is again changed, a work $\frac{1}{2}kx_0^2$ is done on the spring, and the potential energy becomes $\frac{1}{2}kx_0^2$. Now the spring pushes the block and work is done on it by the spring. The block gains energy till its velocity becomes v_0 (kinetic energy $\frac{1}{2}mv_0^2$), and there is a complete transfer of the potential energy $\frac{1}{2}kx_0^2$ to the kinetic energy $\frac{1}{2}mv_0^2$ [Fig. 5.11(e)]. The mass overshoots the mean position and stops at $x = x_0$ [Fig. 5.11(f)]. The above cycle now starts again. This experiment is a good example of transfer of one type of energy into another.

Example 5.6 Show that in Fig. 5.11:

$$\frac{1}{2}kx^2 + \frac{1}{2}mv^2 = \text{constant}$$

where x is the position of the mass m at any time, and v is its velocity at that time.

Let us take two points x_1 and x_2 very close to each other [Fig. 5.11(c)]. After the block is released [Fig. 5.11(b)], let the velocity be v_2 at x_2 and v_1 at x_1. Average force between x_2 and x_1 is $F = \frac{1}{2}(kx_2 + kx_1) = \frac{1}{2}k(x_2 + x_1)$, directed towards $x = 0$. Therefore average acceleration:

$$a = \frac{1}{2m}k(x_2 + x_1)$$

From equations of motion:

$$v_1^2 - v_2^2 = 2as$$

$$= 2 \times \frac{k}{2m}(x_2 + x_1)(x_2 - x_1)$$

$$= \frac{k}{m}(x_2^2 - x_1^2),$$

or

$$\frac{m}{2}(v_1^2 - v_2^2) = \frac{k}{2}(x_2^2 - x_1^2),$$

or

$$\frac{1}{2}mv_1^2 + \frac{1}{2}kx_1^2 = \frac{1}{2}mv_2^2 + \frac{1}{2}kx_2^2,$$

i.e.

$$\frac{1}{2}mv^2 + \frac{1}{2}kx^2 = \text{constant}$$

or the total mechanical energy remains constant.

Example 5.6 also shows that $\frac{1}{2}kx^2$ is the potential energy of the system when m is at x. From the expression:

$$\frac{1}{2}(mv_1^2 - mv_2^2) = \frac{1}{2}(kx_2^2 - kx_1^2) \qquad (5.25)$$

we see that *the change in kinetic energy is equal and opposite to the change in potential energy of the system.*

5.6 Collision of Two Particles

It has been seen in Section 4.2 that the linear momentum of two particles before and after a collision is the same. Not only this, the momentum is conserved even during a collision. But similar statements may or may not hold good for the kinetic energy. Details of the energy aspects can be worked out if the nature of the forces which play their part during a collision is precisely known. Though we often do not know about these forces, the effects of the collisions can be studied in some cases. The collisions can belong to one of the two categories: (1) elastic, and (2) inelastic. Elastic collisions are those in which the total kinetic energy before and after the collision remains unchanged. If the kinetic energy is not conserved, the collision is termed as inelastic.

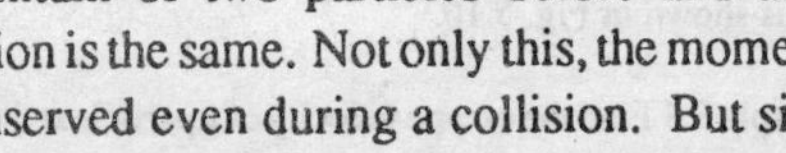

Fig. 5.12 *Head-on collision of two bodies*

Let us study one-dimensional elastic collision of two balls A and B (Fig. 5.12). Collisions of this type are called head-on collisions. In such a collision the line joining the centres of mass of the two balls coincides with their velocities. Here, the balls move along the line xx' both before and after the collision. If v_A' and v_B' are the velocities of A and B after the collision, then from the law of conservation of momentum:

$$\Delta p_A + \Delta p_B = 0,$$

or $$m_A(v_A' - v_A) + m_B(v_B' - v_B) = 0 \quad (5.26)$$

From the law of conservation of kinetic energy:

$$\Delta T_A + \Delta T_B = 0,$$

or $$\frac{1}{2}m_A(v_A'^2 - v_A^2) + \frac{1}{2}m_B(v_B'^2 - v_B^2) = 0, \text{ or}$$

$$m_A(v_A' + v_A)(v_A' - v_A) + m_B(v_B' + v_B)(v_B' - v_B) = 0 \quad (5.27)$$

From Eqs. (5.26) and (5.27), we get:

$$(v_A' + v_A) - (v_B' + v_B) = 0 \quad (5.28)$$

Substituting the value of v_B' from this equation in Eq. (5.26):

$$v_A' = \frac{v_A(m_A - m_B) + 2m_B v_B}{m_A + m_B} \quad (5.29a)$$

Similarly,

$$v_B' = \frac{v_B(m_B - m_A) + 2m_A v_A}{m_A + m_B} \quad (5.29b)$$

From Eq. (5.29) we see that if the bodies have equal masses, then after an elastic collision their velocities will interchange. The discussion of other cases has been left for you.

Example 5.7 **A ball of mass m_A moving with velocity v_A strikes another ball of mass m_B which is stationary. If the collision is elastic and head-on, calculate the fraction of the kinetic energy transferred to the second ball.**

From Eq. (5.29):

$$v_A' = \frac{v_A(m_A - m_B)}{m_A + m_B},$$

and $$v_B' = \frac{2m_A v_A}{m_A + m_B}$$

Initial kinetic energy of ball $A = \frac{1}{2}m_A v_A^2$

Kinetic energy transferred to ball $B = \frac{1}{2}m_B v_B'^2$

Therefore the fraction of kinetic energy transferred to ball B

$$= m_B\left(\frac{2m_A v_A}{m_A + m_B}\right)^2 \times \frac{1}{m_A v_A^2}$$

$$= \frac{4m_A m_B}{(m_A + m_B)^2} \quad (5.30)$$

Find out this fraction for various mass ratios (m_A/m_B) and plot a graph. From the graph find out the value of m_A/m_B for which the energy transfer is maximum.

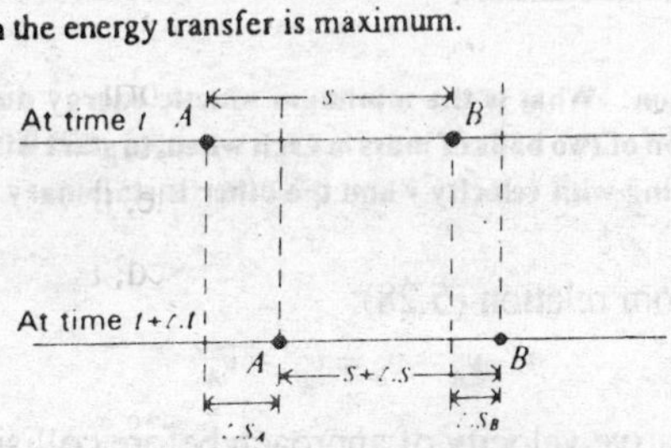

Fig. 5.13 *Mechanism of an elastic collision*

Let us now investigate the condition under which the collision between two bodies A and B will be elastic. Let the distance between the two, when A approaches B, at time t be s and at time $t + \Delta t$ be $s + \Delta s$ (Fig. 5.13). Let the force of repulsion be $F(s)$. Therefore, gain in the kinetic energy of $B = F(s)\Delta s_B$, and loss in the kinetic energy of $A = F(s)\Delta s_A$.

Therefore, total gain in kinetic energy is:

$$= F(s)(\Delta s_B - \Delta s_A) = F(s)\Delta s$$

At the beginning as A approaches B, $\Delta s_A > \Delta s_B$. Thus $F(s)\Delta s$ will be negative and there will be a net loss of kinetic energy. The total loss in kinetic energy up to the distance of minimum approach (zero for contact collisions) is given by $\Sigma F(s)\Delta s$, i.e. by the area under the $F(s) - s$ curve. After this, A starts trailing behind B and soon the collision is over. In this period Δs will be positive. If, now, the repulsive force at distance s is $F'(s)$, the total gain in kinetic energy will be given by $\Sigma F'(s)\Delta s$, i.e. by the area under the $F'(s) - s$ curve. If $F(s) = F'(s)$ for all values of s, then these two areas will be equal and the initial loss in kinetic energy will be completely compensated afterwards.

Hence, *if the force between the colliding particles is the same at a particular distance whether the particles are approaching each other or are receding, i.e. if the force is a function of separation only, then the collision will be elastic.* If it is not so, then the initial loss in kinetic energy will not be equal to the gain afterwards, and thus the kinetic energy will not be conserved. The collision will then be inelastic.

Question: What is the minimum kinetic energy during a collision of two balls of mass m each when, to start with, one is moving with velocity v and the other is stationary ?

From relation (5.28):

$$v_A - v_B = v_B' - v_A'$$

$\therefore$ relative velocity of approach before collision = relative velocity of separation after collision, or

relative velocity after collision = − (relative velocity before collision) (5.31)

In general, the collisions are not elastic, and we define a new quantity, the *coefficient of restitution* e, to explain the inelastic collisions as:

relative velocity after collision

= − e (relative velocity before collision),

i.e. $$v_A' - v_B' = -e(v_A - v_B),$$

or $$e = -(v_A' - v_B')/(v_A - v_B) \qquad (5.32)$$

For $e = 1$, the collision is perfectly elastic, and when $e = 0$, (i.e. when balls stick together after the collision) the collision is perfectly inelastic. These are the two extreme cases. In general the value of the coefficient of restitution is between 0 and 1*.

---

*In atomic collisions when one (or both) of the atoms is in the excited state, the kinetic energy after collision may be greater than the kinetic energy before collision.

Example 5.8 Find the total distance travelled by a ball dropped on a floor from a height h, before it comes to rest. The coefficient of restitution is e.

When a ball drops from a height h it hits the floor with a velocity $v = \sqrt{2gh}$. The velocity after collision is $v' = -ev$, and the ball rises to a height h_1, given by:

$$ev = \sqrt{2gh_1}, \quad \text{or} \quad e = \sqrt{h_1/h}, \quad \text{or} \quad h_1 = e^2 h$$

After going upto a height h_1, the ball will again hit the floor with a velocity v' when it drops back and will rebound with a velocity $v'' = -ev'$, and we again have $e = \sqrt{h_2/h_1}$. Thus for successive heights:

$$e = \sqrt{h_1/h} = \sqrt{h_2/h_1} = \sqrt{h_3/h_2} = \ldots\ldots$$

Total distance travelled

$$= h + 2h_1 + 2h_2 + 2h_3 + \ldots\ldots$$

$$= h[1 + 2(e^2 + e^4 + e^6 + \ldots\ldots)]$$

$$= h\left(1 + \frac{2e^2}{1-e^2}\right) = h\frac{1+e^2}{1-e^2}$$

Let us consider what happens when one particle is projected at a target particle which is at rest. We can always choose a frame in which one particle is at rest. If the direction of projection is along the line joining the centres of the two particles, the particles will move on the same line even after collision. This type of collision has already been discussed. What will happen if the perpendicular distance between the direction of motion of the projectile and the line parallel to it through the centre of mass of the target is b (Fig. 5.14), and not zero ? The particles will move in different directions after collision. These directions will depend on the *impact parameter* b, initial velocities, masses of the particles and the interaction force between them. The particles need not *touch* each other for a collision (interaction) to take place if we consider gravitational, electromagnetic, or nuclear interactions. *The interaction between the particles deflects them from their initial courses.* From conservation of momentum we have:

$$m_A \mathbf{v}_A = m_A \mathbf{v}_A' + m_B \mathbf{v}_B'$$

Taking the initial line of motion as the x-axis and the directions of motion of m_A and m_B to be inclined to it at θ_1 and θ_2 respectively, we can write:

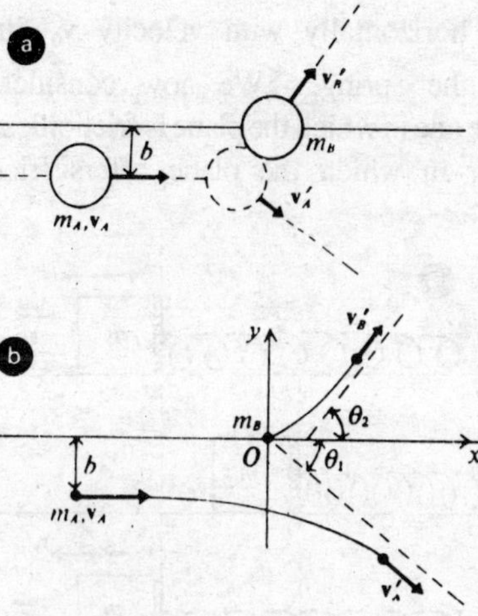

Fig. 5.14 *Oblique collision of two bodies*

$$m_A v_A \mathbf{i} = m_A(v_A' \cos\theta_1 \mathbf{i} - v_A' \sin\theta_1 \mathbf{j})$$

$$+ m_B(v_B' \cos\theta_2 \mathbf{i} + v_B' \sin\theta_2 \mathbf{j})$$

where **i** and **j** are unit vectors along the *x*-axis and *y*-axis respectively. Thus, equating components along **i** and **j** we get:

$$m_A v_A = m_A v_A' \cos\theta_1 + m_B v_B' \cos\theta_2,$$

and $\quad 0 = m_A v_A' \sin\theta_1 - m_B v_B' \sin\theta_2, \quad (5.33)$

Also we have, $\frac{1}{2} m_A v_A^2 = \frac{1}{2} m_A v_A'^2 + \frac{1}{2} m_B v_B'^2$ for an elastic collision.

If we know the initial conditions (m_A, m_B, v_A), we are left with four unknowns $(v_A', v_B', \theta_1, \theta_2)$ but only three equations relating them. We can determine the motion after collision only if we specify a value for one of these quantities.

Example 5.9 Show that in an elastic collision between two particles *A* and *B* of equal mass, one of which is initially at rest, the recoiling particles either move along the same line, or move off at right-angle to each other.

Let the mass of each particle be *m* and initial velocity of *A* be $\mathbf{v}_A$. After collision, let the velocities be $\mathbf{v}_A'$ and $\mathbf{v}_B'$ at inclinations of θ_1 and θ_2 to the initial direction of $\mathbf{v}_A$. The angle between the directions of the particles after collision is $(\theta_1 + \theta_2)$.

From the conservation of momentum:

$$m\mathbf{v}_A' + m\mathbf{v}_B' = m\mathbf{v}_A,$$

or $\quad \mathbf{v}_A' + \mathbf{v}_B' = \mathbf{v}_A \quad (5.33a)$

From the conservation of kinetic energy, as the collision is elastic:

$$\frac{1}{2} m v_A'^2 + \frac{1}{2} m v_B'^2 = \frac{1}{2} m v_A^2,$$

or $\quad v_A'^2 + v_B'^2 = v_A^2 \quad (5.33b)$

Squaring Eq. (5.33a),

$$(\mathbf{v}_A' + \mathbf{v}_B') \cdot (\mathbf{v}_A' + \mathbf{v}_B') = \mathbf{v}_A \cdot \mathbf{v}_A,$$

or $\quad v_A'^2 + v_B'^2 + 2\mathbf{v}_A' \cdot \mathbf{v}_B' = v_A^2 \quad (5.33c)$

We get from Eqs. (5.33b) and (5.33c)

$$\mathbf{v}_A' \cdot \mathbf{v}_B' = 0,$$

or $\quad v_A' v_B' \cos\alpha = 0 \quad (5.33d)$

where $\alpha (= \theta_1 + \theta_2)$ is the angle between the velocity vectors of the particles after collision. Hence, either $v_A' = 0$, or $\alpha = 90°$. The first condition holds good for a head-on collision, and the second for any oblique collision.

Exercise: Draw the vector diagram for momenta and check the above conclusions from it when read in conjunction with Eq. (5.33b).

5.7 Conservative and Non-conservative Forces

In Section 5.3, we saw that the work done in raising a body of mass *m* through a height $h (= OH$ in Fig. 5.6) is *mgh*, and this work is independent of the path followed by the body. There we have considered frictionless inclined planes only. If the planes are rough such that the coefficient of friction between the body and the surface is μ_k, we can see that the work done in raising the body to the height *h* will be different along different paths.

Let us again consider the motion along the plane *AH* (Fig. 5.15). The forces *N* and $mg \cos\theta$ have no contribution to the work when the body moves along the plane. In moving the body up the plane work is done against a total force of $(mg \sin\theta + \mu_k mg \cos\theta)$. Hence, the work done on the body in raising it through the height $h (= OH)$ along *AH* is:

$$W = (mg \sin\theta + \mu_k mg \cos\theta) AH$$

$$= mg(1 + \mu_k \cot\theta) OH \quad (5.34)$$

Thus W depends upon the angle of inclination of the plane.

In the first case the only force acting is that of gravity and the work on the body depends only upon the positions of the points between which the body moves. In the second case, in addition to the gravitational force, the force of friction is also acting and the work in moving between two points depends also upon the path along which the particle moves.

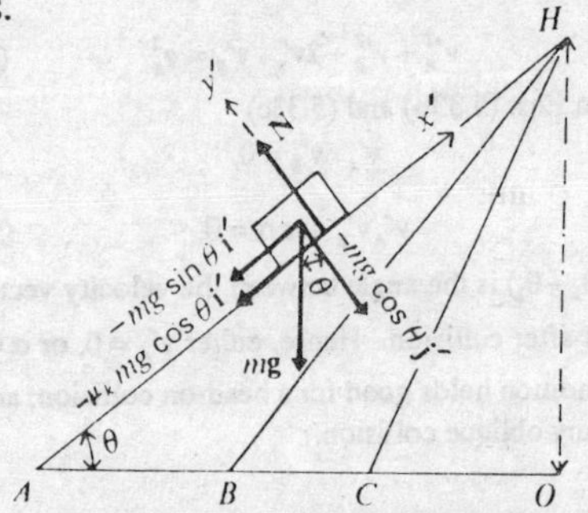

Fig. 5.15 *In the presence of the frictional force the work done in the gravitational field in taking a body from one point to another does depend upon the path taken.*

Question: Will the concept of a potential function, a function of space only, have any meaning here?

A force is called a *conservative force* if the work done by it on a particle that moves between two points depends only on the positions of these points and not on the path followed. A force is called *non-conservative* if the work done by that force on a particle that moves between two points depends on the path followed by the particle between these points. Thus the gravitational force is conservative and the frictional force is non-conservative.

Let us take another example. From Eqs. (5.8) and (5.9) we can see that the work done on a spring to stretch it from one point to another depends only upon the positions of the two points $[W = k(x_2^2 - x_1^2)/2]$. It has nothing to do with the manner in which the spring has been stretched. Thus, the elastic force of a spring is also conservative.

Imagine a massless spring of a force constant k with one end fixed to a wall. Let us fire a block of mass m horizontally with velocity $\mathbf{v}_0$ directly towards the spring. We now consider two cases–the one in which the plane is frictionless, and the other in which the plane offers frictional resistance.

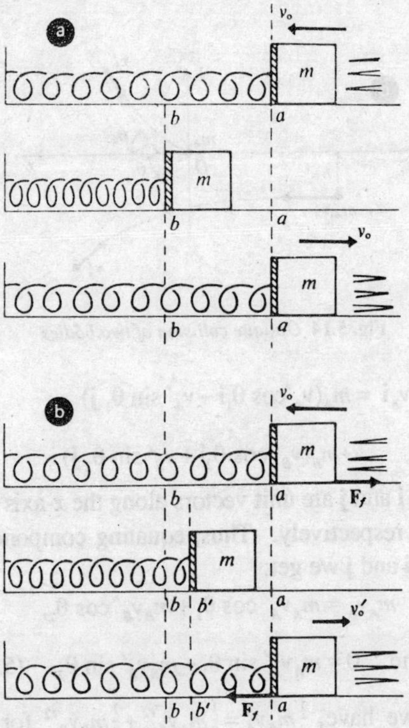

Fig. 5.16 **(a)** *Total mechanical energy is conserved in the presence of conservative forces (here the elastic force of the spring), and work done along a closed path is zero.* **(b)** *In the presence of non-conservative forces the mechanical energy is not conserved and the work done in a closed path is not zero.*

In the first case [Fig. 5.16(a)], after the block touches the spring, the speed and hence the kinetic energy of the block decreases until finally the block is brought to rest by the action of the elastic force of the spring. The kinetic energy $mv_0^2/2$ is converted into the potential energy $kx_0^2/2$ of the spring. The block now reverses its motion as the compressed spring expands. It gains speed and kinetic energy, and when it comes back to its initial position, it has the same speed but in opposite direction. The potential energy $kx_0^2/2$ is converted back into kinetic energy $mv_0^2/2$. This is because the area under the $F(x)-x$ curve is the same in both cases. We can show (see Example 5.6) that in the process of compression and expansion the value of $\left(\frac{1}{2}mv^2+\frac{1}{2}kx^2\right)$ remains constant at any instant. Thus, the total mechanical energy of the system is always conserved. We also see that the work done on the block, or the spring, when the spr`g moves from x to x_0 and then from x_0 to x is zero. Hence, *a force is conservative if the total mechanical energy of the system is conserved, or the work done by the force on a particle that moves through any round trip (or along a closed path) is zero.* When a stone is thrown vertically upward, it comes back to the ground with the speed with which it was thrown. The work done on the stone by the gravitational force during the round trip is zero. Also at any time during the trip the total mechanical energy $\left(\frac{1}{2}mv^2+mgh\right)$ remains constant (Eq. 5.16).

In the second case [Fig. 5.16(b)], a force of friction F_f is continuously acting on the block in a direction opposite to its direction of motion. Here, the area under the $F(x)-x$ curve in the onward journey will be greater than the area in the backward journey because the retarding force in the onward journey is $(kx+F_f)$, while the accelerating force in the backward journey is only $(kx-F_f)$. Thus the negative work on the block is greater than the positive work, and the result is the loss in the energy of the block. Therefore, a force is non-conservative if the work done by the force on a particle that moves through a round trip is not zero. Also the kinetic energy is not completely converted into potential energy, and later the potential energy is not converted into kinetic energy completely*.

*Frictional force is also called a dissipative force because there is a decrease in the mechanical energy. In some other cases there is an increase in mechanical energy (see Sec. 26.8).

Thus the total mechanical energy $\left(\frac{1}{2}mv^2+\frac{1}{2}kx^2\right)$ is not conserved in this case. *The mechanical energy of a body is conserved only when no non-conservative force acts on it.*

Question: Is the force applied by the air to a moving body a conservative force?

5.8 Power

Let us now consider the time involved in doing work. If we have to lift M kg of water out of a well of depth h, the amount of work to be done is Mgh joules. One person may use a bucket of 5 kg capacity and repeat the process $M/5$ times. Another person may use a bucket of 10 kg capacity and repeat the process $M/10$ times only. Thus the second person finishes the job earlier and we term him more powerful. In physics also, an agency which does the work faster is said to be more powerful. *The rate at which the work is done by an agency is called the power developed by that agency.* We define the average power delivered by an agent as:

$$\langle P\rangle=\frac{\text{total work done}}{\text{time taken}}=\frac{W}{t} \qquad (5.35)$$

The instantaneous power delivered by an agent is given by:

$$P=\lim_{\Delta t\to 0}\left(\frac{\Delta W}{\Delta t}\right) \qquad (5.36)$$

where ΔW is the work done in a small interval of time Δt. If the power is constant in time, then $P=\langle P\rangle$. In general, we talk of average power. In SI the unit of power is 1 watt = 1 joule/sec. As a watt is a small quantity, we generally use in practice a unit of kilowatt (=1000 watt). In the British Engineering system the power is given by horse-power (hp), where 1 hp = 746 watt.

When a force **F** acts on a body so as to produce in it a displacement $\Delta\mathbf{x}$ in time Δt, we can see that:

$$P=\frac{\mathbf{F}\cdot\Delta\mathbf{x}}{\Delta t}=\mathbf{F}\cdot\mathbf{v} \qquad (5.37)$$

where $v = \Delta x/\Delta t$ is the velocity of the body. A new unit of work can also be defined from Eq. (5.35). The work done by an agent working constantly at the rate of 1 kW in 1 hour is called a kilowatt-hour.

$$1\ \text{kWh} = 60\times 60\times 10^3\ \text{J} = 3.6\times 10^6\ \text{J}$$

One kilowatt-hour is the basic unit of energy used in consuming electricity in India.

Example 5.10 A boy whose mass is 48 kg climbs a vertical rope of length 7.5 m in 12 s. What is the power output of the boy during the climb?

Work done by the boy $= Mgh$

$$= 48\times 9.8\times 7.5\ \text{J} = 3528\ \text{J}$$

Power output during the climb $= \dfrac{3528}{12}\ \text{Js}^{-1}$

$$= 294\ \text{W} = 0.294\ \text{kW}$$

5.9 Mass and Energy

In Section 4.5 we have studied a perfectly elastic collision. Let us now consider a perfectly inelastic collision. Two balls in S' frame, each of rest mass m_0, collide head on with equal speeds v' and stick to each other [Fig. 5.17(a)]. From the law of conservation of momentum it is clear that the compound ball will be at rest. Let the total mass (rest) of the compound ball be M_0.

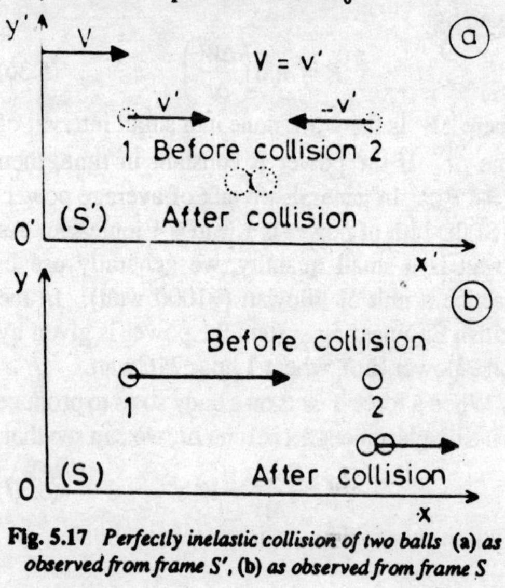

Fig. 5.17 *Perfectly inelastic collision of two balls* **(a)** *as observed from frame S',* **(b)** *as observed from frame S*

From the point of view of an observer in S the velocities of balls 1 and 2 along x-x' axis will be (by using the velocity addition theorem),

$$v_1 = \frac{v'+V}{1+\frac{v'V}{c^2}} = \frac{2v'}{1+\frac{v'^2}{c^2}}, \tag{5.38}$$

and

$$v_2 = \frac{-v'+V}{1-\frac{v'V}{c^2}} = 0, \tag{5.39}$$

respectively. The velocity of the compound ball along x-x' direction in this frame will be

$$v = V = v' \tag{5.40}$$

Applying the law conservation of momentum in frame S along x-x' direction, we get

$$m_1 v_1 + 0 = Mv \tag{5.41}$$

where m_1 and M are the masses of ball 1 and the compound ball, respectively, in frame S. Using Eqs. (5.38), (5.40) and the law of variation of mass with velocity, Eq. (5.41) becomes:

$$\frac{m_0}{\sqrt{1-\frac{v_1^2}{c^2}}} v_1 = \frac{M_0}{\sqrt{1-\frac{v^2}{c^2}}} v,$$

or

$$\frac{m_0}{\sqrt{1-\frac{(2v'/c)^2}{(1+(v'^2/c^2))^2}}} \times \frac{2v'}{\left(1+\frac{v'^2}{c^2}\right)} = \frac{M_0}{\sqrt{1-\frac{v'^2}{c^2}}} \times v',$$

or

$$\frac{m_0\left(1+\frac{v'^2}{c^2}\right)}{\left(1-\frac{v'^2}{c^2}\right)} \times \frac{2v'}{\left(1+\frac{v'^2}{c^2}\right)} = \frac{M_0 v'}{\sqrt{1-\frac{v'^2}{c^2}}},$$

or

$$M_0 = \frac{2m_0}{\sqrt{1-\frac{v'^2}{c^2}}} \tag{5.42}$$

Thus the rest mass of the compound ball is not equal to the sum of the rest masses of balls 1 and 2, but is greater than this. The kinetic energy of the balls (before collision) has been converted into mass. For $v' \ll c$,

$$M_0 = 2m_0\left(1-\frac{v'^2}{c^2}\right)^{-1/2}$$

$$\cong 2m_0\left(1+\frac{v'^2}{2c^2}\right),$$

or $$M_0 \cong 2m_0 + \frac{\text{kinetic energy}}{c^2} \qquad (5.43)$$

Thus the rest mass M_0 of the compound ball is equal to the sum of the rest masses of the individual balls plus a contribution proportional to their kinetic energy. It can be shown that Eq. (5.43) holds good for high velocities also.* As the kinetic energy of the compound ball (in S' frame) is zero, the decrease in kinetic energy equals the increase in mass multiplied by c^2. Or:

$$\Delta T = \Delta mc^2 \qquad (5.44)$$

Now, quite many things may happen in an inelastic collision. The constituent particles may keep on moving after sticking together, like a proton and a neutron. They may get excited to higher energy states. They might also be having some potential energy. If the balls were initially macroscopic particles, then the thermal energy will also be produced, which of course is the random motion of the constituent molecules. Therefore, we do not talk about various forms of rest energy. We only say that when a particle has rest mass m_0, its total rest energy is m_0c^2. If it is moving, its energy is mc^2. Hence the total energy of an object of mass m and moving with speed v is:

$$E = mc^2 = \frac{m_0c^2}{\sqrt{1-\frac{v^2}{c^2}}} \qquad (5.45)$$

Question: Show that mc^2 has the dimensions of energy.

If mass m_0 is converted into energy we can get work (equal to m_0c^2) out of the released energy. This is the basic source of atomic energy. In other words, we can say that the mass m_0 has a potentiality to produce an energy $E_0 = m_0c^2$. This energy may therefore be called the *potential energy of the mass content.*

From Eq. (5.45) we see that when the speed of a particle reaches c *(or* equals the speed of light) the mass and energy become infinite. Hence, no particle can attain the speed of light if its rest mass is non-zero. In other words, particles moving with the velocity of light have zero rest mass. Such particles are photons (light particles) and neutrinos.

The amount of energy which can be produced by conversion of mass into energy can be estimated very easily. If one gram of a substance is converted into energy, we get an amount of 9×10^{13} joules of energy, which is sufficient to evaporate about thirty thousand tons of ice into steam.

Example 5.11 A 10,000 metric ton block of copper is heated so that its temperature increases by 900°C. What is the increase in the mass of the block?

Specific heat of copper can approximately be taken as 0.1 cal/g-°C

$$Q = 10000\times100\times10\times1000\times0.1\times900 \text{ cal}$$
$$= 9\times10^{11} \text{ cal}$$

This heat has been absorbed by the copper block. As heat is energy and we know that energy and mass are related by the relation $m = E/c^2$, there will be an increase in mass. The energy absorbed by the copper block in joules is:

$$E = 9\times10^{11}\times4.2 \text{ J}$$

Hence, increase in mass is:

$$\Delta m = \frac{9\times4.2\times10^{11}}{9\times10^{16}} \text{ kg} = 4.2\times10^{-5} \text{ kg}$$
$$= 4.2\times10^{-2} \text{ g} = 42 \text{ mg}$$

$$*T = \int_0^r \mathbf{F}\cdot d\mathbf{l}$$

Let us take the motion along the x-axis. Then:

$$T = \int_0^r \frac{d(mv)}{dt}dx = \int_0^r d(mv)\frac{dx}{dt}$$

$$= \int_0^r (mdv + vdm)v = \int_0^r (mvdv + v^2dm)$$

From Eq. (4.15):

$$m^2c^2 - m^2v^2 = m_0^2c^2$$

$$\therefore \qquad 2mdmc^2 - 2mv^2dm - 2m^2vdv = 0,$$

or $$mvdv + v^2dm = c^2dm$$

$$\therefore \qquad T = \int_0^r c^2dm = c^2(m - m_0)$$

5.10 Binding Energy

The gravitational potential energy of a body in the earth's gravitational field is [Eq. (5.21)]:

$$U(r) = -GMm/r$$

A body of mass m at a distance r from the centre of earth ($r >$ radius of the earth) can remain in equilibrium state with it only when it is orbiting around the earth. Otherwise, it will fall on the earth and will not remain at a distance r. We have already seen that the velocity of a satellite at a distance r is given by:

$$v^2 = GM/r$$

The kinetic energy of the mass m orbiting at a distance r from the centre of the earth is:

$$T = \frac{1}{2}mv^2 = \frac{1}{2}(GMm/r) \qquad (5.46)$$

Hence, the total mechanical energy of the earth-satellite system is*:

$$E = T + U = -\frac{1}{2}(GMm/r) \qquad (5.47)$$

If we supply an energy equal to $(GMm/2r)$ the mechanical energy of the system will be zero, i.e. the two bodies will be at rest and separated by an infinite distance. In other words, the quantity $(GMm/2r)$ can be termed as the *binding energy* between the body and the earth. The binding energy between two bodies is defined as the *energy required just to separate them to infinite distance* (at infinite distance the mutual interaction and, hence, the potential energy is taken as zero). Thus it is numerically equal to E in Eq. (5.47) given above.

The planets are bound in the solar system under the influence of the gravitational field of the sun. (There is a small binding due to mutual attraction of planets, but this effect is very small in comparison with the effect of the sun's gravitational field.) The binding energy of planets is given by:

$$GMM_p/2R \qquad (5.48)$$

where M and M_p are the masses of the sun and the planet respectively, and R is the distance between them.

Binding energy of atoms and molecules: An atom consists of a positively charged nucleus surrounded by electrons. The electrostatic potential at any point due to a charge is also given in the same manner as the gravitational potential (see Section 31.3). If we consider a hydrogen atom, its nucleus of positively charged particle, proton, having a charge $+e = 1.602 \times 10^{-19}$ coulomb, and an orbiting electron of charge $-e$, we can show (on similar lines as in gravitational binding energy) that the binding energy of the electron and the proton is $e^2/8\pi\varepsilon_0 r$. The average separation between the electron and the proton (i.e. r) remains at about 5×10^{-11} m. Hence, the binding energy of the electron is about 22×10^{-19} J or 13.6 electron volts (eV). This new unit of energy, mostly used in atomic and nuclear physics, is given by:

$$1 \text{ eV} = 1.6 \times 10^{-19} \text{ J (see Section 19.6)}$$

The atoms do not in general remain independent. Different atoms bind with one another to form molecules. The molecular binding and chemical reactions involve the outer electrons of the atoms. Hence, molecular binding energies are smaller than the B.E. of the electron-nucleus system. When two or more atoms or molecules are combined to form a molecule, energy equal to their binding energy is released in the process. From exothermic chemical reactions we can easily calculate the binding energy of a molecule. Thus:

$$2H_2 + O_2 = 2H_2O + 116 \text{ kcal}$$

means that 116 kcal energy has been shared by $2N_A$ water molecules (where N_A is Avogadro number = 6.02×10^{23} mole^{-1}). Hence, energy released per molecule is:

$$E = \frac{4.2 \times 10^7 \times 1.16 \times 10^5}{2 \times 6.02 \times 10^{23}} = 4 \times 10^{-12} \text{ erg}$$

$$= 2.5 \text{ eV}$$

Therefore, the binding energy of a water molecule is 2.5 eV.

*The earth being very heavy as compared with the satellite, its velocity and hence its kinetic energy can be neglected.

Binding energy of nuclear particles: Nucleus of an atom, in a broad sense, is supposed to be constituted of two types of particles, the protons and neutrons. The simplest nucleus is deuteron consisting of one neutron and one proton. Now:

m_p = mass of proton = 1.00783 amu
m_n = mass of neutron = 1.00867 amu
m_d = mass of deuteron = 2.01410 amu

where 1 amu $\cong 1.66 \times 10^{-27}$ kg. We see that:

$$m_p + m_n (= 2.01650 \text{ amu}) > m_d$$

The difference of the mass is equivalent to an energy:

$$E = 0.00240 \times 1.66 \times 10^{-27} \times 9 \times 10^{16} \text{ J}$$

$$= 3.59 \times 10^{-13} \text{ J}$$

$$= 2.24 \times 10^{6} \text{ eV}$$

When a neutron and a proton combine to form a deuteron, this amount of energy is given out in the form of γ-radiation. If we want to disintegrate a deuteron into a proton and a neutron, we must supply an energy 2.2 MeV (1 MeV = 10^6 eV) to a deuteron nucleus. In general, for most of the nuclei the binding energy per nucleon is about 8 MeV. We see that the nuclear binding energies are much larger (10^6 times) as compared with the atomic binding energies. The nuclear binding energy is different for lighter nuclei and heavier nuclei. The difference in binding energies of various nuclei is the source of energy for an atom bomb and a hydrogen bomb. (The principles of atom bomb and hydrogen-bomb will be discussed later in Chapter 34.)

Discussion and Problems

5.1 In nuclear reactors, the moderators used for slowing down neutrons are hydrogenous matter. Why?

5.2 Labourers taking heavy materials to the top of a building under construction generally follow a path of lesser slope. Why?

5.3 A projectile of mass m has a speed v_1 when at a height h_1. What will be its speed when at a height h_2?

5.4 A linear elastic spring is compressed by 15 cm by a force of 15 N. (i) What is the spring constant? (ii) What will be the potential energy stored in it when compressed by 10 cm? (iii) What will be the spring constant of two such springs put in series, and put in parallel?

5.5 A 1.5 kg block is dropped from a height of 0.6 m onto a spring of force constant k = 1764 N/m. Find the maximum compression of the spring (neglect friction).

5.6 A linear spring, whose force constant k is 50 N/m, hangs vertically and supports a 1.0 kg mass at rest. The mass is now pulled down a distance 10 cm and released. (i) How high will it rise from the point of release? (ii) What will be its maximum velocity and at what point this will be attained?

5.7 A car of mass m is moving with a speed v on a level surface and brakes are suddenly applied. Show that the car will stop after covering a distance greater than $v^2/2g\mu_s$, where μ_s is the coefficient of static friction.

5.8 A block of 10 kg accelerates uniformly from rest to a speed of 2 ms^{-1} in 30 s. What is the work done on the block at the end of first 10 s? Neglect friction. What is the instantaneous power delivered at the end of the 10th second?

5.9 A block of mass 4 kg is dragged at a constant speed a distance 4 m along a horizontal plane with the help of a rope exerting a constant force of 8.4 N and making an angle of 60° with the horizontal. Compute: (i) the total work done on the block, (ii) the work done by the rope on the block, (iii) the work done by the friction on the block, (iv) the coefficient of kinetic friction between the block and the plane.

5.10 A car can move up the road having a slope of 1.0 ft rise for every 60 ft with a speed of 40 mph. The resisting force (due to friction and air) is equal to 1/30 of the weight of the car. How fast will the same car move down the hill with the same horse-power? Assume the resisting force to be independent of the speed of the car.

5.11 Two perfectly elastic balls of equal kinetic energy have head-on collision. Find the velocities after collision if one ball is four times heavier than the other.

5.12 A 1 m long spring with a force constant of 2000 N/m is compressed to half its size and tied to remain as such. If the spring breaks up into two equal parts, find the kinetic energy of each part.

5.13 A gas molecule having a speed of 400 ms^{-1} collides elastically with a similar molecule at rest and gets deflected through an angle 45°. Find the speed of each molecule, and the angle the second molecule makes with the original direction of motion.

5.14 Ballistic pendulums are used to find the velocities of bullets. A bullet of mass m and having horizontal velocity v strikes the bob of a pendulum and gets stuck into it. If the mass of the bob is M, length of the thread is l, and the pendulum rises to a height h, express the velocity of the bullet in terms of other known quantities. Is the kinetic energy conserved?

5.15 An earth satellite in a circular orbit has a kinetic energy T. How much more energy would be needed to make it escape from the earth?

5.16 Show that the gravitational potential due to a uniform solid sphere at a point inside it and at a distance r from its centre is given by:

$$U(r) = -\frac{GM}{2R^3}(3R^2 - r^2)$$

where R is the radius of the sphere and M is the mass. [Hint: Divide the sphere into concentric shells.]

5.17 Show that the kinetic energies T and T' of a mechanical system as measured in two inertial frames S and S', respectively (S' moving with velocity $\mathbf{V}$ relative to S), are related by:

$$T = T' + \frac{1}{2}MV^2 + \mathbf{V}\cdot\mathbf{P}'$$

where $\mathbf{P}'$ is the linear momentum of the system in S', and M is the total mass of the system. Take non-relativistic case.

5.18 (a) Show that for an object the velocity v, the momentum p, and total energy E are related as:

$$pc = Ev/c,$$

and

$$E^2 = p^2c^2 + m_0^2c^4$$

(b) Show that the momentum of a particle having zero rest mass is E/c, and its speed is c.

5.19 An electron is accelerated to have an energy ten times its rest energy. What is the velocity of the electron? What is the ratio of the relativistic kinetic energy and the kinetic energy evaluated using the classical expression $\frac{1}{2}m_0v^2$?

5.20 An elementary particle Σ^-, at rest in a certain frame of reference, decays spontaneously into two particles according to the equation $\Sigma^- \rightarrow \pi^- + n$. The rest masses are:

$$m_{\Sigma^-} = 2340m_e, \quad m_{\pi^-} = 273.2m_e,$$

$$m_n = 1838.65m_e, \text{ where } m_e \text{ is the mass of the electron}$$

(i) How much kinetic energy is generated in the process?
(ii) What are the velocities of π^- and n?
(Hint: $E^2 = m_0^2c^4 + p^2c^2$)

5.21 In the following reaction

$$p + {}^7_3Li \rightarrow \text{compound nucleus} \rightarrow \alpha + \alpha$$

$$\text{if } m_p = 1.6726\times10^{-27}\text{ kg}, \; m_{Li} = 11.6399\times10^{-27}\text{ kg}$$

and

$$m_\alpha = 6.6404\times10^{-27}\text{ kg}$$

what is the kinetic energy of each α-particle?

5.22 A ball strikes another ball, which has the same mass and is stationary, with zero error ($b = 0$). Show that as long as both balls are in motion, the kinetic energy will be less than the initial kinetic energy. When is the kinetic energy minimum?

5.23 Gravitational potential difference between any two points is equal to the work done in carrying a unit mass from one point to the other. Will this definition also hold good for work done in non-conservative force fields?

5.24 (i) When a man on skates pushes against a wall he starts moving. Where from has this kinetic energy come? Which forces have worked on him? Does it violate the law of conservation of momentum?
(ii) Engine of a moving car is stopped, and it comes to rest without slipping. Where has its kinetic energy gone?

5.25 Show that the form of the equations:

$$\mathbf{F} = m\mathbf{a}$$

$$\sum_{t=t_1}^{t_2} \mathbf{F}\Delta t = m\mathbf{u}_2 - m\mathbf{u}_1,$$

and

$$\sum_{t=t_1}^{t_2} \mathbf{F}\cdot\Delta\mathbf{r} = \frac{1}{2}mu_2^2 - \frac{1}{2}mu_1^2$$

remains invariant under Galilean transformation.

Chapter 6

Rotational Dynamics

6.1 Torque

Let us consider a plane sheet hinged at a point O (Fig. 6.1). If we want to rotate this sheet about an axis through the point O and perpendicular to it, what should we do? The obvious answer is to apply a force. Can we apply the force at any point, and in any direction? If we apply the force at the fixed point O, nothing will happen. The plane will not rotate even if we pull it at some other point, provided the direction in which we pull passes through O. As soon as the force is applied at a point P shifted from O, in a direction other than OP, rotation takes place. If we move farther from O the rotation becomes easier. At a given point P, the rotation is easiest if the force lies in the plane and is perpendicular to OP. Thus, in rotating a body the force itself is not the only important factor, but the direction of force and the distance at which it is acting from the axis are also of vital importance. Actually, the important quantity is the product of the force and the perpendicular distance from the axis of rotation. This product is called the moment of the force.

Let us find the moment of a force $\mathbf{F}$ (or torque) acting at a point P about a point O fixed in an inertial frame. Let the radius vector $\mathbf{r}(= OP)$ be inclined to $\mathbf{F}$ at an angle θ [Fig. 6.2(a)]. The magnitude of the moment of the force is, by definition:

$$\left.\begin{aligned} \tau = |\tau| &\equiv F \times OQ = rF \sin\theta \\ &= |\mathbf{r} \times \mathbf{F}| \end{aligned}\right\} \qquad (6.1)$$

Suppose OP is a weightless rod hinged at O. We can very easily see that the rod OP will rotate about

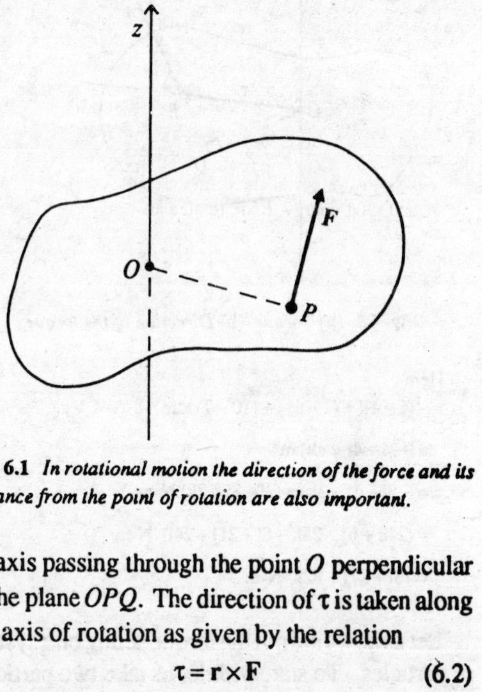

Fig. 6.1 *In rotational motion the direction of the force and its distance from the point of rotation are also important.*

an axis passing through the point O perpendicular to the plane OPQ. The direction of τ is taken along the axis of rotation as given by the relation

$$\tau = \mathbf{r} \times \mathbf{F} \qquad (6.2)$$

In Fig. 6.2(b) the force $\mathbf{F}$ is acting in such a manner that the rotation produced will move a right-handed screw in the positive z-direction, i.e., the rotation will be positive.

Question: If r and F lie in the same plane, will τ have any component in that plane?

Example 6.1 A force $\mathbf{F} = (7\mathbf{i} - 8\mathbf{j} + 2\mathbf{k})$ N is passing through a point (1, 2, 10) m. Calculate its torque about the point (4, 6, 7) m.

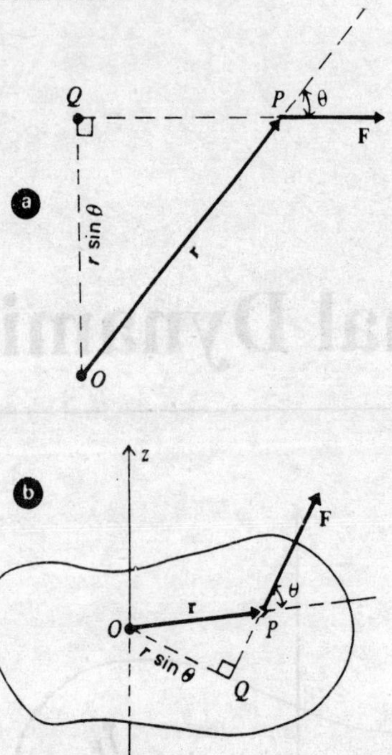

Fig. 6.2 (a) *Torque* (b) *Direction of the torque*

Here

$$\mathbf{r} = (1-4)\mathbf{i} + (2-6)\,\mathbf{j} + (10-7)\,\mathbf{k}\ \text{m}$$

$$= (-3\mathbf{i} - 4\mathbf{j} + 3\mathbf{k})\ \text{m}$$

$$\therefore\quad \tau = (-3\mathbf{i} - 4\mathbf{j} + 3\mathbf{k}) \times (7\mathbf{i} - 8\mathbf{j} + 2\mathbf{k})\ \text{N-m}$$

$$= (24\mathbf{k} + 6\mathbf{j} + 28\mathbf{k} - 8\mathbf{i} + 21\mathbf{j} + 24\mathbf{i})\ \text{N-m}$$

$$= 16\mathbf{i} + 27\mathbf{j} + 52\mathbf{k}\ \text{N-m}$$

Let us now find out the torque acting on a system of particles. To start with let us take two particles of masses M_1 and M_2 and assume that the external forces acting on them are $\mathbf{F}_{1e}$ and $\mathbf{F}_{2e}$, respectively (Fig. 6.3). The mutual interaction forces are also shown in the figure. The torque acting on M_1 is:

$$\tau_1 = \mathbf{r}_1 \times (\mathbf{F}_{1e} + \mathbf{F}_{12}),$$

and the torque acting on M_2 is:

$$\tau_2 = \mathbf{r}_2 \times (\mathbf{F}_{2e} + \mathbf{F}_{21})$$

Therefore, the torque acting on the system of the two particles is:

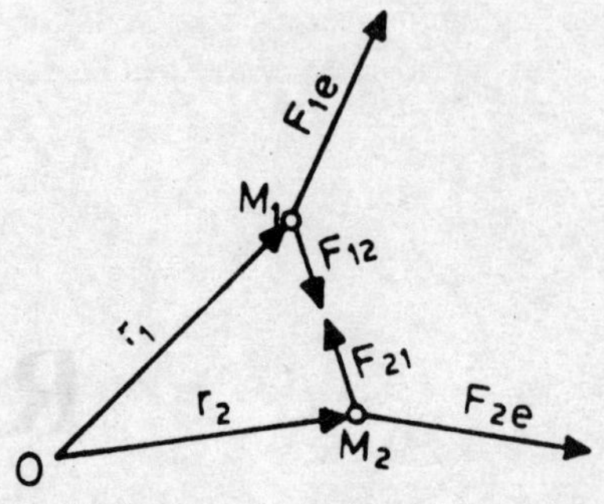

Fig. 6.3 *Torque acting on a system of two particles*

$$\tau = \tau_1 + \tau_2$$

$$= \mathbf{r}_1 \times \mathbf{F}_{1e} + \mathbf{r}_1 \times \mathbf{F}_{12} + \mathbf{r}_2 \times \mathbf{F}_{2e} + \mathbf{r}_2 \times \mathbf{F}_{21}$$

If Newton's third law holds good,

$$\mathbf{F}_{12} = -\mathbf{F}_{21}$$

$$\therefore\quad \tau = \mathbf{r}_1 \times \mathbf{F}_{1e} + \mathbf{r}_2 \times \mathbf{F}_{2e} + (\mathbf{r}_2 - \mathbf{r}_1) \times \mathbf{F}_{21}$$

The last term on the right-hand side is zero because the vectors $(\mathbf{r}_2 - \mathbf{r}_1)$ and $\mathbf{F}_{21}$ are in the same line. Therefore,

$$\tau = \mathbf{r}_1 \times \mathbf{F}_{1e} + \mathbf{r}_2 \times \mathbf{F}_{2e}$$

$$= \tau_{1,ext} + \tau_{2,ext} \qquad (6.3)$$

The above treatment can be extended to a system of more than two particles. Hence, *the total torque acting on a system of particles is equal to the sum of the external torques acting on the individual particles.*

6.2 Angular Momentum

Let us take a particle of mass M moving in the xy plane (Fig. 6.4) under the influence of a force **F**. Let at a certain instant the particle be at P. The

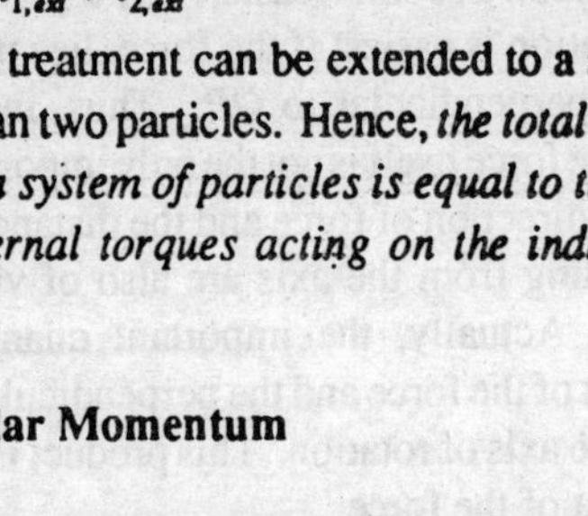

Fig. 6.4 *Angular impulse equals change in angular momentum*

torque acting on the particle relative to the origin O will be in the z-direction, and will be given by:

$$\tau_z = xF_y - yF_x$$

$$= xMa_y - yMa_x$$

$$= xM\frac{\Delta v_y}{\Delta t} - yM\frac{\Delta v_x}{\Delta t}$$

$$\therefore\ \tau_z \Delta t = xM\Delta v_y - yM\Delta v_x$$

$$= xM\Delta v_y + \frac{M\Delta x\Delta y}{\Delta t} - \frac{M\Delta x\Delta y}{\Delta t} - yM\Delta v_x$$

$$= Mx\Delta v_y + M\Delta x v_y - Mv_x\Delta y - My\Delta v_x$$

$$= \Delta(Mxv_y) - \Delta(Myv_x)^*,$$

$$\text{or}\quad \tau_z \Delta t = \Delta(xp_y - yp_x)^{**} \qquad (6.4)$$

Thus the quantity $\tau_z \Delta t$, called the angular impulse (see impulse in Section 4.1), is equal to the change in a quantity $(xp_y - yp_x)$. Drawing the parallelism between Eqs. (6.4) and (4.4), the quantity $(xp_y - yp_x)$ is called the angular momentum about point O, of the particle at P which is moving in the xy-plane. This quantity is denoted by L_z. Or:

$$L_z = xp_y - yp_x \qquad (6.5)$$

This L_z is the z-component of the three dimensional angular momentum vector defined, in general, by the relation:

$$\mathbf{L} = \mathbf{r} \times \mathbf{p} \qquad (6.6)$$

We get from Eq. (6.4),

$$\tau_z = \frac{\Delta L_z}{\Delta t},$$

and similar equations for motion in the other two planes. Therefore, in general,

$$\tau = \frac{\Delta \mathbf{L}}{\Delta t} \qquad (6.7)$$

Hence, *the torque acting on a particle about a reference point is equal to the time rate of change of angular momentum of the particle about that point.*

If we have more than one particles having angular momenta $\mathbf{L}_1, \mathbf{L}_2, \ldots$, etc., relative to a point, and are acted upon by torques $\tau_1, \tau_2, \ldots\ldots$, etc., relative to that point, then:

$$\frac{\Delta \mathbf{L}_1}{\Delta t} + \frac{\Delta \mathbf{L}_2}{\Delta t} + \ldots = \tau_1 + \tau_2 + \ldots.$$

Using Eq. (6.3),

$$\frac{\Delta(\mathbf{L}_1 + \mathbf{L}_2 + \ldots)}{\Delta t} = \tau_{1,ext} + \tau_{2,ext} + \ldots,$$

$$\text{or}\quad \frac{\Delta \mathbf{L}}{\Delta t} = \tau_{ext} \qquad (6.8)$$

Here $\mathbf{L} = \sum_i \mathbf{L}_i$ is the total angular momentum of the particles, and $\tau_{ext} = \sum_i \tau_{i,\,ext}$ is the total torque exerted on the system by external forces only. It may be noted that Eq. (6.8) is valid for rotation about a point fixed in an inertial frame.

Equation (6.8) states that *the time rate of change of the total angular momentum of a system of particles relative to a point is equal to the total torque relative to that point, of the external forces acting on the system.* This is subject to the condition that Newton's third law holds good.

---

*Here the following formula has been used. If

$$F = xy$$

$$F + \Delta F = (x + \Delta x)(y + \Delta y)$$

$$= xy + x\Delta y + y\Delta x + \Delta x\Delta y$$

$$\therefore\quad \Delta F = x\Delta y + y\Delta x + \Delta x\Delta y$$

For very small increments the term $\Delta x\Delta y$ can be neglected, and we get:

$$\Delta F = x\Delta y + y\Delta x$$

**Using calculus:

$$\tau_z = xF_y - yF_x$$

$$= xM\frac{d^2y}{dt^2} - yM\frac{d^2x}{dt^2}$$

$$= xM\frac{d^2y}{dt^2} + M\frac{dx}{dt}\frac{dy}{dt} - M\frac{dx}{dt}\frac{dy}{dt} - yM\frac{d^2x}{dt^2}$$

$$= M\frac{d}{dt}\left(x\frac{dy}{dt}\right) - M\frac{d}{dt}\left(y\frac{dx}{dt}\right)$$

$$= \frac{d}{dt}(xp_y - yp_x)$$

Example 6.2 Show that for a collection of particles the total angular momentum about a given point is equal to the angular momentum about that point of the system concentrated at the centre of mass plus the angular momentum of the constituent particles about the centre of mass.

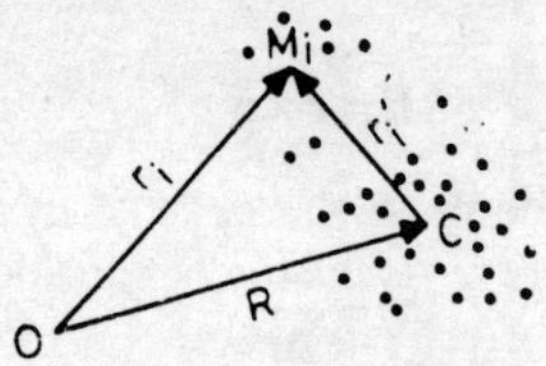

Fig. 6.5 *For Example 6.2*

In Fig. 6.5, if $\mathbf{r}_i$ and $\mathbf{r}_i'$ are the position vectors of the *i*th particle of mass M_i with respect to the origin O and the centre of mass C, respectively, and if $\mathbf{R}$ is the position vector of the centre of mass with respect to the origin, then:

$$\mathbf{r}_i = \mathbf{r}_i' + \mathbf{R}$$

$$\therefore \quad \mathbf{v}_i = \mathbf{v}_i' + \mathbf{v}$$

Here $\mathbf{v}_i$ and $\mathbf{v}_i'$ are the velocities of M_i with respect to O and C, respectively, and $\mathbf{v}$ is the velocity of the centre of mass with respect to O.

If $\mathbf{L}$ is the angular momentum of the system of particles about O, then:

$$\mathbf{L} = \sum_i \mathbf{r}_i \times \mathbf{p}_i$$

$$= \sum_i \mathbf{r}_i \times M_i \mathbf{v}_i$$

$$= \sum_i (\mathbf{r}_i' + \mathbf{R}) \times M_i(\mathbf{v}_i' + \mathbf{v})$$

$$= \sum_i (\mathbf{r}_i' \times M_i \mathbf{v}_i' + \mathbf{r}_i' \times M_i \mathbf{v} + \mathbf{R} \times M_i \mathbf{v}_i' + \mathbf{R} \times M_i \mathbf{v})$$

$$= \sum_i (\mathbf{r}_i' \times M_i \mathbf{v}_i') + \left(\sum_i M_i \mathbf{r}_i'\right) \times \mathbf{v} + \mathbf{R} \times \sum_i (M_i \mathbf{v}_i') + \mathbf{R} \times \sum_i M_i \mathbf{v}$$

But, $\sum_i M_i \mathbf{r}_i' = 0$, because $\sum_i M_i \mathbf{r}_i' / \sum_i M_i$ defines the radius vector of the centre of mass relative to the centre of mass. Also $\sum_i M_i \mathbf{v}_i' = 0$, because $\sum_i M_i \mathbf{v}_i' / \sum_i M_i$ gives the velocity of the centre of mass relative to the centre of mass. Hence,

$$\mathbf{L} = \sum_i (\mathbf{r}_i' \times \mathbf{p}_i') + \mathbf{R} \times M\mathbf{v}$$

$$= \mathbf{L}' + \mathbf{R} \times \mathbf{P} \qquad (6.9)$$

Here, $\mathbf{L}'$ is the angular momentum of the system about the centre of mass. As M is the total mass of the system, $\mathbf{P}(= M\mathbf{v})$ gives the linear momentum of the system concentrated at the centre of mass and moving along with it. Hence, $\mathbf{R} \times \mathbf{P}$ is the angular momentum about O of the total mass kept at the centre of mass.

Angular momentum relative to the centre of mass (i.e., in the C-frame) is called internal angular momentum. In the case of a rigid body, the internal angular momentum is also called spin. The term $\mathbf{R} \times \mathbf{P}$, that is, the angular momentum in the L-frame of the mass concentrated at the centre of mass, is called the orbital angular momentum. Thus, Eq. (6.9) states that the total angular momentum of a system of particles is equal to the vector sum of the internal and orbital angular momenta of the system.

An interesting result follows from Eq. (6.9). If the centre of mass is at rest, then $\mathbf{L} = \mathbf{L}'$. Hence, in this case, the angular momentum is independent of the point of reference. It depends only on the motion around the centre of mass.

6.3 Conservation of Angular Momentum

If there are no external forces acting on a system of particles, then, $\tau_{ext} = 0$. We then get from Eq. (6.8):

$$\mathbf{L} = \sum_i \mathbf{L}_i = \mathbf{L}_1 + \mathbf{L}_2 + \ldots\ldots = \text{constant} \qquad (6.10)$$

This gives the law of conservation of angular momentum – *the total angular momentum of a system in the absence of external torques is constant.* So far this is one of the sacred laws of physics – extending from elementary particles up to the galaxies.

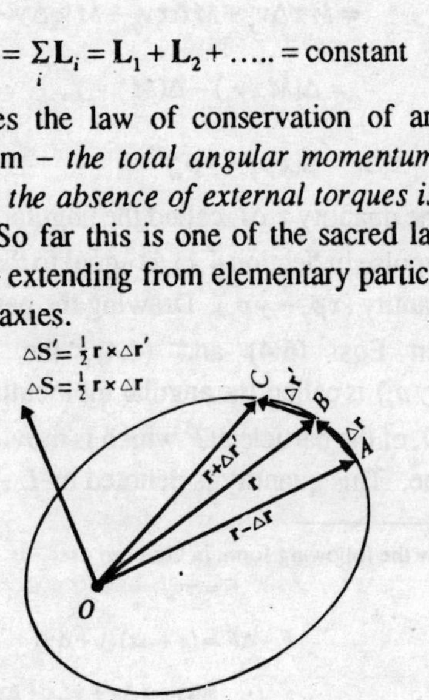

Fig. 6.6 *When a body rotates about another body in the absence of any external torque, the radius vector sweeps equal area in equal intervals of time*

Let us consider the motion of a particle of mass M around a fixed point O (Fig. 6.6) under the action of a central force. A central force is a force whose direction always passes through a fixed point – point O in our case. It could be the motion of a planet around a star, say, the motion of the earth around the sun. In this case, if O is taken as origin, $\mathbf{r} \times \mathbf{F} = 0$. Therefore, the angular momentum of our particle relative to O will remain constant. As angular momentum is a vector, its direction will also remain fixed. Hence, the particle will move in a plane.

Let the particle move from A to B, and from B to C in equal time intervals Δt. Considering OB as vector **r**, we see that:

$$\text{area } OAB = \Delta \mathbf{S} = \frac{1}{2}\mathbf{r} \times \Delta \mathbf{r},$$

and $$\text{area } OBC = \Delta \mathbf{S}' = \frac{1}{2}\mathbf{r} \times \Delta \mathbf{r}',$$

or $$\frac{\Delta \mathbf{S}}{\Delta t} = \frac{1}{2}\mathbf{r} \times \frac{\Delta \mathbf{r}}{\Delta t}; \quad \frac{\Delta \mathbf{S}'}{\Delta t} = \frac{1}{2}\mathbf{r} \times \frac{\Delta \mathbf{r}'}{\Delta t},$$

or $$\frac{\Delta \mathbf{S}}{\Delta t} = \frac{1}{2M}\mathbf{r} \times \mathbf{p}; \quad \frac{\Delta \mathbf{S}'}{\Delta t} = \frac{1}{2M}\mathbf{r} \times \mathbf{p}',$$

where $$\mathbf{p} = M\frac{\Delta \mathbf{r}}{\Delta t}; \ \mathbf{p}' = M\frac{\Delta \mathbf{r}'}{\Delta t}$$

This gives:

$$\frac{\Delta \mathbf{S}}{\Delta t} = \frac{1}{2M}\mathbf{L}; \quad \frac{\Delta \mathbf{S}'}{\Delta t} = \frac{1}{2M}\mathbf{L}'$$

As there is no torque acting on the particle relative to O:

$$\mathbf{L} = \mathbf{L}',$$

or $$\frac{\Delta \mathbf{S}}{\Delta t} = \frac{\Delta \mathbf{S}'}{\Delta t}$$

Thus, *in equal intervals of time the area swept by the radius vector of the particle remains constant. This is Kepler's second law of planetary motion.*

6.4 Equilibrium of Rigid Bodies

A rigid body is a body which has fixed interparticle distances, and these distances do not change under the application of a force or a torque. A rigid body, therefore, maintains its shape and size during the motion. If such a body is constrained to move *about a fixed axis,* a given particle of the body will move in a plane perpendicular to this axis. Such rotation of a rigid body about a fixed axis is called plane rotation or rotation in two dimensions.

Changes in angular momentum can take place only along the axis of rotation. Therefore, the torque components in the direction of the axis of rotation only will be of interest to us. The torque components perpendicular to the axis of rotation will be balanced out by the reactions from the hinge. This is the rotational analogue of motion along a straight line. Think of a bead constrained to move along a straight wire. Therefore, torques (or angular momenta) can be added by the ordinary laws of algebra. As the torques are along the axis of rotation, force components lying in planes perpendicular to the axis of rotation will be effective. Further, the distances of these forces from the axis of rotation will be only those measured in these planes. Hence, our approach will be to divide the body into a very large number of thin slices (laminas) perpendicular to the axis of rotation. Motions of particles in a particular slice will be considered, and then the effects of all slices in the body will be added up.

In the translatory motion of a rigid body (say a thin lamina) we can just consider the motion of the body as the motion of a point particle – the centre of mass (Section 4.4). Let us consider a number of forces acting on the body and in its plane in such a manner that the centre of mass remains at rest, or continues to move with a constant velocity. The motion of the centre of mass in such a case is under no net force. The system of all the forces is said to be in equilibrium. How is this possible? When we have a number of forces acting on a rigid body [Fig. 6.7(a)] such that the total force acting at the centre of mass is zero, we can write:

$$\mathbf{F}_1 + \mathbf{F}_2 + \ldots. = 0,$$

or $$\Sigma \mathbf{F} = 0 \tag{6.11}$$

Only translational equilibrium, but not the complete equilibrium, is assured by these conditions as can be seen in Fig. 6.7(b). Here, two equal and opposite forces are acting in a plane at two points P and Q of the rigid body. Therefore, though $\Sigma\mathbf{F} = 0$, yet we can very easily see that the system of these forces will try to rotate the body in anticlockwise direction. Now the body is not in equilibrium because of the couple Fd acting on the body. Hence, besides $\Sigma\mathbf{F} = 0$, for equilibrium we must have a further condition that:

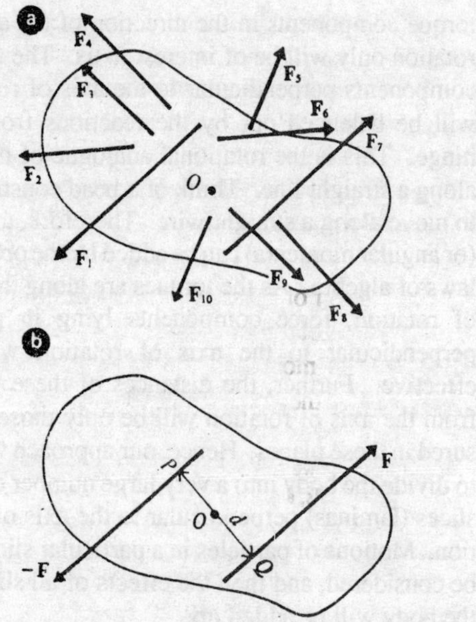

Fig. 6.7 **(a)** *Vector sum of the forces is zero. Therefore there will be translational equilibrium.* (b) *If torque is not zero, rotational equilibrium will not be there.*

$$\Sigma \mathbf{r} \times \mathbf{F} = \tau = 0 \qquad (6.12)$$

i.e. the algebraic sum of the moments of all the forces about any axis perpendicular to the plane of the forces must be zero.*

Example 6.3 Show that if a body (say, a plane sheet) is suspended from a point, the body will rest in such a manner that the point of suspension and the centre of mass fall in a vertical line.

Let the coordinate system be as shown in Fig. 6.8 with P as origin. Here, the y-axis is vertical and the x-axis is horizontal. Let us consider the body to be divided into a number of sub-units with masses $M_1, M_2, \ldots, M_N$ situated at distances $x_1, x_2, \ldots, x_N$ from the y-axis. In other words, $x_1, x_2, \ldots, x_N$ are the abscissae of these particles. The total moment of the gravitational forces about an axis passing through P and perpendicular to the plane of the figure, i.e. the z-axis, is:

$$\sum_{n=1}^{N} M_n g x_n = g \sum_{n=1}^{N} M_n x_n$$

*A couple is a pair of equal but opposite forces not acting along the same line, and its magnitude is the product of one force and the perpendicular distance between the forces. It is also equal to the sum of moments of these forces about any point in the body.

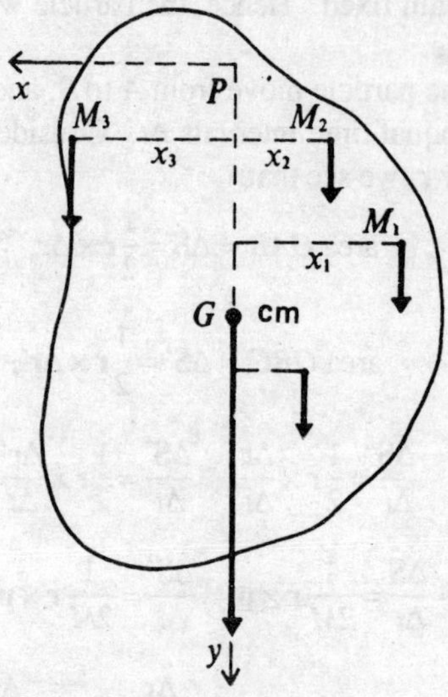

Fig. 6.8 *For Example 6.3*

As the body is not rotating about P:

$$g \sum_{n=1}^{N} M_n x_n = 0$$

Now, the abscissa of the centre of mass is given by

$$x = \frac{\sum_{n=1}^{N} M_n x_n}{M}$$

where M is the total mass of the sheet. Therefore, the x-coordinate of the centre of mass is zero. Thus, the centre of mass lies on the y-axis, i.e. vertically below P. Because the moments of the gravitational forces balance about the centre of mass, it is also called the *centre of gravity*.

6.5 Moment of Inertia

In the last section we have seen that when a force is applied at a distance from a fixed axis, it is possible to produce rotational motion. Also the body is in rotational equilibrium (either no rotation or rotation with uniform angular speed) if the total torque $\tau = \Sigma \mathbf{r} \times \mathbf{F} = 0$. Let us find the amount of torque required to move a body with a constant angular acceleration α. As the body is rigid, all particles will have the same angular acceleration. Otherwise interparticle distances will change. Let the axis of rotation pass through a point O and let us suppose the body is constrained to move about this axis. We consider the body to be constituted of small elementary masses $M_1, M_2, \ldots, M_N$ situated from the axis at distances $r_1, r_2, \ldots, r_N$ respectively (Fig. 6.9). The distances $r_1, r_2, \ldots, r_N$, can be found by drawing planes perpendicular to the fixed axis through the masses $M_1, M_2, \ldots, M_N$. We have seen in Section 2.3 that linear acceleration and angular acceleration are related as:

$$a = \alpha r \tag{6.13}$$

where r is the distance of M from the axis. Thus, linear accelerations of the constituent masses are $\alpha r_1, \alpha r_2, \ldots, \alpha r_N$, giving rise to forces $\alpha M_1 r_1, \alpha M_2 r_2, \ldots, \alpha M_N r_N$ acting at distances $r_1, r_2, \ldots, r_N$ from the fixed axis of rotation. Total moment of all these forces is:

$$\tau = \alpha M_1 r_1 r_1 + \alpha M_2 r_2 r_2 + \ldots + \alpha M_N r_N r_N$$

$$= \alpha (M_1 r_1^2 + M_2 r_2^2 + \ldots + M_N r_N^2)$$

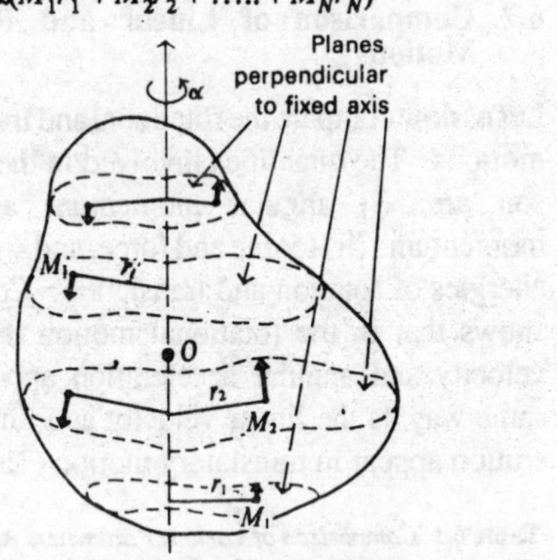

Fig. 6.9 ***When a torque acts on a body, angular acceleration is produced in it***

$$= \alpha \sum_{n=1}^{N} M_n r_n^2, \tag{6.14}$$

or $$\tau = I\alpha, \tag{6.15}$$

where $$I = \sum_{n=1}^{N} M_n r_n^2 \tag{6.16}$$

is called the *moment of inertia* of the body about the axis of rotation. Thus, the torque is given by the product of moment of inertia and the angular acceleration. In rotational motion, therefore, not only the mass but its distribution about the axis of rotation is also important. If we apply the same torque on two bodies of equal masses but with different distributions of mass about their axes of rotation (Fig. 6.10), we will produce different angular accelerations. If more mass is situated near the axis of rotation [Fig. 6.10(a)], the moment of inertia will be less, and the angular acceleration produced by the given torque will be more. If the concentration of mass is more at positions away from the axis of rotation [Fig. 6.10(b)], the moment of inertia is more and the angular acceleration produced by the given torque will be less.

Example 6.4 Two rings of equal mass and thickness, but of different materials, are acted upon by the same torque about an axis passing through their centres and perpendicular to the planes of the rings. If the densities of the materials are in the ratio of 4 : 1, find the ratio of their angular accelerations.

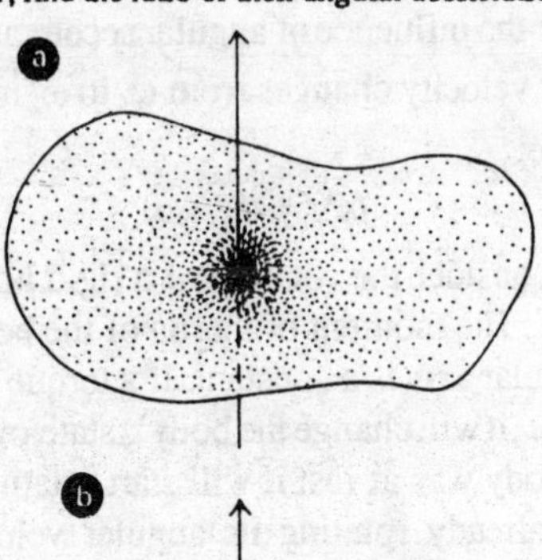

Fig. 6.10 ***Angular acceleration produced by a torque acting on a body of given mass depends upon the distribution of the mass.* (a) *More acceleration,* (b) *less acceleration***

Because the rings are of equal thickness and equal mass, their circumferences will be inversely proportional to their densities. Therefore:

$$2\pi r_1 \rho_1 = 2\pi r_2 \rho_2$$

where r_1 and r_2 are the radii of the rings, and ρ_1 and ρ_2 are the densities of their materials. Therefore $r_1/r_2 = 1/4$. For a ring

$I = mr^2$, because the whole mass of the ring is at a distance r from the centre. Therefore:

$$\frac{mr_1^2}{mr_2^2} = \frac{I_1}{I_2} = \frac{1}{16}$$

Thus the moments of inertia are in the ratio of 1 : 16. If the same torque is applied, then we have:

$$\tau = I_1\alpha_1 = I_2\alpha_2,$$

therefore $$\frac{\alpha_1}{\alpha_2} = \frac{I_2}{I_1} = \frac{16}{1}$$

Hence, the angular accelerations are in the ratio of 16 : 1.

6.6 Angular Momentum of a Rigid Body

We know that (from Section 2.3):

$$\text{angular acceleration} = \frac{\text{change in angular velocity}}{\text{time of change of angular velocity}}$$

If under the influence of angular acceleration α the angular velocity changes from ω_1 to ω_2 in time Δt, we have:

$$\alpha\Delta t = \omega_2 - \omega_1 \tag{6.17}$$

Let us consider the rotation of a rigid body about an axis. The moment of inertia of the body about a particular axis is a constant. If a torque τ acts for a time Δt, it will change the body's state of rotation. If the body was at rest it will start rotating, and if it was already rotating its angular velocity will change. We can write from Eq. (6.15):

$$\tau\Delta t = I\alpha\Delta t$$

$$= I(\omega_2 - \omega_1) = I\omega_2 - I\omega_1 \tag{6.18}$$

Comparing Eqs. (6.18) and (6.8) we note that $I\omega$ is the angular momentum of the rigid body about the axis of rotation. Or,

$$L = I\omega^* \tag{6.19}$$

Example 6.5 Prove that the kinetic energy of rotation of a rigid body is $\frac{1}{2}I\omega^2$.

The total kinetic energy is the sum of the kinetic energies of the constituent masses. Therefore:

$$T = T_1 + T_2 + \ldots.. + T_N.$$

Now $$T_n = \frac{1}{2}M_n v_n^2$$

$$= \frac{1}{2}M_n(\omega r_n)^2$$

$$= \frac{1}{2}M_n r_n^2 \omega^2.$$

Hence $$T = \frac{1}{2}M_1 r_1^2\omega^2 + \frac{1}{2}M_2 r_2^2\omega^2 + \ldots.. + \frac{1}{2}M_N r_N^2\omega^2$$

$$= \frac{1}{2}\omega^2 \sum_{n=1}^{N} M_n r_n^2$$

$$= \frac{1}{2}I\omega^2 \tag{6.20}$$

Thus the rotational kinetic energy is equal to half the product of moment of inertia and square of angular velocity.

6.7 Comparison of Linear and Rotational Motions

Let us now compare the rotational and translational motions. The quantities involved in the comparison are: (i) angular momentum and linear momentum, (ii) torque and force, and (iii) kinetic energies of rotation and translation. Comparison shows that in the rotational motion the angular velocity and angular acceleration appear in the same way as the linear velocity and linear acceleration appear in translatory motion (Table 6.1).

Table 6.1 *Comparison of Various Parameters Appearing in Rotational and Translational Motions*

Rotational motion	*Translational motion*
(i) Angular displacement θ	Displacement s
(ii) Angular velocity ω	Velocity **v**,
(iii) Angular acceleration α	Linear acceleration **a**
(iv) Angular momentum **L** (The change in **L** gives the total angular impulse) $\mathbf{L} = I\omega\,(= \mathbf{r}\times\mathbf{p})$	Linear momentum **p** (The change in **p** gives the total linear impulse) $\mathbf{p} = M\mathbf{v}$
(v) Torque: $\tau = \Delta\mathbf{L}/\Delta t$ (The torque produces a change in angular momentum) $\tau = I\alpha (= \mathbf{r}\times\mathbf{F})$	Force: $\mathbf{F} = \Delta\mathbf{p}/\Delta t$ (The force produces a change in linear momentum) $\mathbf{F} = M\mathbf{a}$
(vi) Kinetic energy of rotation $= \frac{1}{2}I\omega^2$	Kinetic energy of translation $= \frac{1}{2}Mv^2$

*Actually: $\mathbf{L} = \hat{\mathbf{I}}\omega$, where **L** and ω are vectors and $\hat{\mathbf{I}}$ is a dyadic or a tensor of second rank. But here we have considered the rotation constrained about a fixed axis. This way we have avoided the vector notation with moment of inertia. Refer to: C. Kittel, W. D. Knight and M. A. Ruderman, Mechanics--Berkeley Physics Course, Vol. I (McGraw-Hill), 1965.

The moment of inertia has the same important role to play in rotational motion as that played by mass in translatory motion. From Eq. (6.16):

$$I = \sum_{n=1}^{N} M_n r_n^2 = MK^2 \qquad (6.21)$$

where K is known as the radius of gyration defined as: *The radius of gyration of a body about an axis is the distance from the axis at which the whole mass of the body is imagined to be concentrated for the purpose of obtaining moment of inertia about that axis.* The radii of gyration of some common geometrical shapes are given in Table 6.2.

Table 6.2 *Radii of Gyration of Some Uniform Common Geometrical Forms*

S.No.	Body	Axis	K^2
1.	Slender rod of length L	(i) Through the centre and perpendicular to the rod	$\frac{1}{12}L^2$
		(ii) Through the end and perpendicular to the rod	$\frac{1}{3}L^2$
2.	Ring of radius R	Through the centre and perpendicular to the plane of the ring	R^2
3.	Thin walled cylinder of radius R	Through the centre and parallel to the length of the cylinder	R^2
4.	Thin circular solid plate of uniform thickness and radius R	Through the centre and perpendicular to the plane of the plate	$\frac{1}{2}R^2$
5.	Solid cylinder of radius R and length L	(i) Axis of the cylinder	$\frac{1}{2}R^2$
		(ii) Through the centre and perpendicular to the axis of the cylinder	$\frac{R^2}{4}+\frac{L^2}{12}$
6.	Hollow thin sphere of radius R	Diameter	$\frac{2}{3}R^2$
7.	Solid sphere of radius R	Diameter	$\frac{2}{5}R^2$
8.	Rectangular lamina of length L and breadth B	Through the centre perpendicular to the plane of the lamina	$\frac{1}{12}(L^2+B^2)$

From the knowledge of moment of inertia and mass we can directly write:

$$K = \sqrt{I/M} \qquad (6.22)$$

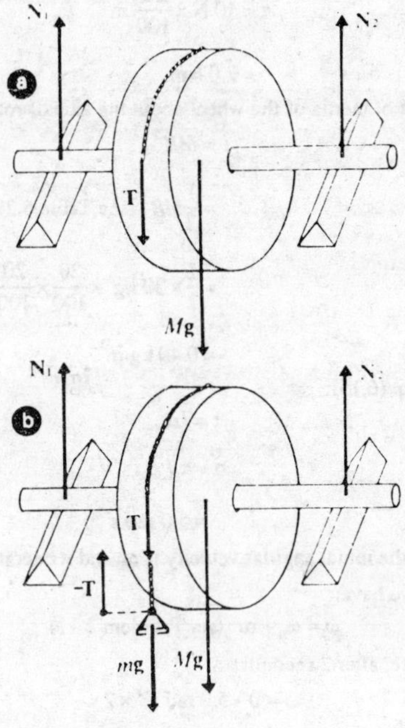

Fig. 6.11 *For Example 6.6*

which may also be considered as the definition of radius of gyration. Lastly, we may write $\tau = I\alpha$. When $\alpha = 1, \tau = I$. Thus, *the moment of inertia of a body about an axis is equal in magnitude to a torque which produces a unit angular acceleration about that axis.*

Question: From the equivalence of quantities appearing in rotational and translational motions (Table 6.1), it can be inferred that work done in rotating a body about an axis by an angle $\Delta\theta$ is given by $\tau\Delta\theta$. Prove this.

Example 6.6 A fly-wheel of mass 20 kg and diameter 40 cm can rotate without friction about a horizontal axis passing through its centre and perpendicular to its face [Fig. 6.11(a)]. A cord is wound around its circumference and is pulled downward with a force of 10 N. Find the angular acceleration of the wheel, and its angular velocity after 2 s. Neglect the moment of inertia of the axle.

The forces acting on the system are shown in Fig. 6.11(a). The weight of the wheel Mg and the pull T are acting downward, while the reaction forces N_1 and N_2 at the supports act upward. As the wheel rotates about its centre (which is also its centre of mass), the torques of Mg, N_1 and N_2 will be zero. Therefore, the torque is:

$$\tau = 10\text{ N} \times \frac{20}{100}\text{ m}$$

$$= 2.0\text{ Nm}$$

Moment of inertia of the wheel about the axis of rotation is:

$$I = MK^2$$

$$= \frac{1}{2}MR^2 \text{ (see Table 6.2)}$$

$$= \frac{1}{2} \times 20\text{ kg} \times \frac{20}{100} \times \frac{20}{100}\text{ m}^2$$

$$= 0.40\text{ kg m}^2$$

From Eq. (6.15):

$$\tau = I\alpha$$

$$\therefore \quad \alpha = \tau/I$$

$$= 5.0\text{ rad s}^{-2}$$

Now, if the initial angular velocity is ω_0 and it becomes ω after time t, we have:

$$\omega = \omega_0 + \alpha t \quad \text{(see Problem 2.11)}$$

Therefore, after 2 seconds:

$$\omega = 0 + 5.0\text{ rad s}^{-2} \times 2\text{ s}$$

$$= 10\text{ rad s}^{-1}$$

Question: If the supports in Fig. 6.11(a) are symmetrically placed about the wheel, what are the magnitudes of N_1 and N_2? Assume $g = 10\text{ ms}^{-2}$.

If instead of pulling the cord with a force of 10 N we suspend a mass of 1.0 kg by it [Fig. 6.11(b)], the situation becomes quite different. Now the force of 10 N is acting on the system mass plus wheel. If a is the acceleration of the suspended mass, the tension in the string is given by:

$$mg - T = ma,$$

or

$$T = (10 - a)\text{ N}$$

$$= \left(10 - \alpha \times \frac{20}{100}\right)\text{N}$$

where α is the angular acceleration of the wheel. Therefore, the equation of motion for the wheel becomes:

$$\left(10 - \frac{\alpha}{5}\right) \times \frac{1}{5} = \frac{1}{2} \times 20 \times \left(\frac{1}{5}\right)^2 \alpha$$

$$\therefore \quad 10 - \frac{\alpha}{5} = 2\alpha,$$

or

$$\alpha = \frac{50}{11} = 4.5\text{ rad s}^{-2}$$

which is less than the value calculated above. The acceleration of the falling mass is:

$$a = R\alpha$$

$$= \frac{1}{5} \times 4.5$$

$$= 0.9\text{ ms}^{-2}$$

which is much smaller than the value 10 ms^{-2}, its acceleration in free fall.

Question: What are the magnitudes of N_1 and N_2 now?

In general, if M is the mass of the wheel, R is its radius and m is the mass of the hanging object, the angular acceleration of the wheel is given by:

$$\alpha = \frac{mg}{\left(m + \frac{1}{2}M\right)R}$$

The derivation has been left as an exercise for the reader.

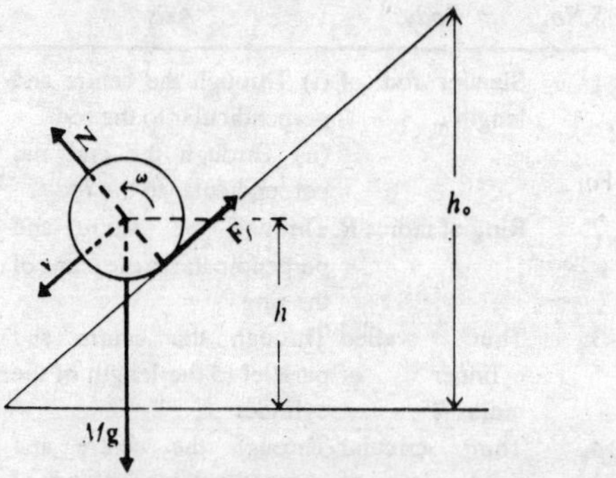

Fig. 6.12 *For Example 6.7*

Example 6.7 A sphere and a cylinder start from rest at the same position and roll down the same inclined plane. Show that the sphere will reach the bottom before the cylinder, irrespective of their masses and radii.

Let us take a body rolling down an incline. The forces acting on it are shown in Fig. 6.12. If the body does not slide, the work done by the frictional force will be zero. Under these conditions the mechanical energy of the body will remain unchanged. Let the body start rolling from the top, which is at a height h_0. When it is at any height h, let the translational velocity of the centre of mass be v, and let the body be rotating about the centre of mass with angular velocity ω.

When a point at the circumference of the body is in touch with the incline, it is momentarily at rest. Therefore, the velocity of the centre of mass must be equal to the linear velocity of the point. Further, the velocity of the point of contact is equal to the linear velocity of a point on the circumference of a circle of radius R (equal to the radius of the sphere or the cylinder). Let us illustrate this point. Take a disc of radius R rolling on a surface without slipping, and complete one rotation. The distance covered by the point of contact is $2\pi R$. The same is the distance covered by a point on the circumference around the centre of the disc. The centre of the disc will also move a distance $2\pi R$. Hence:

$$v = R\omega$$

where R is the radius of the rolling body. Now the motional energy of a body is equal to the sum of the kinetic energy due to the motion of the centre of mass and that due to the rotation of the body about the centre of mass (see Example 5.2). Thus, the total kinetic energy at the height h is:

$$\frac{1}{2}Mv^2 + \frac{1}{2}I\omega^2$$

where M is the mass of the object and I is its moment of inertia about the axis of rotation passing through the centre of mass. As there is no slipping, work done by the frictional force will be zero. From the law of conservation of energy:

$$\frac{1}{2}Mv^2 + \frac{1}{2}I\omega^2 + Mgh = Mgh_0$$

or $$\frac{1}{2}Mv^2 + \frac{1}{2}MK^2 \times \frac{v^2}{R^2} = Mg(h_0 - h),$$

or $$v^2 = \frac{2g(h_0 - h)}{(1 + K^2/R^2)}$$

For a sphere: $K^2 = \frac{2}{5}R^2$, and for a cylinder: $K^2 = \frac{1}{2}R^2$.

Therefore, for a sphere:

$$v_s = \sqrt{1.43g(h_0 - h)}$$

and for a cylinder

$$v_c = \sqrt{1.33g(h_0 - h)}$$

Hence, the sphere translates faster than the cylinder and, therefore, will come down earlier.

As v_c and v_s are independent of the masses and the radii of the objects, this conclusion will hold good for spheres and cylinders of any size and made of any material provided these are solid from inside.

Question: Will a ring come down faster or slower as compared with: (i) a cylinder, (ii) a sphere? As compared with: (i) a hollow thin sphere, (ii) a thin walled cylinder?

We conclude from the above that the speed of a body rolling down an incline does not depend on its mass and size, but depends only on the geometrical distribution of the mass. It can also be noted that the translational speed is less than the speed of free fall. This is because the loss in potential energy is shared by the rotational motion also.

6.8 Theorems of Parallel and Perpendicular Axes

Suppose that there is a body whose moment of inertia about an axis passing through its centre of gravity is known, and we want to find the moment of inertia about an axis parallel to this axis. We can cut the body into a number of thin plane sheets which are perpendicular to these axes. The axis passing through the centre of gravity cuts one such sheet 1 at G and the parallel axis cuts it at O (Fig. 6.13). Let this sheet be divided into a large number of constituent masses and let a representative mass at P be M_n. Then the moment of inertia about the axis through G is:

$$I_{G1} = \sum_n M_n |GP|^2 = \sum_n M_n |\mathbf{r}_n|^2, \qquad (6.23)$$

where $\mathbf{r}_n = GP$

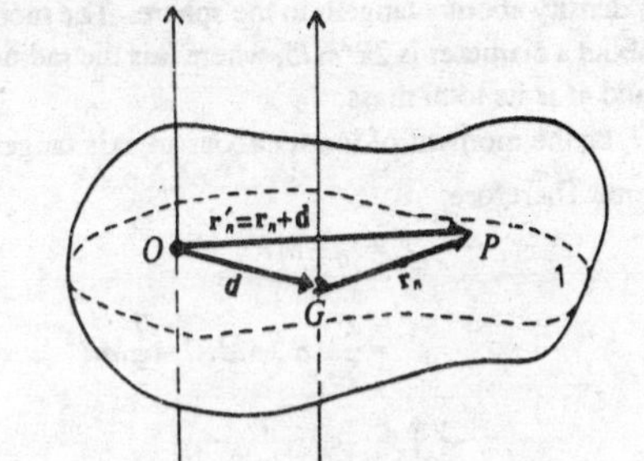

Fig. 6.13 *Parallel axes theorem*

The moment of inertia about the axis through O is:

$$I_{O1} = \sum_n M_n |OP|^2 = \sum_n M_n |\mathbf{r}_n'|^2, \qquad (6.24)$$

where $\mathbf{r}_n' = OP$

But $\mathbf{r}_n' = \mathbf{r}_n + \mathbf{d}$, where $\mathbf{d} = OG$

Therefore $|\mathbf{r}_n'|^2 = \mathbf{r}_n' \cdot \mathbf{r}_n' = (\mathbf{r}_n + \mathbf{d}) \cdot (\mathbf{r}_n + \mathbf{d})$

$$= |\mathbf{r}_n|^2 + |\mathbf{d}|^2 + 2\mathbf{r}_n \cdot \mathbf{d} \qquad (6.25)$$

Hence:

$$I_{O1} = \sum_n M_n |\mathbf{r}_n|^2 + \sum_n M_n |\mathbf{d}|^2 + 2\sum_n M_n \mathbf{r}_n \cdot \mathbf{d}$$

But d is constant for all masses M_n, giving:

$$I_{O1} = I_{G1} + M_1 d^2 + 2\mathbf{d} \cdot \sum_n M_n \mathbf{r}_n \qquad (6.26)$$

Here M_1 is the mass of this sheet.

We can treat all other parallel sheets similarly and write equations similar to Eq. (6.26). As the axes of rotation are fixed in space, moments of inertia of different sheets will add arithmetically. Thus, for the whole body:

$$I_O = I_G + Md^2 \qquad (6.27)$$

because for the whole body* :

$$\Sigma M_n \mathbf{r}_n = 0 \qquad (6.28)$$

Here M is the mass of the body.

Eq. (6.27) states that moment of inertia about a given axis is equal to the moment of inertia about a parallel axis passing through the centre of gravity plus the product of the mass of the body and the square of the distance between the two axes. This is *the parallel axes theorem* or *Steiner's theorem.*

Example 6.8 Find out the moment of inertia of a sphere of uniform density about a tangent to the sphere. The moment of inertia about a diameter is $2R^2M/5$, where R is the radius of the sphere and M is its total mass.

Let I_t be the moment of inertia about an axis tangential to the sphere. Therefore:

$$I_t = I_G + MR^2$$

$$= \frac{2}{5}MR^2 + MR^2 = \frac{7}{5}MR^2$$

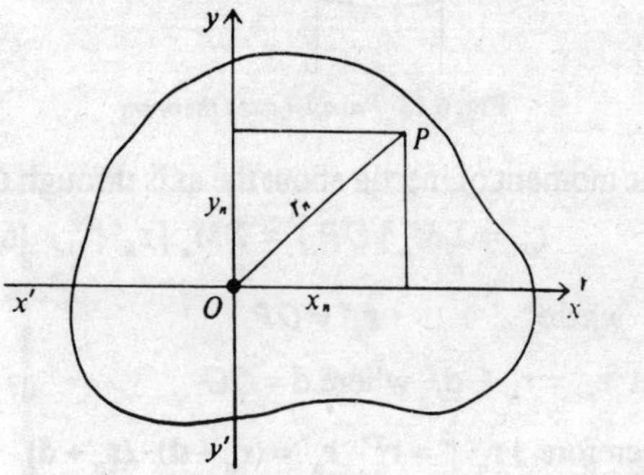

Fig. 6.14 *Perpendicular axes theorem*

Next, let us consider a thin plane sheet (Fig. 6.14). The moments of inertia about the x-axis and the y-axis respectively are:

$$\left.\begin{aligned} I_x &= \sum_n M_n y_n^2 \\ I_y &= \sum_n M_n x_n^2 \end{aligned}\right\} \qquad (6.29)$$

Hence

$$I_x + I_y = \sum_n M_n (x_n^2 + y_n^2)$$

$$= \sum_n M_n r_n^2 \qquad (6.30)$$

Now, $\sum_n M_n r_n^2$ is the moment of inertia about an axis passing through O and perpendicular to the sheet. Thus, we conclude that for a thin plane sheet the sum of moments of inertia about two mutually perpendicular axes lying in the sheet is equal to the moment of inertia about an axis perpendicular to the sheet and passing through the point of intersection of the two axes. This is *the theorem of perpendicular axes.* As an example, let us calculate the moment of inertia of a thin circular disc about its diameter. The moment of inertia of a disc of mass M and radius R about an axis passing through its centre (intersection of two diameters) and perpendicular to its plane is $\frac{1}{2}MR^2$ (Table 6.2).

Let us take two mutually perpendicular diameters. If I_d is the moment of inertia about each one of these diameters:

$$I_d + I_d = \frac{1}{2}MR^2,$$

or

$$I_d = \frac{1}{4}MR^2$$

*Take the axis through the centre of gravity as the z-axis. Now $\mathbf{r}_n$ will be in the xy-plane Therefore,

$$\mathbf{r}_n = x_n\mathbf{i} + y_n\mathbf{j},$$

and

$$\Sigma M_n \mathbf{r}_n = \Sigma M_n x_n \mathbf{i} + \Sigma M_n y_n \mathbf{j}$$

Here the summation is over the whole body.

Now, $\Sigma M_n x_n/\Sigma M_n$ and $\Sigma M_n y_n/\Sigma M_n$ give the x and y coordinates, respectively, of the centre of mass. These coordinates will be zero because the centre of mass is on the z-axis.

Discussion and Problems

6.1 A projectile does not have any angular momentum about the point of projection in the beginning. But it acquires one as the time passes. Does it violate the law of conservation of angular momentum?

6.2 A boy standing on a turntable holds two heavy dumbbells with arms stretched. The turntable is given a push and it starts rotating at the rate of one revolution per second. The boy pulls the dumbbells towards his chest. What happens to the speed of rotation? Calculate the angular speed if the dumbbells have been pulled from a distance of 50 cm from the axis of rotation to a distance of 10 cm. Neglect the angular momentum of the boy as compared with that of the dumbbells. What happens if lets the dumbbells go?

6.3 A ring is rolling on an inclined plane without slipping. What fraction of its total kinetic energy is due to translation? Due to rotation?

6.4 Two space ships 1 and 2 of masses 1000 kg and 2000 kg are travelling in a gravity-free intersteller space with speeds 100 m/s and 150 m/s respectively (Fig. 6.15). A shell of mass 10 kg is fired from ship 2 with a speed 5000 m/s. The shell strikes the ship 1 at P and gets embedded in it. Calculate the angular momentum of each ship and the total angular momentum with respect to the point O: (i) before the gun is fired; (ii) after the gun is fired but the shell is on its way to ship 1; and (iii) after the shell has struck ship 1.

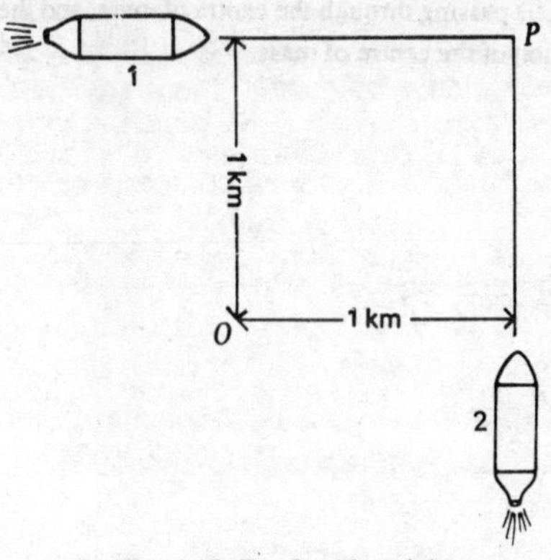

Fig. 6.15 *For Problem 6.4*

6.5 A 25 kg ladder is standing at 45° angle against a frictionless wall. Calculate the force of the ladder against the wall. Can this ladder stand in this position if there is no friction at the floor? Discuss.

6.6 Assuming the moment of inertia to be $\approx 10^{38}$ mks units, find the rotational kinetic energy of the earth. What average torque is required to stop the rotation of the earth in a week. If the kinetic energy can be converted into matter, how much mass is created ? For how long could the earth supply 1 kW power to each of the 5×10^9 persons on the earth?

6.7 A body of mass M and radius R is rolling horizontally without slipping with speed v. It then rolls up a hill to a maximum height h. If $h = 5v^2/6g$: (i) What is the body's rotational inertia? (ii) What might be the shape of the body ?

6.8 (a) Show that the work done by a torque τ is given by:

$$\Delta W = \tau \Delta\theta,$$

and the power developed by it is given by:

$$P = \tau\omega$$

(b) An engine develops 120 hp when rotating at a speed of 2000 rpm. What torque does it deliver?

6.9 In Atwood's machine (see Problem 3.9) one block has a mass of 2.5 kg and the other a mass of 2.0 kg. The pulley, which is mounted on frictionless bearings, has a radius of 6.0 cm. When released from rest, the heavier block is observed to fall 1.2 m in 6.0 sec. What is the rotational inertia of the pulley?

6.10 A man wants to come down from a building. He has a rope and a very heavy pulley. He ties the rope around the pulley and jumps. Will the man attain an unaccelerated motion? Discuss.

6.11 In a water molecule, the O–H distance is 0.91 Å, and the angle between the two O–H bonds is 105°. Calculate the moment of inertia of the molecule about an axis passing through its centre of mass and perpendicular to the plane containing the atoms.

6.12 In Fig. 6.16, when mass m is let go it gets detached from the fly-wheel after falling through a height h. Show that if the wheel rotates frictionlessly and makes N revolutions per unit time, then:

$$N = \frac{1}{2\pi}\sqrt{\frac{2mgh}{mr^2 + I}}$$

where r is the radius of the axle and I is the moment of inertia of the fly-wheel about the axis of rotation.

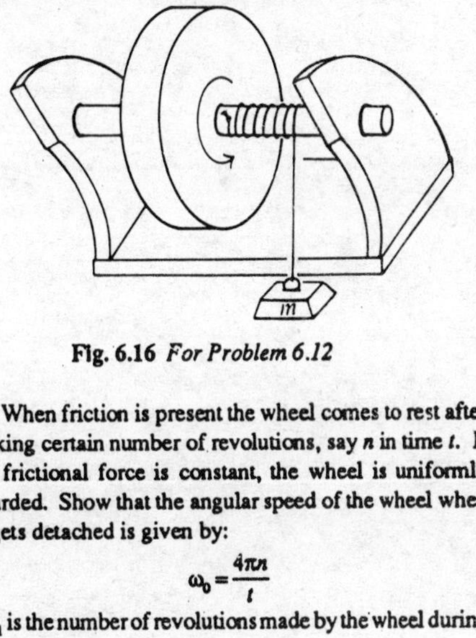

Fig. 6.16 *For Problem 6.12*

When friction is present the wheel comes to rest after making certain number of revolutions, say n in time t. If the frictional force is constant, the wheel is uniformly retarded. Show that the angular speed of the wheel when m gets detached is given by:

$$\omega_0 = \frac{4\pi n}{t}$$

If n_1 is the number of revolutions made by the wheel during the descent of m, show that:

$$mgh = \frac{1}{2}mr^2\omega_0^2 + \frac{1}{2}I\omega_0^2\left(1 + \frac{n_1}{n}\right)$$

[Hint: Assume that the energy expended against friction in one revolution is E_f].

This is a laboratory method to determine the moment of inertia of a fly-wheel.

6.13 Show that the angular momenta $\mathbf{L}_1$ and $\mathbf{L}_2$ of a system of particles relative to points O_1 and O_2 are related as:

$$\mathbf{L}_2 = \mathbf{L}_1 + \mathbf{r} \times \mathbf{P}$$

where $\mathbf{P}$ is the linear momentum of the system, and $\mathbf{r}$ is the position vector of O_2 relative to O_1.

6.14 Show that the angular momenta $\mathbf{L}$ and $\mathbf{L}'$ of a system of particles as measured in two inertial frames S and S', respectively, about their origins at an instant when they coincide are related as:

$$\mathbf{L} = \mathbf{L}' + M\mathbf{R} \times \mathbf{V}$$

Here M is the total mass of the system, $\mathbf{R}$ is the radius vector of centre of mass, and $\mathbf{V}$ is the velocity of S' relative to S.

6.15 Two flywheels A and B are mounted side by side on a common shaft. Their moments of inertia about the shaft are 5.0 kg-m^2 and 20.0 kg-m^2, respectively. A is made to rotate at the rate of 10.0 rev/s and left free. B is stationary. They are now coupled with the help of a clutch, and rotate together. Assume bearings to be frictionless. (i) What is the speed of revolution of the wheels? (ii) Compare the rotational energies before and after coupling. (iii) During coupling A makes 10 revolutions more than B. In how much time the complete coupling has taken place? (iv) How much torque is transmitted by the clutch?

6.16 How can a space vehicle be rotated about its centre of mass?

6.17 In Example 6.7 find out the acceleration of the centre of mass. (Hint: $v \sin\theta = -\Delta h/\Delta t$). Also find this acceleration by: (i) taking torque and angular momentum about an axis parallel to the axis of rotation and passing through the instantaneous point of contact; (ii) taking rotation about an axis passing through the centre of mass, and the linear motion of the centre of mass.

Chapter 7

Simple Harmonic Motion

7.1 Introduction

Two important types of motion in nature are the nonperiodic motion and the periodic motion. Examples of the nonperiodic motion are the motion of a car, a bullet, etc. Circular motion, beating of heart, etc., are examples of the periodic motion. Some of these have been dealt with in earlier chapters. In this chapter an important kind of periodic motion called *simple harmonic motion* (SHM) will be studied. A large number of phenomena like the motion of a simple pendulum, vibration of a loaded spring, discharging of a capacitor through an inductor involve simple harmonic oscillations. At times the oscillations occurring at a point spread out to the neighbouring points and give rise to a wave motion. This will form the subject matter of Chapter 10. Here we will study oscillations confined about a point.

7.2 Characteristics of Simple Harmonic Motion

An oscillation means a repetitive variation of a parameter of a physical system about a mean value. For a vibrating system the parameter is displacement. A body is said to oscillate when it repetitively moves to and fro about a point of equilibrium. In electromagnetic oscillations the electric and magnetic fields undergo vibrations about a mean value of the field (see Section 27.10). The measurable quantities associated with an oscillation are: *amplitude, periodic time, frequency and angular frequency.* These are discussed below.

Amplitude: This is the maximum variation of the variable parameter from its mean value in one direction. We will denote it by A. It has the same units as the variable parameter.

Periodic time: It is the time taken between the successive attainments of a reference value of the parameter during variation in the same direction. It is denoted by T and its unit is second.

Frequency: It is the number of oscillations completed in a second and is, therefore, equal to $1/T$. We denote it by ν. The unit for frequency is cycles/sec. In SI it is denoted by *hertz.*

Angular frequency: It is the angle described by the oscillating system in a second, and is denoted by ω. In one oscillation the system moves through 2π radians. For ν oscillations the angle moved will be $2\pi\nu$. Therefore, $\omega = 2\pi\nu = 2\pi/T$. Though called angular frequency here, it is actually the angular velocity.

On analysing an oscillatory motion, under certain restrictions, the instantaneous value of the variable turns out to be the sum of sine and cosine functions of time (see Sec. 10.8). These functions are called *harmonic functions.* Therefore, such oscillations are called harmonic oscillations. In general, the oscillatory motion is complex in the sense that the analysis may reveal presence of several motions of different periods at any instant. If it turns out to be of only one period, we call it *simple harmonic motion*, abbreviated as SHM.

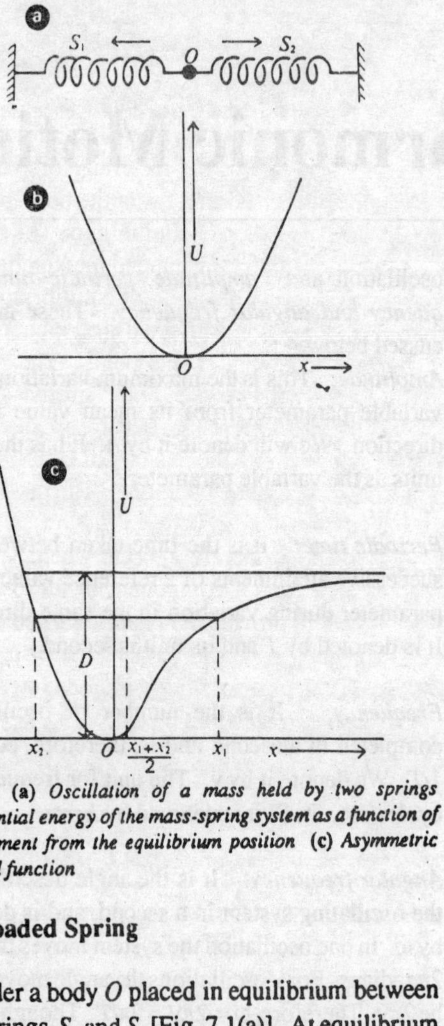

Fig. 7.1 (a) *Oscillation of a mass held by two springs* (b) *Potential energy of the mass-spring system as a function of displacement from the equilibrium position* (c) *Asymmetric potential function*

7.3 Loaded Spring

Consider a body O placed in equilibrium between two springs S_1 and S_2 [Fig. 7.1(a)]. At equilibrium the body has minimum energy. Let us now displace it along the x-axis through a distance x. This disturbs the state of equilibrium of the system. The energy of the body will now be higher than that at the equilibrium. If a curve between the potential energy and the displacement is plotted, it will show a minimum at the centre with a rise for the displacements both towards right and left. Thus the potential energy is a function of displacement. Denoting potential energy by U, we can write:

$$U = f(x) \tag{7.1}$$

There is a theorem in mathematics, called Taylor's theorem, which states that a function can be expanded in a series in powers of the variable. Eq. (7.1), therefore, becomes:

$$U = ax + \frac{1}{2}kx^2 + bx^3 + cx^4 + \ldots\text{ to } \infty \tag{7.2}$$

where a, $k/2$, b,....., etc. are constant coefficients. The coefficient of x^2 has been deliberately chosen as $\frac{1}{2}k$.

It was stated in Section 7.2 that an oscillatory motion becomes simple harmonic under certain restrictions. To specify it let us assume that O has been given a small displacement x. Under these conditions the terms containing x^3 and higher powers of x become negligibly small and can be dropped. Eq. (7.2) then becomes:

$$U = ax + \frac{1}{2}kx^2 \tag{7.3}$$

The potential energy at displacement $(x + \Delta x)$ then becomes:

$$U + \Delta U = a(x + \Delta x) + \frac{1}{2}k(x + \Delta x)^2$$

$$= a(x + \Delta x) + \frac{1}{2}k(x^2 + 2x\Delta x) \tag{7.4}$$

Here the term $(\Delta x)^2$ on RHS has been neglected because it is very small as compared with other terms. Subtracting Eq. (7.3) from Eq. (7.4), we get:

$$\Delta U = a\Delta x + kx\Delta x \tag{7.5}$$

Now, $F = -\dfrac{\Delta U}{\Delta x}$ (Eq. 5.19), where F is the force due to the springs. Therefore,

$$F = -a - kx \tag{7.6}$$

Because $F = 0$ when $x = 0$, a becomes zero.

Hence, $$U = \frac{1}{2}kx^2 \tag{7.7}$$

and $$F = -kx \tag{7.8}$$

Eq. (7.7) is the equation of a parabola [Fig. 7.1(b)]. The oscillations consistent with Eq. (7.7) are called harmonic. It can also be noted that U is a symmetric function of x. If x is large the higher powers of x will also appear in Eq. (7.3). Then the curve for the potential energy is not parabolic and the oscillations are called *anharmonic*. U will also become asymmetric in this case. There is always some anharmonicity in oscillations occurring in nature. But many of them have very small anharmonicity and can be treated as harmonic oscillations. Atoms in molecules and crystals are held together by electromagnetic forces. They can be pictured as being held by springs. Thermal energy agitates the atoms and sets them into vibration. These vibrations are nearly harmonic.

As stated above the vibrations are only nearly harmonic. To analyse these motions a greater number of terms of Eq. (7.2) have to be considered. The coefficients $a, k, b, \ldots$, etc. are found from the experimental data. Sometimes an arbitrary potential function is chosen, which accounts for the anharmonicity satisfactorily. One such potential function is the Morse potential, stated as :

$$U = D(1 - e^{-px})^2 \tag{7.9}$$

Here, D and p are constants and x is the displacement. D determines the depth of the potential. A plot of Eq. (7.9) [Fig. 7.1 (c)] shows asymmetry of U about the ordinate. In crystals the force between the atoms is highly repulsive if the distance between them is less than the equilibrium value. Therefore, the positive displacement of an atom is greater than the negative displacement as is also apparent in Fig. 7.1 (c). Thermal expansion of the crystals is caused by these anharmonicities.

7.4 Analysis of SHM

We see from Eq. (7.8) that k is the restoring force per unit displacement. For a spring it is called the spring constant. We are now in a position to state the salient features of SHM. These are:

(a) The motion is periodic.

(b) The restoring force acting on an oscillating body is proportional to the displacement from the position of equilibrium.

(c) The restoring force is always directed towards the position of equilibrium.

(d) The plot of instantaneous displacement with time is a sine or a cosine curve (Eq. 7.21).

It has already been seen that the potential energy increases with displacement (Eq. 7.7). Let us now analyse SHM from the point of view of energy. The body is initially supplied with kinetic energy to disturb it. As the potential energy builds up, the kinetic energy decreases. Finally all the kinetic energy is converted into potential energy and the body comes to rest at the point of maximum displacement. Here the restoring force acting is also maximum, and as a result the body begins to return to the position of equilibrium. The potential energy goes on decreasing and the kinetic energy builds up. At the position of equilibrium all the potential energy is converted into kinetic energy which is maximum at this point. The resulting momentum pushes the body in the opposite direction with decreasing velocity, because as soon as the body crosses the position of equilibrium the restoring force starts operating. The process goes on like this.

Oscillations, therefore, occur when:

(a) There is a component in the oscillating system to provide the restoring force.

(b) A component possessing inertia is present.

(c) There is a possibility of interconversion of energy between two forms.

In the example considered in Section 7.3 the elastic force of the spring provides the restoring force. The mass of the body O supplies inertia. Energy can exist in two –: the kinetic and the potential – forms.

Another oscillatory system is the discharge of a capacitor C through an inductor L (Fig. 7.2). Here, C is shown charged with the lower plate carrying maximum negative charge. The key K is pressed. The excess electrons on the lower plate repel one another and are driven out. The flow of charge gives rise to a current. This current through the inductor builds up a magnetic field. After the capacitor has completely discharged the magnetic field in the inductor tends to collapse. An induced current is thereby maintained. This current recharges the capacitor with reversed polarity, i.e. the upper plate getting the negative charge. The repulsive force between the electrons on the upper plate stops the charging at some point. The discharging of the capacitor starts again. In this system the repulsive force between the electrons acts as the restoring force, the inductor plays the role of the component possessing inertia, and the energy oscillates between the electric and the magnetic forms given respectively by $q^2/2C$ (Section 19.12) and $LI^2/2$ (Section 26.5).

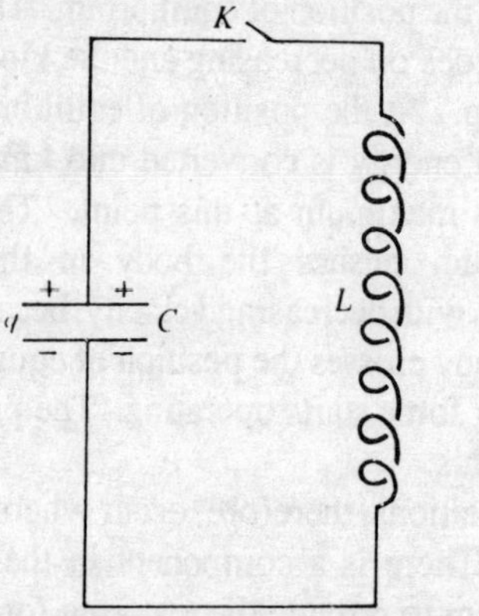

Fig. 7.2 *Electromagnetic oscillations in an LC circuit*

In Fig. 7.3, a spring of negligible mass, attached to a rigid support and resting on a horizontal and frictionless surface holds a block of mass M at O in equilibrium. M is now displaced. The resulting deformation of the spring brings the restoring force $-\mathbf{F}$ into operation. Fig. 7.4 shows the variation of F with displacement x. We know from Example (5.6) that:

$$\frac{1}{2}kx^2+\frac{1}{2}Mv^2=C \tag{7.10}$$

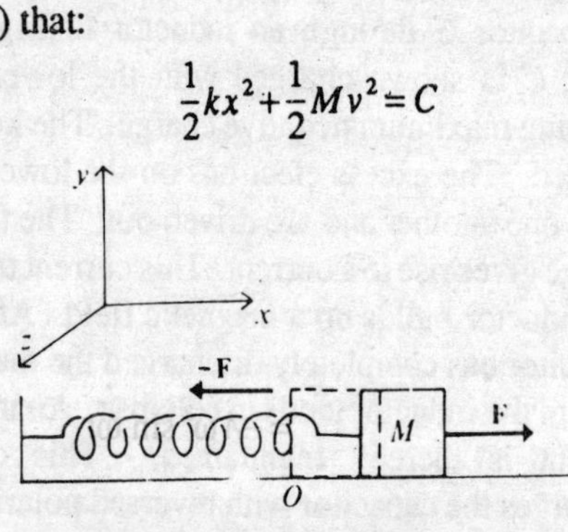

Fig. 7.3 *SHM of a block held by a spring*

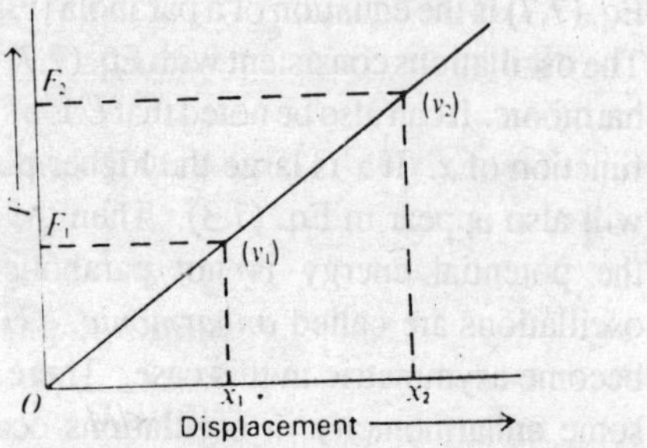

Fig. 7.4 *Spring force as a function of displacement of the block from its equilibrium position*

where C is a constant. At the extreme points where the displacement is A, the kinetic energy is zero.

Therefore, $\frac{1}{2}kA^2=C,$

$$\text{and} \quad \frac{1}{2}kx^2+\frac{1}{2}Mv^2=\frac{1}{2}kA^2 \tag{7.11}$$

$$\text{or,} \quad \frac{v^2}{k/M}+x^2=A^2 \tag{7.12}$$

This is the equation of a circle of radius A, similar to $y^2+x^2=A^2$. Only y has been replaced by $v/\sqrt{k/M}$. Fig. 7.5 is a plot of Eq. (7.12), which is nothing but the velocity-displacement curve of the mass M executing a SHM. As point P moves along a circle of radius A, the foot N of the perpendicular from P on COD moves along CD in the sequence $OCODO$. Thus, as P moves along the circle, N moves on a straight line and its motion is oscillatory. From Fig. 7.5:

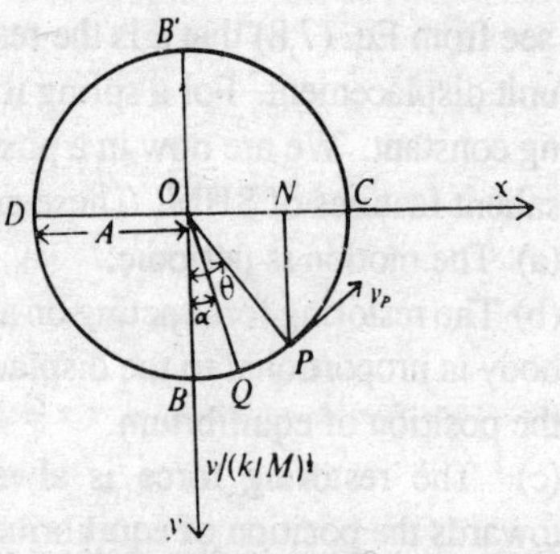

Fig. 7.5 *Reference circle for SHM*

linear velocity of point $P = v_P$

$$= \text{velocity of } N \times \frac{OP}{NP} \quad (7.13)$$

The velocity of N is equal to the ordinate NP on proper scale. Since the ordinate is plotted as $v/\sqrt{k/M}$, velocity at any point is:

$$v = \text{distance along ordinate} \times \sqrt{k/M} \quad (7.14)$$

or,
$$v = NP\sqrt{k/M} \quad (7.15)$$

Substituting the value of v in Eq. (7.13) :

$$v_P = NP\sqrt{k/M} \times \frac{OP}{NP}$$

$$= A\sqrt{k/M} = \text{constant} \quad (7.16)$$

Conversely, if a point moves in a circle with a constant velocity, the foot of the perpendicular from the point to vertical or horizontal diameter executes a SHM. This circle is called the *reference circle*. Note that the motion of M is not the same as the circular motion of point P. The reference circle is a useful representation that helps us in deriving many important relations. As P moves in a circle of radius A with a constant velocity $A\sqrt{k/M}$, the displacement x is along COD, i.e. the movement of M is the same as that of N. Further, the periodic time of N (or M), which is moving simple harmonically, is the same as that of P. Or:

$$T = \frac{2\pi A}{A\sqrt{k/M}} = \frac{2\pi}{\sqrt{k/M}} \quad (7.17)$$

If the reference point P describes an angle θ radians in t seconds:

$$t = \frac{A\theta}{A\sqrt{k/M}}, \text{ or } \theta = t\sqrt{k/M}$$

Therefore, from Fig. 7.5

$$ON = OP \sin\theta = OP \sin(\sqrt{k/M}\, t),$$

or
$$x = A \sin(\sqrt{k/M}\, t) \quad (7.18)$$

We know that $\omega = 2\pi/T$. Using Eq. (7.17):

$$\omega = \sqrt{k/M} \quad (7.19)$$

Thus ω^2 is equal to the restoring force per unit displacement per unit mass.

Hence,
$$x = A \sin\omega t$$

$$= A \sin\frac{2\pi}{T} t \quad (7.20)$$

At time $t = 0, x = 0$; at $t = T/4, x = A$; at $t = T/2$, $x = 0$; at $t = 3T/4, x = -A$.

We obtain Eq. (7.20) if the time is measured from point B of the reference circle. If the SHM starts when P is at point Q on the reference circle, the value of θ at time t is $(\omega t + \alpha)$. Under these conditions we get

$$x = A \sin(\omega t + \alpha) \quad (7.21)$$

Angle $(\omega t + \alpha)$ is called the *phase* of the SHM. The initial phase angle α is also called *epoch* of the oscillation.

To find the instantaneous velocity and acceleration of M we have to find these parameters for N. From Fig. 7.5, the velocity of N is ordinate NP. Measuring time from B, we get:

$$v/\sqrt{k/M} = A\cos\theta,$$

or
$$v = A\omega\cos\omega t \quad (7.22a)$$

On adding the initial phase angle α, Eq. (7.22a) becomes:

$$v = A\omega\cos(\omega t + \alpha), \quad (7.22b)$$

or
$$v = A\omega\sin(\omega t + \alpha + \pi/2) \quad (7.22c)$$

Comparing Eqs. (7.22c) and (7.21), we note that the phase between velocity and displacement differs by $\pi/2$. Velocity and displacement are at quadrature, i.e. when the displacement is maximum the velocity is zero. Similarly, maximum velocity occurs when the displacement is zero, i.e. at the position of equilibrium.

Acceleration of M is the same as that of N. Now

$$a_P = \frac{-v_P^2}{A} = -A\omega^2,$$

therefore
$$a_N = -A\omega^2 \sin\theta \quad (7.23a)$$

$$= -A\omega^2 \sin\omega t \quad (7.23b)$$

$$= -\omega^2 x \quad (7.23c)$$

Therefore $a_N \propto -x$, as required in SHM.

The curves of displacement, velocity and acceleration as a function of time are given in Fig. 7.6 for the case $\alpha = 0$.

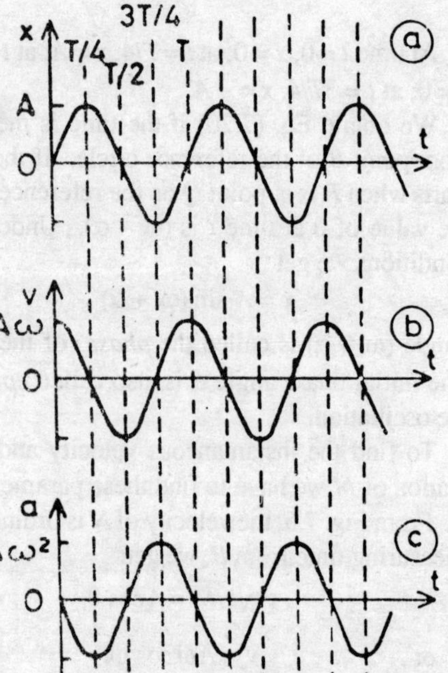

Fig. 7.6 *Variation of* (a) *displacement,* (b) *velocity, and* (c) *acceleration as a function of time.*

Question: A loaded spring executes SHM. At what fraction of the oscillation the kinetic energy and the potential energy are equal? Neglect the mass of the spring.

Example 7.1 Two SHMs are given by the equations:

$$y_1 = 15 \sin 6t, \text{ and } y_2 = 18 \sin (6t + \pi/4)$$

Find: (i) the periodic times of both, (ii) the phase difference between them, and (iii) the velocity amplitude and maximum value of acceleration for both.

(i) Periodic time for both can be found out by using the formula

$$T = 2\pi/\omega = \frac{2 \times 3.14}{6} = 1.05 \text{ s}$$

Note that the periodic time does not change with amplitude and phase angle.

(ii) The phase difference between the two is $\pi/4$. Since 2π radians are completed in one oscillation, $\pi/4$ radian corresponds to 1/8th of an oscillation. Thus y_2 leads y_1 by 1/8th of an oscillation.

(iii) Make use of Eqs. (7.22a) and (7.23a). Since velocity amplitude and maximum value of acceleration are needed, we have to substitute maximum values of cos θ and sin θ, i.e. one, in the above relations.

$$\therefore \quad v_{1max} = A \times \omega = 15 \times 6 = 90 \text{ units},$$

$$v_{2max} = 18 \times 6 = 108 \text{ units},$$

$$a_{1max} = -A \times \omega^2 = -15 \times 36 = -540 \text{ units},$$

$$\text{and } a_{2max} = -18 \times 36 = -648 \text{ units}$$

Example 7.2 Find out the frequency of vibration of an oxygen molecule. Given: $k = 1170 \text{ N m}^{-1}$.

As already pointed out the atoms forming a molecule are held together by electromagnetic forces. The complete interatomic force in oxygen can be expressed in terms of potential energy between the atoms. A Morse potential plot for O_2 is shown in Fig. 7.7 (a). Here OA is the equilibrium distance at which the potential energy is minimum. For large separations the potential energy approaches zero as shown near P. This leads to the dissociation of the molecule, and E_D is the dissociation energy of the molecule.

For small amplitudes of oscillation the Morse plot almost coincides with the parabolic potential energy plot. The oscillations for small amplitudes are, therefore, nearly harmonic. Treating the molecule as two masses m_1 and m_2 held by a spring of spring constant k [Fig. 7.7(b)], we have two equations:

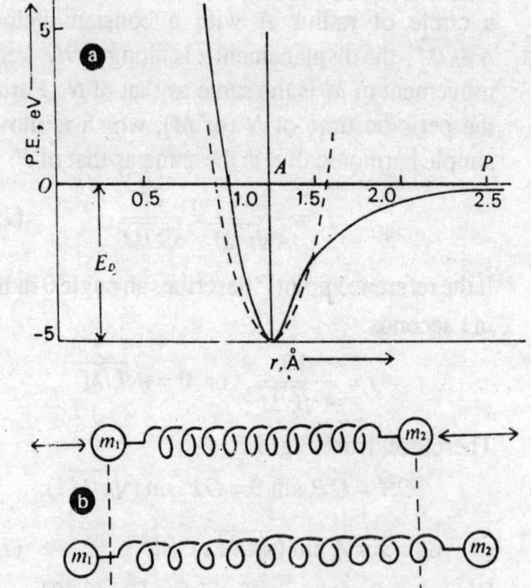

Fig. 7.7 **(a)** ***Morse potential function for an oxygen molecule.*** **(b)** ***Vibration of an oxygen molecule can be visualized as vibration of two masses held by a spring***

$$m_1 a_1 = -kx, \tag{7.24}$$

and

$$m_2 a_2 = -kx \tag{7.25}$$

Multiplying Eq. (7.24) by m_2, Eq. (7.25) by m_1 and adding*.

$$m_1 m_2 (a_1 + a_2) = -(m_1 + m_2)\, kx \tag{7.26}$$

Writing a for $(a_1 + a_2)$:

$$\frac{m_1 m_2}{m_1 + m_2} a = -kx, \tag{7.27}$$

or

$$Ma = -kx, \tag{7.28}$$

where $M = \dfrac{m_1 m_2}{m_1 + m_2}$ is called the reduced mass.

Now, Eq. (7.28) represents a relation for SHM. Therefore using Eq. (7.17):

$$T = 2\pi\sqrt{M/k}, \tag{7.29}$$

or

$$\nu = 1/T = \frac{1}{2\pi}\sqrt{k/M} \tag{7.30}$$

Since for an oxygen molecule $m_1 = m_2 = m_0$

$$M = m_0/2 = 8 \times 1.67 \times 10^{-27} \text{ kg},$$

and

$$k = 1170 \text{ Nm}^{-1},$$

$$\nu = \frac{1}{6.28}\sqrt{\frac{1170 \text{ Nm}^{-1}}{8 \times 1.67 \times 10^{-27} \text{ kg}}}$$

$$= 4.71 \times 10^{13} \text{ Hz}$$

*It is a characteristic of the SHMs that they can be added in this manner. This can happen because the equation of force is linear. If the equation becomes nonlinear as in anharmonic oscillation, i.e. if terms with powers of displacement higher than one occur, then they cannot be added as here. This leads to the breakdown of the superposition principle. For anharmonic oscillations Eq. (7.8) will contain terms of higher powers of x. Suppose x is as large as to justify the inclusion of x^2 term, Eq. (7.8) then becomes:

$$F = -k_1 x - k_2 x^2$$

If $x = A \sin \omega t$ is substituted in this, and as $\sin^2 \omega t$ can be written as $(1 - \cos 2\omega t)/2$, the resulting expression will contain terms of frequency equal to double the fundamental. This is called the second harmonic. Nonlinear oscillations thus generate harmonics of the fundamental. Nonlinear effects have been put to great use in technology. The harmonic generation by non-linear capacitances in the junction diodes (see Section 35.2) have been used to produce low noise microwave frequencies. Very strong electric fields produced by intense laser beams in materials make the dielectric constant nonlinear. Two laser beams then do not combine according to the superposition principle. Operating in nonlinear region they give rise to quadratic effects and light of different frequency is also generated. This is called nonlinear optics.

This frequency lies in the infrared region. Therefore the vibration states of the molecules are studied by infrared spectroscopy.

7.5 Simple Pendulum

A simple pendulum is defined as a point mass suspended from a rigid support by an inextensible weightless suspension (Fig. 7.8). Let the length of the suspension be l. The weight mg of the bob at O is balanced by tension in the suspension. As the bob is displaced sideways the suspension is deviated. The weight mg can then be resolved into two components: one parallel to the suspension balanced by the tension, and the other perpendicular to the suspension acting as restoring force. These components are respectively $mg \cos \theta$ and $mg \sin \theta$ for the angular displacement θ of the suspension. The restoring force $mg \sin \theta$ increases with θ bringing the motion to a halt. The pendulum then swings back with increasing velocity. The inertia of m carries it along to the other side. This way the oscillations continue. At the extreme ends where angular displacement is β:

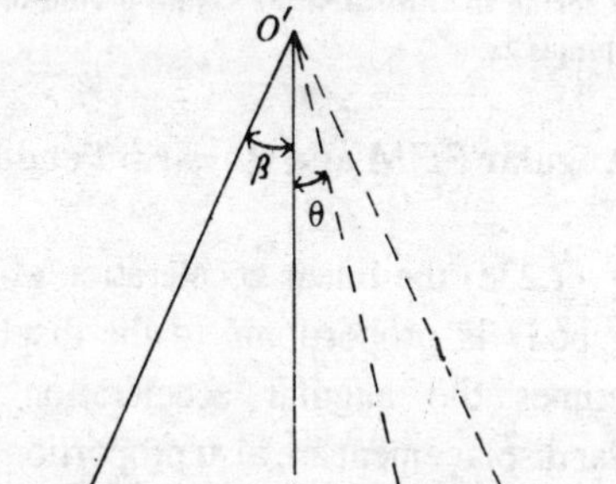

Fig. 7.8 *Simple pendulum*

$$\left.\begin{aligned} mg \cos \beta &= |\mathbf{T}| \\ -mg \sin \beta &= ma \end{aligned}\right\} \tag{7.31}$$

where a is the linear acceleration of m. Therefore,

$$a = -g \sin \beta \tag{7.32}$$

At any other point: $|\mathbf{T}| = mg\cos\theta + mv^2/l$, where v is the speed of the bob. But under the approximation of very small amplitude and large l the term mv^2/l can be neglected as compared with $mg\cos\theta$. Therefore, for a small angular displacement, $\sin\theta = \tan\theta$, or,

$$a = -g\tan\theta \qquad (7.33)$$

$$= -g\frac{x}{l} \qquad (7.34)$$

Hence $\qquad a \infty - x \qquad (7.35)$

The motion of a pendulum is, therefore, simple harmonic. Comparing Eq. (7.23c) with Eq. (7.34), we get:

$$\omega^2 = g/l, \qquad (7.36)$$

or $$T = 2\pi\sqrt{l/g} \qquad (7.37)$$

Question : (a) A seconds pendulum is carried in a satellite revolving around the earth at a height of 1000 km. What is its periodic time? (b) What will be the time period of a loaded spring in this satellite? Time period of a seconds pendulum is 2s.

7.6 Angular SHM and Physical Pendulum

In Eq. (7.23c) the linear acceleration of an oscillating body is proportional to the displacement. Sometimes the angular acceleration and the angular displacement are also proportional to each other. The motion is then called angular SHM.

The simple pendulum discussed above is an idealization. In practice pendulums, like the pendulum of a wall clock, are massive bodies. We call these pendulums physical pendulums or compound pendulums. In Fig. 7.9, a physical pendulum is suspended from O. The weight Mg acts at the centre of gravity G. As the body is displaced G moves through an arc to G' in a rotatory motion. The equation of motion is then:

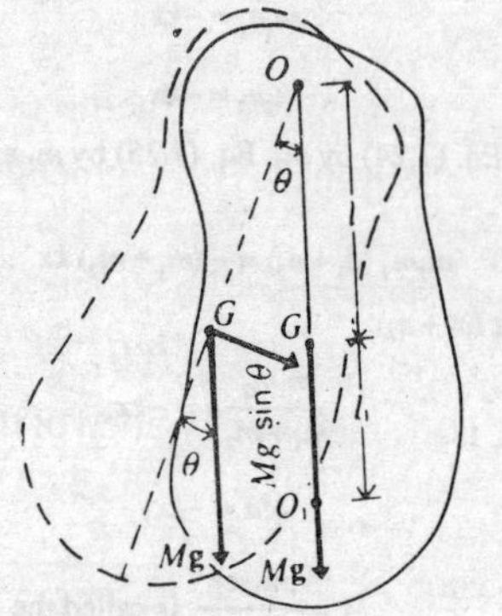

Fig. 7.9 ***Compound pendulum***

$$I\alpha = -Mg\sin\theta \times G'O \qquad (7.38)$$

where α is the angular acceleration, I is the moment of inertia of the pendulum about an axis passing through O and perpendicular to the plane of the pendulum, called the axis of suspension. If $GO = G'O = l$, Eq. (7.38) becomes:

$$I\alpha = -Mgl\sin\theta \qquad (7.39)$$

For small values of θ

$$I\alpha = -Mgl\theta, \qquad (7.40)$$

or $$\alpha = -\frac{Mgl}{I}\theta \qquad (7.41)$$

Therefore, $\qquad \alpha \infty - \theta \qquad (7.42)$

This is the condition of angular simple harmonic motion. Comparing Eq. (7.41) with Eq. (7.23c):

$$\omega^2 = Mgl/I, \qquad (7.43)$$

or $$T = 2\pi\sqrt{I/Mgl} \qquad (7.44)$$

Eq. (7.44) is true for pendulums in general. For a simple pendulum total mass M is due to point mass at the bottom of suspension. The value of I is,

$$I = Ml^2,$$

and $$T = 2\pi\sqrt{Ml^2/Mgl} = 2\pi\sqrt{l/g} \qquad (7.45)$$

This is the expression which we have already obtained for a simple pendulum [Eq. (7.37)].

If I_G is the moment of inertia of the physical pendulum about an axis perpendicular to the plane of the paper and passing through G, then using the theorem of parallel axes (Section 6.8)

$$I = Ml^2 + I_G = M(l^2 + K^2)$$

where K is the radius of gyration of the body about the axis through G. Eq. (7.44) then becomes:

$$T_1 = 2\pi\sqrt{\frac{K^2+l^2}{gl}} \tag{7.46}$$

On inverting the physical pendulum and suspending it from a point O_1 such that $GO_1 = l_1$, Eq. (7.44) becomes (see Fig. 7.9):

$$T_2 = 2\pi\sqrt{\frac{K^2+l_1^2}{gl_1}} \tag{7.47}$$

If T_1 and T_2 are equal:

$$\frac{l^2+K^2}{lg} = \frac{l_1^2+K^2}{l_1g}, \tag{7.48}$$

or

$$l_1l^2 - l_1^2l = K^2(l - l_1),$$

or

$$K^2 = l_1 l \tag{7.49}$$

Therefore,

$$T_1 = T_2 = T = 2\pi\sqrt{\frac{l_1+l}{g}}$$

$$= 2\pi\sqrt{L/g} \tag{7.50}$$

Thus the physical pendulum can be treated as equivalent to a simple pendulum of length $(l + l_1)$. Here, l_1 and l are the lengths from those points of suspension to the centre of gravity for which the physical pendulum has the same periodic time. For given l, l_1 can be obtained from Eq. (7.49). O_1 is the point of oscillation corresponding to the point of suspension O.

Example 7.3 Find the length of an equivalent simple pendulum for an oscillating hoop, supported from a point on its periphery.

The moment of inertia of the hoop about the axis through the center G normal to the plane of the hoop is:

$$I_G = mR^2 \tag{7.51}$$

where m is the mass of the hoop, and R is its radius.

To find MI about an axis AB through point of suspension (Fig. 7.10) and normal to the plane of the hoop, we use the theorem of parallel axes. Therefore:

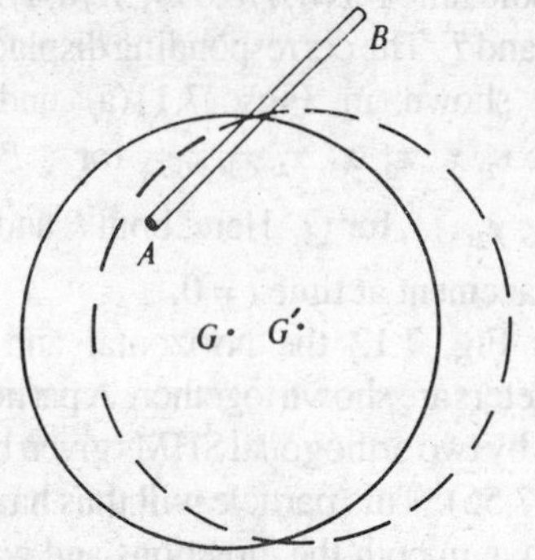

Fig. 7.10 *For Example 7.3*

$$I_s = 2mR^2 \tag{7.52}$$

Using Eq. (7.44)

$$T = 2\pi\sqrt{\frac{2mR^2}{mgR}} = 2\pi\sqrt{\frac{2R}{g}} \tag{7.53}$$

The length of the equivalent pendulum is therefore $2R$.

7.7 Lissajous Figures

Lissajous figures arise by the combination of two orthogonal simple harmonic motions acting on a particle at the same time.

In Section 7.4, it was pointed out that if a point P moves on a reference circle the foot of the perpendicular from P on the horizontal diameter, shown as N, then executes a SHM. The equation of motion is:

$$x = A \sin \omega t \tag{7.54}$$

This is shown again in Fig. 7.11(a). In Fig. 7.11(b) the path of a point M on a vertical diameter is considered. This is the foot of perpendicular from Q which is also moving along the same reference circle as P. The equation of motion of M is:

$$y = A \sin \omega t \tag{7.55}$$

Here, the amplitudes and periods of the two motions have been assumed to be equal. These SHMs are confined along definite directions–first along the x-axis and second along the y-axis. Such vibrations are also called linearly polarised vibrations, since a particular vibration is confined along a straight line in a fixed direction.

To study their combined effect, it will be necessary to know the displacements of both the SHMs at various instants. For convenience we divide the periodic time T as 0, $T/8$, $T/4$, $3T/8$, $T/2$, $5T/8$, $3T/4$, $7T/8$ and T. The corresponding displacements have been shown in Figs. 7.11(a) and 7.11(b) as $x_0, x_1, x_2, x_1, x_0, x_3, x_4, x_3, x_0$ for P; and as $y_0, y_1, y_2, \ldots,$ for Q. Here, both P and Q have zero displacement at time $t = 0$.

In Fig. 7.12 the horizontal and the vertical diameters are shown together. A particle O is acted upon by two orthogonal SHMs given by Eqs. (7.54) and (7.55). This particle will thus have a tendency to move in both the directions and will, therefore, choose the resultant path. This is due to the vector addition of two orthogonal displacements. At a particular time, the point for the resultant displacement, shown as ×, has been plotted after measuring x and y along the x- and y-axes at that time (Fig. 7.12). The values of x's and y's can be taken from Figs. 7.11 (a) and 7.11 (b). The resultant on joining these points in Fig. 7.12 gives a straight line. The resultant motion starts from O and follows a straight line path in the order $OAOBO$. Thus, it is also a periodic motion of the same period as its components.

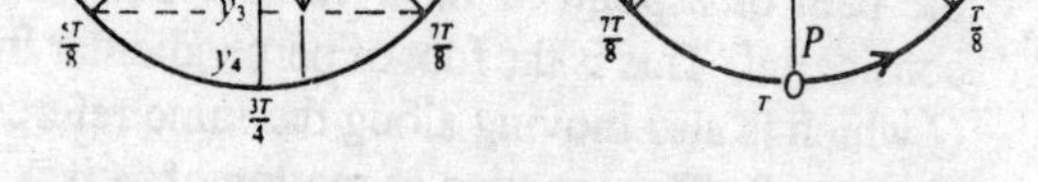

Fig. 7.11 *Representation of two SHMs with the help of reference circles* : **(a)** *along the x-axis* **(b)** *along the y-axis*

As a second example take the following two SHMs:

$$x = A \sin \omega t, \tag{7.56}$$

and $$y = A \cos \omega t,$$

$$= A \sin (\omega t + \pi/2) \tag{7.57}$$

Here, y leads x by a phase angle of $\pi/2$. It means y has completed 1/4th of an oscillation already when x starts moving. Since y has completed 1/4th of an oscillation, it has already gone from y_0 to y_2 [Fig. 7.11(b)] and is on its way back when the motion of the particle along the x-axis just starts [from x_0 in Fig. 7.11(a)]. The resultants of the displacements y and x at $t = 0, T/8, T/4$, etc., are plotted in Fig. 7.13. One can see that the resulting curve is a circle. The point now moves in a circle of radius A in a clockwise direction. The vibration is then said to be circularly polarised. If, instead, we had chosen:

$$x = A \cos \omega t, \tag{7.58}$$

and $$y = A \sin \omega t \tag{7.59}$$

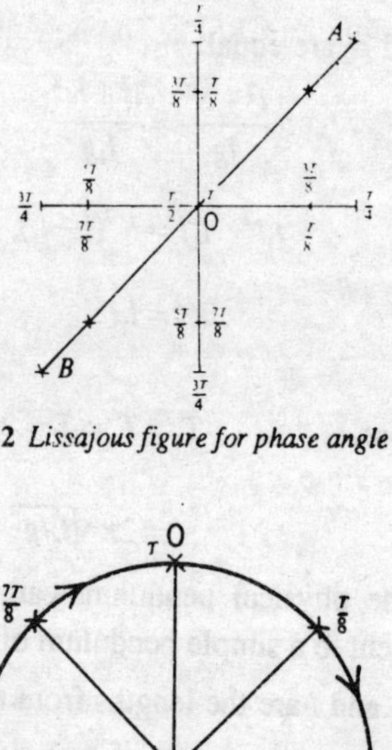

Fig. 7.12 *Lissajous figure for phase angle zero*

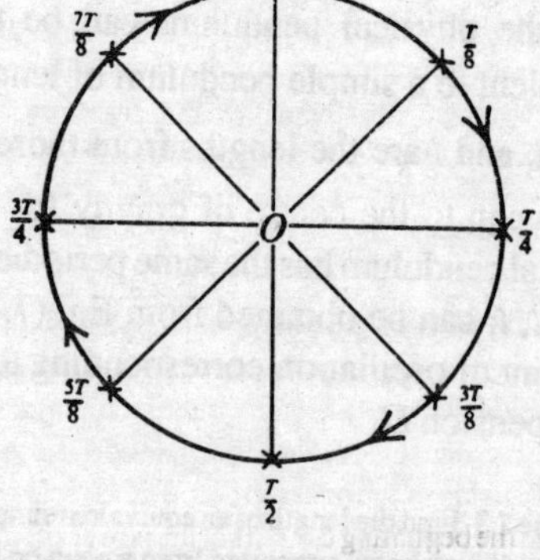

Fig. 7.13 *Lissajous figure for phase angle of π/2*

a circle with anticlockwise rotation would have resulted. If the amplitudes in Eqs. (7.56) to (7.59) differ, an ellipse will be formed. Thus, from the shape of the curve obtained by the vectorial addition, the resultant motion due to two orthogonal SHMs acting simultaneously on a particle can be obtained. The Lissajous patterns for various phase differences between x and y are shown in Fig. 7.14.

These patterns can be observed experimentally with the help of a CRO*. The ac voltages across the passive reactive and resistive elements in a circuit differ in phase (see Chapter 27). The phase difference can be adjusted by adjusting these passive elements. The voltages are then fed to the horizontal and the vertical plates of the CRO. The attenuator points for both the channels are adjusted at the same value so that the signals are equally amplified. The resultant pattern on the screen shows the Lissajous figure.

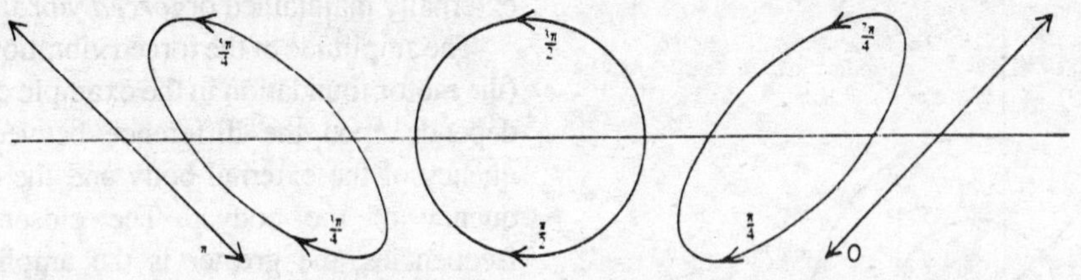

Fig. 7.14 *Lissajous diagrams for various values of phase angles. The direction in which the figure will be traced for a particular phase difference is shown*

*CRO or Cathode Ray Oscilloscope is an electronic instrument for visual examination of electrical signals. The instrument has two main components—a cathode ray tube, and the associated circuitry. The tube has a cathode which emits thermally excited electrons. These electrons pass through a system of electrodes. The field distribution between the electrodes focuses the stream of electrons into a narrow pencil and imparts a high velocity to them. This pencil of fast-moving electrons passes through two pairs of plates which are at right angles to each other. The stream then impinges on a fluorescent screen producing a bright spot there. The signals to be tested are fed into one of the two pairs of plates. This produces an electric field between the plates which deflects the electron beam.

If both the pairs of the plates are supplied with electrical signals the electron beam is influenced by two orthogonal fields. The resulting pattern on the screen is thus a Lissajous diagram. If the available signal is feeble and cannot deflect the electron stream appreciably, it can be amplified. The associated circuitry referred to in the beginning consists of the power supply circuits, the amplifying circuits, and a network to supply calibrated horizontal deflection. The latter network is called *time base* or *sweep circuit*.

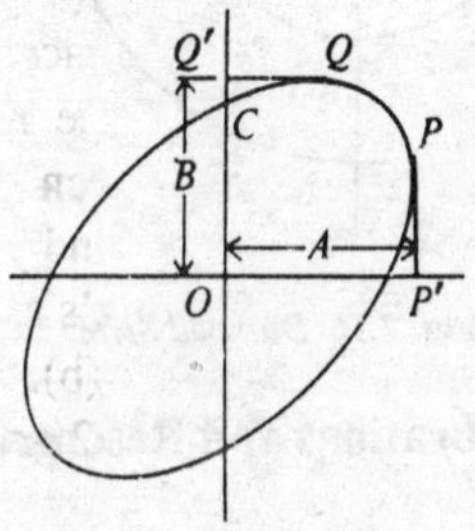

Fig. 7.15 *For Example 7.4*

Example 7.4 Lissajous figure for two orthogonal SHMs of the same frequency is given in Fig. 7.15. What is the phase angle between the two SHMs ?

Let the required phase angle be ϕ. Therefore, the two SHMs are:

$$x = A \sin \omega t \quad (7.60)$$

$$y = B \sin (\omega t + \phi) \quad (7.61)$$

In Fig. 7.15 draw two tangents PP' and QQ' on the curve parallel to the y-axis and the x-axis respectively. Obviously, $OP' = A$, and $OQ' = B$. When the resultant is OC, $x = 0$. It occurs at time $t = 0$ [Eq. (7.60)]. Therefore, from Eq. (7.61):

$$OC = B \sin \phi,$$

or

$$\phi = \sin^{-1}(OC/OQ') \quad (7.62)$$

7.8 Damped SHM

There cannot be any SHM which of itself will keep on vibrating for ever. Oscillations of a simple pendulum die off after some time. The reason is the friction between the moving system and the air. In other words, the oscillating systems are never conservative in practice. Their energy is dissipated by friction into useless energy. The loss in energy decreases the amplitude of the oscillating body. This is called a damped SHM. The reference diagram of a damped SHM is shown in Fig. 7.16.

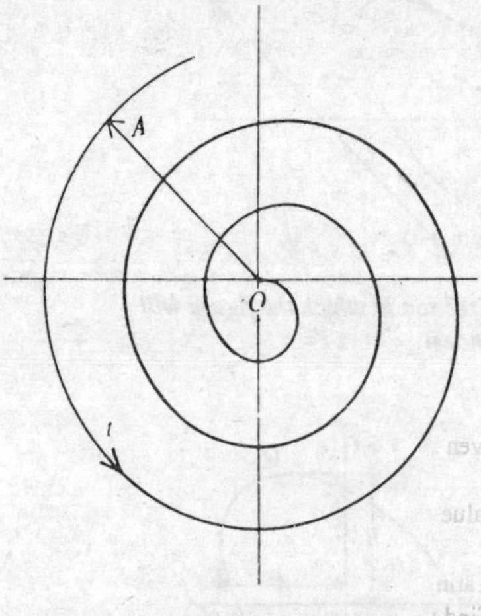

Fig. 7.16 *Damped SHM*

7.9 Forced Vibrations and Resonance

Every system has a natural frequency of vibration. If made to vibrate by applying an impulse, then in the beginning the motion is very complex due to the presence of several frequencies at the same time. Gradually all the frequencies except the natural frequency of the body die off, and then the body is said to be vibrating freely. The free vibrations also cannot continue for long. This is because of friction between the vibrating body and the medium. In this sense the vibrations are not entirely free. Accurately stated, the body is executing a damped oscillation with the natural frequency. If a vibrating body is put in contact with a body at rest, then the vibrations of the former are communicated to the latter. For example, an electric motor while working may vibrate at its natural frequency. The foundation on which the motor is mounted also starts vibrating. In the beginning all sorts of frequencies exist in the foundation. Slowly all but one of the frequencies die off. The frequency which is maintained is the frequency with which the electric motor vibrates. If it is not so, the phase difference between the motor and the foundation will keep on changing, and the steady state will not be reached. The energy required to maintain these vibrations comes from the electric motor. Such vibrations are called externally maintained or *forced vibrations.*

The amplitude of the forced vibrations of a body (the motor foundation in the example cited above) depends upon the difference between the frequency of the external body and the natural frequency of the body. The closer are these frequencies, the greater is the amplitude. The amplitude also depends upon the damping factor arising out of the dissipative mechanism. If the frequency of the two bodies is the same the amplitude rises sharply, and the magnitude is restricted by damping only. This phenomenon is called *resonance.* Sometimes during resonance the amplitude of the forced vibration becomes so large as to damage the body. Electrical machines have been reported to have cracked due to the resonance phenomenon. Suspension bridges have collapsed when the frequency of oscillations caused by wind equalled the natural frequency of the bridge.

It is clear from the above discussion that the effective way of maintaining the natural vibration of a body is to apply a forced vibration of the same frequency. Consider now a simple pendulum vibrating at its natural frequency. Left to itself the vibrations of the pendulum would die off. To maintain the vibrations, energy has to be fed to the pendulum periodically. But at which instant should the energy be supplied? In Fig. 7.17(a) the energy is supplied by giving a push to the right every time the pendulum, A, reaches the extreme left and is starting its journey back. To supply the necessary energy to this pendulum the kinetic energy of the second pendulum, B, should be maximum. It means that the two pendulums are having a phase difference of $\pi/2$. This is because whereas the first pendulum, A, completes a quarter of vibration the second pendulum, B, completes one half. To maintain this phase relation between them, the frequencies of the two pendulums should either be equal or the frequency of the second should be a submultiple of the first.

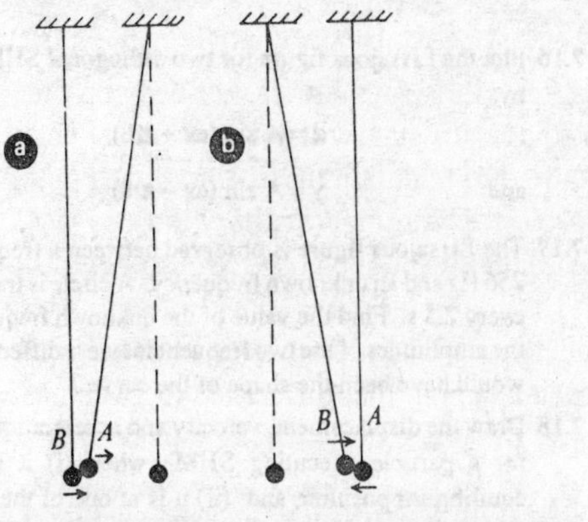

Fig. 7.17 *Vibrations of one pendulum as maintained by another pendulum called the driver.* (a) *Energy transfer is such as to help build up the vibrations of the driven.* (b) *Energy transfer hinders the motion of the driven*

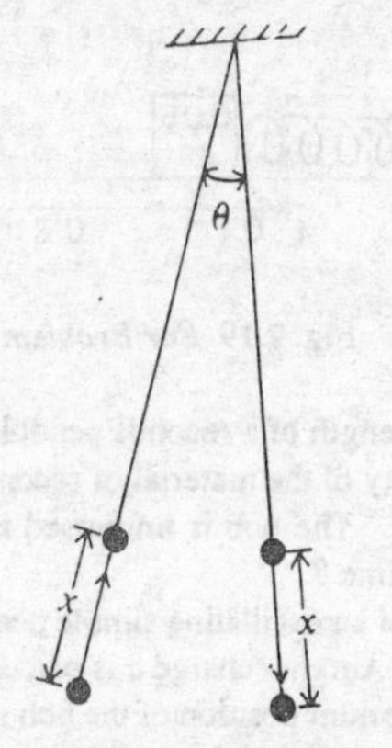

Fig. 7.18 *Parametric oscillations*

In Fig. 7.17(b), the second pendulum feeds the energy to the first when it is crossing the position of equilibrium to the left. The push from the second is now opposite to the motion of the first. The oscillations in this case get suppressed. It is, therefore, very important to supply the maintenance energy at the right phase and frequency.

The above examples deal with externally maintained oscillations. Sometimes the oscillations can be self maintained. Consider the simple pendulum in Fig. 7.18. The length of the pendulum is increased by an amount x every time it is at the position of equilibrium. The mass m loses potential energy equal to mgx. This energy is delivered to the vibrations of the pendulum. The length of the pendulum is decreased at the extreme ends. The mass m now gains potential energy equal to $mgx \cos \theta$. Since $mgx \cos \theta < mgx$, there is a net loss $mgx (1 - \cos \theta)$ in the potential energy of m. This amount of energy is pumped into the simple pendulum in every beat of the oscillation. The oscillation can thus be maintained by a constant supply of energy. In this type of self maintenance a parameter l of the pendulum is being changed. Therefore, the arrangement is called a *parametric amplifier*. The word amplifier is added because the supply of energy can also amplify the oscillations. The principle of electronic amplifiers is similar to the above example. One of the parameters of the device is changed in such a way that the potential energy due to a dc electric field is pumped into the R.F. signal with a consequent amplification.

Discussion and Problems

7.1 A SHM is given by the equation:

$$x = 5 \sin (2\pi t + \pi/4)$$

What is the value of x at $t = 0$? When is the acceleration maximum?

7.2 A body is vibrating with a period of 10 s and an amplitude of 0.025 m. Find the maximum value of the acceleration.

7.3 A point is executing SHM with a frequency of 20 Hz. It has an initial displacement of 0.01 m and velocity of 0.4π m s^{-1}. Write equations for the displacement, velocity and acceleration for all times.

7.4 A light spring loaded with a mass M is hung vertically. Addition of mass m extends the spring by a distance x. It is now pulled with a force F and released. Derive an expression for the periodic time in terms of m, M, g and x. What is the mean position of the oscillation?

7.5 The mass m in Problem (7.4) is removed. Do the periodic time and the mean position of oscillation change? Do the periodic time and mean position of oscillation change if g is changed? Explain.

7.6 A spring carrying a load is cut into two at the middle. The load is now hung from the two halves joined in parallel. If the spring constant of the original spring is k and mass of the load is M, find the periodic time.

7.7 Two springs with spring constants k_1 and k_2 are placed in series. The combined system is loaded with mass M. Find the periodic time.

7.8 A mass of 0.05 kg is placed in contact with a horizontally placed spring of $k = 10$ N/m. The spring is compressed through 0.1 m (Fig. 7.19). It is then released. The mass is thrown off and strikes another spring of the same spring constant placed 0.2 m away. Calculate (i) the time taken by the mass to cover the distance AB, and (ii) the time interval between the two positions of rest of the mass.

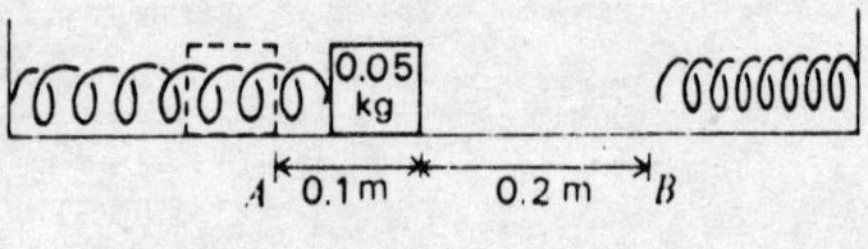

Fig. 7.19 *For Problem 7.8*

7.9 Find the length of a seconds pendulum at the moon.

7.10 The density of the material of seconds pendulum is 8.9×10^3 kg/m^3. The bob is immersed in water. What is its periodic time ?

7.11 The bob of an oscillating simple pendulum is given some charge q. Another charge q is placed at a distance d from the equilibrium position of the bob so that $d \gg A$, where A is the amplitude of the oscillating pendulum. Will the motion be simple harmonic ?

7.12 A hole is dug through the centre of the earth along a diameter. A mass m is gently released into it. Will the mass (i) come out from the other end ? (ii) not go down ? (iii) come to rest at the centre ? (iv) oscillate for ever ? (v) come to rest at the centre after executing some oscillations of decreasing amplitude ?

7.13 A block of mass m rests on a platform. The platform is given up and down SHM with an amplitude d. What can be the maximum frequency so that the block never leaves the platform ?

7.14 The speed of a bullet can be measured by a ballistic pendulum. Derive an expression for the velocity of the bullet in terms of the periodic time, the amplitude of the pendulum, the mass of the bullet, and the mass of the pendulum bob.

7.15 Find out the frequency of vibration of a HCl molecule. The force constant is 514 N/m.

7.16 Plot the Lissajous figure for two orthogonal SHMs given by:

$$x = A \sin (\omega t + \pi/8),$$

and $$y = A \sin (\omega t - \pi/8)$$

7.17 The Lissajous figure is observed between a frequency of 256 Hz and an unknown frequency. A circle is traced after every 2.5 s. Find the value of the unknown frequency. If the amplitudes of the two frequencies were different, what would have been the shape of the curve ?

7.18 Draw the displacement, velocity and acceleration vectors for a particle executing SHM when (i) it is at the equilibrium position, and (ii) it is at one of the extreme ends.

7.19 A U-tube of uniform cross-section contains mercury column of length l. The column is pressed a little on one side and let go. Show that the frequency of oscillation of the mercury column is:

$$\frac{1}{2\pi}\sqrt{\frac{2g}{l}}$$

7.20 A straight tunnel is dug through the earth which may or may not pass through its centre. If a particle is let go at one end of the tunnel show that it will execute SHM, and its period will be $2\pi\sqrt{3/4\pi\rho G}$. Assume the earth to be a sphere of uniform density ρ.

7.21 By eliminating t in Eqs. (7.60) and (7.61) show that the general equation for the Lissajous curve is:

$$\frac{x^2}{A^2} + \frac{y^2}{B^2} - \frac{2xy}{AB} \cos \phi = \sin^2 \phi$$

Starting from this relation derive Eq. (7.62).

Chapter 8

Fluid Mechanics

8.1 Fluids and Pressure—Pascal's Law

A fluid can be broadly defined as a substance that can flow. Thus the term fluid, in general, includes liquids and gases. Unlike solids the fluids do not have any regular shape or size. The shape of a fluid depends upon the container in which it is kept. The two important parameters frequently used in the study of fluids are density and pressure.

The *density of a fluid, ρ, is defined as its mass per unit volume.* This may depend upon many factors, such as the fluid's temperature and pressure. Mathematically, we write:

$$\rho = \text{mass/volume} = M/V \qquad (8.1)$$

If we exert a force on a solid along its surface, the solid can withstand the force. But this is not possible in the case of fluids. The fluid layers slip over one another just like a pack of cards. The force on a fluid at rest can only be normal to its surface. The effect of such forces depends on the area of the surface of the fluid. To express it quantitatively we introduce a parameter p, called pressure, which is defined as the magnitude of the normal force per unit surface area. Or:

$$\mathbf{F} = p\mathbf{A} \qquad (8.2)$$

As usual, the direction of **A** is normal to the surface, and outwards from the surface. Pressure p is a scalar whose unit in the SI is N m^{-2} or pascal (Pa). Other common units are dyn/cm^2, lb/in^2, atmosphere, mm-Hg, and bar (1 bar = 10^6 dyn/cm^2).

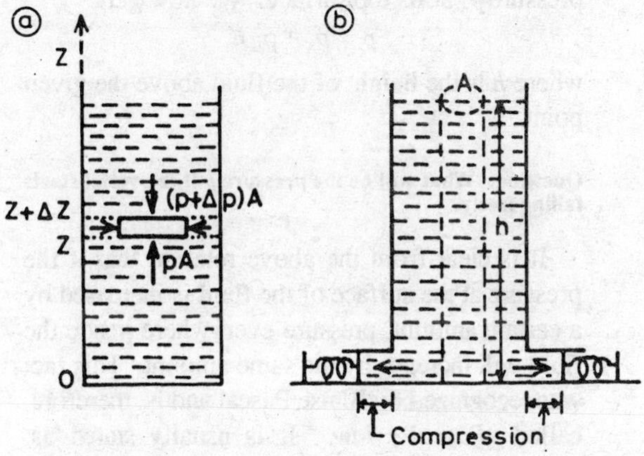

Fig. 8.1 **(a)** *Pressure inside a fluid* **(b)** *Fluids exert pressure in all directions*

Take a fluid in a vessel [Fig. 8.1(a)]. When the vessel is in equilibrium, every element of the fluid will also be in equilibrium. Take a slab of the fluid of thickness Δz at a height z from the bottom. Let its area of cross-section be A. Force exerted on this element by the surrounding fluid is everywhere normal to its surface. As there is no external force in the horizontal direction, horizontal forces exerted by the surrounding fluid on its vertical sides will be equal, and their sum will be zero. Let us now consider horizontal faces. Since there is equilibrium in the vertical direction also,

$$\Sigma \mathbf{F}_z = 0,$$

or $(p + \Delta p)A$ + weight of the fluid in the element = pA,

or $(p + \Delta p)A + A\Delta z \rho g = pA$,

or $$\Delta p = -\rho g \Delta z,$$

or $$\frac{\Delta p}{\Delta z} = -\rho g$$

As ρ and g are both positive, Δp and Δz will have opposite signs. Therefore, as we go up, pressure inside the fluid decreases. If p_1 and p_2 are pressures at heights z_1 and z_2, respectively, then,

$$p_2 - p_1 = -\rho g (z_2 - z_1)$$

Let p_1 now represent pressure p at any point inside the fluid, and p_2 represent atmospheric pressure p_A at its top surface. We now get

$$p = p_A + \rho g h$$

where h is the height of the fluid above the given point.

Question: What will be the pressure *p* if the vessel starts falling freely?

It is clear from the above relation that if the pressure at the surface of the fluid is increased by a certain amount, pressure everywhere inside the fluid will increase by the same amount. This fact was recognized by Blaise Pascal and is, therefore, called *Pascal's law*. It is usually stated as: 'Pressure applied to an enclosed fluid is transmitted undiminished to every portion of the fluid and the walls of the containing vessel'.

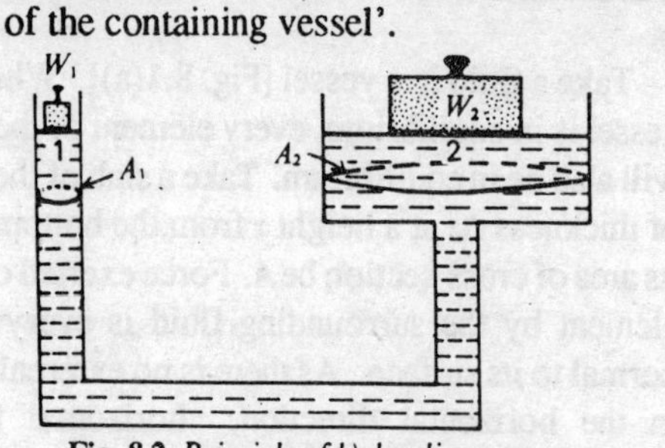

Fig. 8.2 *Principle of hydraulic press*

If we fit pistons of unit area of cross-section attached with springs of known force constant as in Fig. 8.1(b), we can verify that the compression produced on the springs shows a force equal to $\rho g h$. This shows that the fluid pressure is acting in all directions and depends only on the depth at which measurements are made.

About the fluid pressure we thus know that: (i) it is the normal force exerted by the fluid per unit area, and (ii) pressure at one point is equal to the pressure at any other point at the same level.

Let us consider two tubes of different areas of cross-section A_1 and A_2, connected with each other (Fig. 8.2). If the surface at 1 is compressed by a weight W_1, then the surface at 2 is capable of withstanding a weight $W_2 = W_1 A_2/A_1$ because:

pressure at 1 = pressure at 2 (both being at the same level),

or $$\frac{W_1}{A_1} = \frac{W_2}{A_2},$$

or $$W_2 = \frac{A_2}{A_1} W_1 \tag{8.4}$$

If $A_1 = 5\ \text{cm}^2$ and $A_2 = 500\ \text{cm}^2$, we see that by using a force of 1 N at A_1 we can produce a total force of 100 N at A_2. This device is therefore a force transformer. This principle is used in constructing a hydraulic press.

Question: Why does the atmospheric pressure fall as we go up a mountain or any elevated place?

Example 8.1 Archimedes' principle states that a body immersed, wholly or partly, in a fluid is buoyed up with a force equal to the weight of the fluid displaced by the body. Explain.

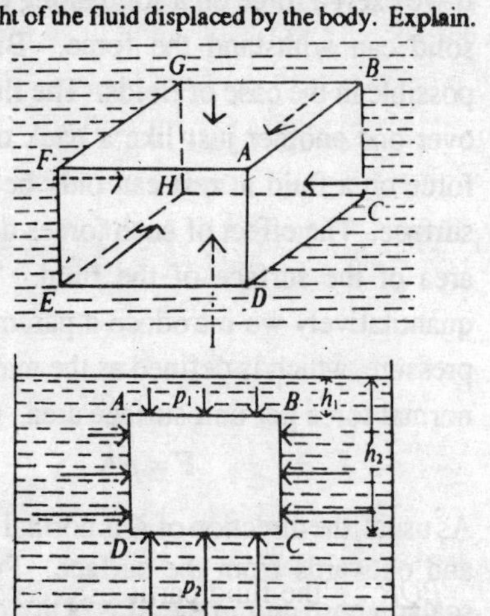

Fig. 8.3 *Archimedes' principle*

Let us consider a rectangular block $ABCDEFGH$ immersed in a liquid of density ρ (Fig. 8.3). Inside the liquid the forces due to liquid pressure balance out on areas $ADEF$, $BCHG$, $ABCD$ and $EFGH$. But the pressure on $ABGF$ is only $\rho g h_1$ in downward direction while the pressure on $CDEH$ is $\rho g h_2$ in upward direction. Therefore, the net upward force on the block is:

$$\rho g h_2 \times \text{area } CDEH - \rho g h_1 \times \text{area } ABGF$$
$$= \rho g (h_2 - h_1) \times \text{area } ABGF$$
$$= \rho g V$$

where V is the volume of the block. Hence, the force of buoyancy equals the weight of the liquid displaced.

Question: What is the force on face ABCD?

8.2 Streamline and Turbulent Motions

In the previous section we have considered only the stationary state of a fluid. Let now the particles of the fluid be in motion. When the fluid velocity at any given point is constant of time, the fluid motion is said to be steady. In Fig. 8.4, whichever fluid particle passes through P it has a velocity $\mathbf{v}_P$ in a steady flow. When one fluid particle leaves P with velocity $\mathbf{v}_P$, it is followed by another fluid particle. When this second particle leaves P it also has a velocity $\mathbf{v}_P$. This process will continue and every subsequent particle will leave P with the same velocity $\mathbf{v}_P$. This thing must hold good for all other points in the fluid, i.e. the fluid particles should always leave Q with velocity $\mathbf{v}_Q$, R with $\mathbf{v}_R$, A with $\mathbf{v}_A$, B with $\mathbf{v}_B$ and so on. It is not necessary for these velocities to be equal.

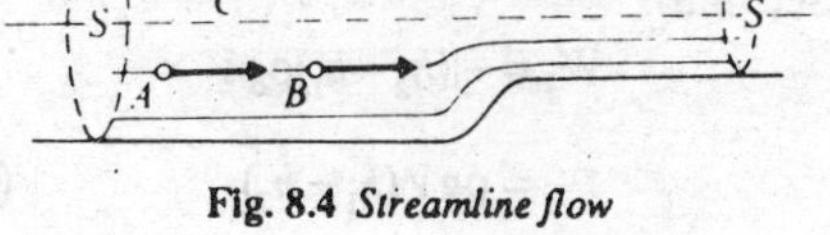

Fig. 8.4 *Streamline flow*

In a steady flow, let a fluid particle follow the path PQR. If the fluid particles preceding it and following it also follow the same path PQR, the path is called a *line of flow* or a *streamline*. A streamline is parallel to the velocity of the fluid particles. Tangent at a point on the line of flow gives the direction of velocity of the particles at that point.

Question: Can two steamlines cross ?

The streamline motion is possible only when the fluid velocity is low and the velocity along a path does not change too abruptly and too frequently.

If the velocity is high, or the change in velocity is frequent and very abrupt, the motion ceases to be a steamline flow. This will happen in a tube having area of cross-section too large at some places, too small at some other places, and sharp bends. Such a motion is called a turbulent motion. If an element of a fluid at each point has no net angular velocity about that point, the fluid flow is irrotational. If there is a net angular momentum, the flow will be obstructed by the formation of whirlpools (fluid particles start moving in circular paths during their flow). Irrotational steamline motion can quite easily be treated mathematically.

The motion of a fluid is obstructed by its *viscosity*, just as the motion of solids is obstructed by friction. The viscous force acts between various layers of the fluid. When we neglect viscous forces, the flow is called nonviscous. First of all we will discuss the fluid motion which is steady, irrotational and nonviscous. We will also assume that the density of the fluid remains constant throughout the fluid, i.e. the fluid is non-compressible.

A collection of streamlines is called a tube of flow. There will be lines of flow at its surface also. As two lines of flow do not cross each other, the fluid will neither go out of the tube nor will it enter into the tube from outside. The fluid in a tube of flow will remain in the space enclosed by the lines of flow at its surface. This way a tube of flow can be compared with an ordinary tube of variable cross-section. Let the area of cross-section of a thin tube of flow be S at one place and S' at another (Fig. 8.4).

Since the flow is irrotational and fluid is non-viscous, the velocity for all fluid particles entering S is the same, say v. Let the velocity of all fluid particles leaving at S' be v'. Because the fluid is non-compressible, at steady state, the quantity of fluid that enters the tube in time Δt through S is equal to the quantity of fluid that leaves the tube at S'. Now:

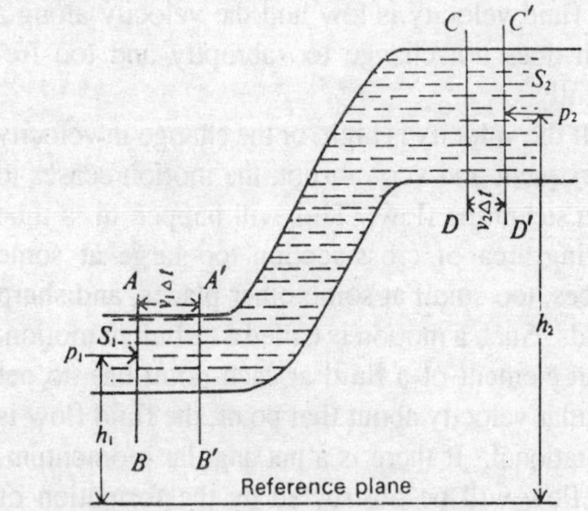

Fig. 8.5 *A portion of fluid moves through a pipe from position 1 to position 2*

The volume of fluid that enters S in time Δt

$= $ area of cross-section $\times$ distance travelled

$$= S \times v \Delta t \qquad (8.5)$$

The volume of the fluid that leaves S' in time

$$\Delta t = S'v'\Delta t$$

Hence $\qquad S v \Delta t = S'v'\Delta t,$

or $$Sv = S'v' \qquad (8.6)$$

There should be no source or sink of the fluid in between S and S'. Eq. (8.6) is called the *equation of continuity*. It states that the velocity of flow through a tube is inversely proportional to the area of cross-section of the tube.

8.3 Bernoulli's Equation

The fundamental equation of hydrodynamics (the mechanics of fluid flow) is Bernoulli's equation. This equation can be derived from Newton's laws, and connects pressure, velocity and height of a fluid at various places.

A fluid is flowing through a tube (Fig. 8.5). Consider the fluid between cross-sections AB and CD. The height of AB from a reference level (say, the surface of earth) is h_1, and its cross-section is S_1. The corresponding parameters of CD are h_2 and S_2. The cross-section of the pipe may vary between AB and CD. Let at S_1 the fluid pressure be p_1 and the fluid velocity be v_1. In a small time interval Δt the layer of the fluid at AB is pushed to $A'B'$ under the influence of force p_1S_1. Thus the displacement of AB is $l_1 = v_1\Delta t$. Hence, the work done on the system is

$$W_1 = \text{Force} \times \text{displacement}$$

$$= p_1 S_1 l_1,$$

or $$W_1 = p_1 S_1 v_1 \Delta t \qquad (8.7)$$

The layer at CD advances to $C'D'$ by a distance $l_2 = v_2\Delta t$. If the fluid pressure and the velocity are p_2 and v_2 at S_2, the work done by the system is:

$$W_p = p_2 S_2 l_2 = p_2 S_2 v_2 \Delta t \qquad (8.8)$$

Hence, the net work done by the fluid pressure is

$$W_p = W_1 - W_2 = p_1 S_1 l_1 - p_2 S_2 l_2$$

$$= p_1 V - p_2 V$$

$$= (p_1 - p_2)V \qquad (8.9)$$

Here, $V = S_1 l_1 = S_2 l_2$ is the volume of the fluid that flows through the tube in time Δt.

During the time Δt, the mass ρV of the fluid has been lifted from height h_1 to h_2. The work done on the fluid by the gravitational field can be written as (Sec. 5.3):

$$W_g = -(h_2 - h_1)\rho g V$$

$$= \rho g V (h_1 - h_2) \qquad (8.10)$$

Eqs. (8.9) and (8.10) give the total work done on the system as:

$$W = W_p + W_g$$

$$= (p_1 - p_2)V + \rho g V (h_1 - h_2) \qquad (8.11)$$

The increase in kinetic energy of the fluid element (ρV) is:

$$\frac{1}{2}(\rho V)(v_2^2 - v_1^2)$$

$$\therefore \quad \frac{1}{2}(\rho V)(v_2^2 - v_1^2) = (p_1 - p_2)V + \rho g V(h_1 - h_2)$$

or, $$\frac{p_1}{\rho g} + \frac{v_1^2}{2g} + h_1 = \frac{p_2}{\rho g} + \frac{v_2^2}{2g} + h_2 \qquad (8.12)$$

or, $$\frac{p}{\rho g} + \frac{v^2}{2g} + h = \text{constant} \qquad (8.13)$$

Eq. (8.13) is the Bernoulli's equation for streamline flow without resistance. Every term in this equation has dimensions of length and is called a *head*. The term $p/\rho g$ is called the pressure head, $v^2/2g$ the velocity head, and h the elevation head or gravitational head.

Question: From Bernoulli's theorem, show that the pressure at a depth h in a liquid is $p + \rho g h$, where p is the atmospheric pressure.

Example 8.2 Discuss the flow of water through a horizontal tube at the bottom of a water tank shown in Fig. 8.6.

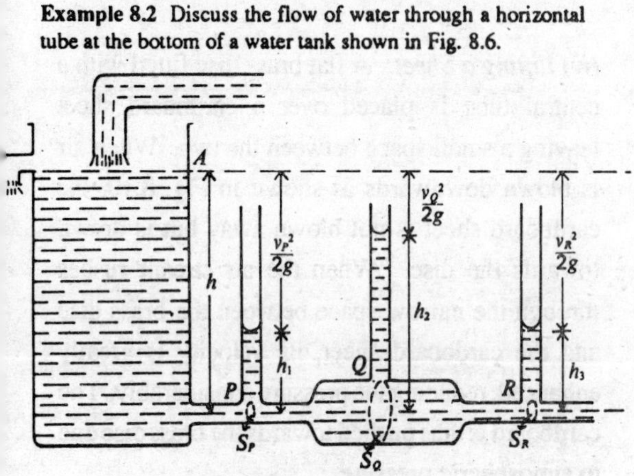

Fig. 8.6 *For Example 8.2*

The water level in the tank is kept fixed. This ensures a constant pressure in the tank. The height of water at the points P, Q and R represents the pressure of water inside the tube at these points. The level of A is higher than the level of P, Q and R. From Bernoulli's theorem, at A, P and Q, we have:

$$\frac{p_A}{\rho g} + \frac{v_A^2}{2g} + h_A = \frac{p_P}{\rho g} + \frac{v_P^2}{2g} + h_P = \frac{p_Q}{\rho g} + \frac{v_Q^2}{2g} + h_Q$$

Now, $p_A =$ the atmospheric pressure p, $v_A = 0$, $h_A = h$,

$$p_P = \rho g h_1 + p, \quad h_P = 0,$$

$$p_Q = \rho g h_2 + p, \quad h_Q = 0,$$

Therefore

$$\frac{p}{\rho g} + h = \frac{p + \rho g h_1}{\rho g} + \frac{v_P^2}{2g} = \frac{p + \rho g h_2}{\rho g} + \frac{v_Q^2}{2g},$$

or $$h = h_1 + \frac{v_P^2}{2g} = h_2 + \frac{v_Q^2}{2g}$$

The heights $(h - h_1)$ and $(h - h_2)$ appear at P and Q as velocity heads. From continuity condition: $S_P v_P = S_Q v_Q$. As $S_P < S_Q$, $v_P > v_Q$. Hence, $h_1 < h_2$. This principle is used in a venturimeter, an instrument used for measuring the velocity of a fluid in a pipe. It can be easily deduced that:

$$v_Q = S_P \sqrt{\frac{2g(h_2 - h_1)}{S_Q^2 - S_P^2}}$$

and the volume of the fluid passing through the pipe in one second is:

$$V = S_Q S_P \sqrt{\frac{2g(h_2 - h_1)}{S_Q^2 - S_P^2}}$$

A few illustrative examples of Bernoulli's theorem are now discussed.

(i) Velocity of Efflux of a Liquid: Let a liquid of density ρ filled in a vessel be allowed to escape through a narrow orifice (Fig. 8.7). At the free surface the pressure is the atmospheric pressure p, the velocity of the liquid is zero, and the height is (taking orifice as reference) h, while at the orifice the pressure is again p, the height is zero, and the velocity is v (say). From Bernoulli's theorem we get:

$$p/\rho g + h = p/\rho g + v^2/2g,$$

or $$v = \sqrt{2gh}$$

This is the velocity of efflux of the liquid at the orifice. This is also the velocity acquired by a body falling freely under gravity through a height h. Hence, *the velocity of efflux of a liquid through an orifice is equal to that which a body would attain in falling freely from the free surface to the orifice.* This statement is also known as Torricelli's theorem.

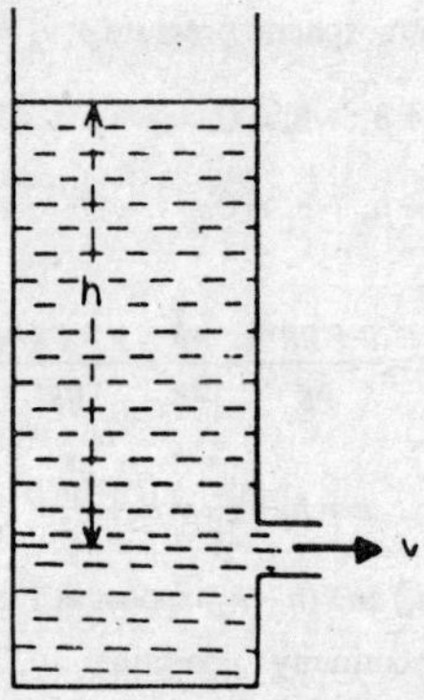

Fig. 8.7 *Velocity of efflux v of the liquid is the same as if the liquid has a free fall from height h*

(ii) Filter Pump : This is a device for reducing pressure in a vessel (Fig. 8.8). Water is made to flow very rapidly through a tube A having a narrow orifice O. Tube A opens into a wide tube B which is connected through a tube C to the vessel to be evacuated. Water coming out at extremely high velocity at O reduces the pressure in B. The air is sucked out of the vessel and is carried away by the water stream as it flows down the outlet. In a good pump the pressure inside the vessel soon reduces to a value slightly higher than the vapour pressure of water.

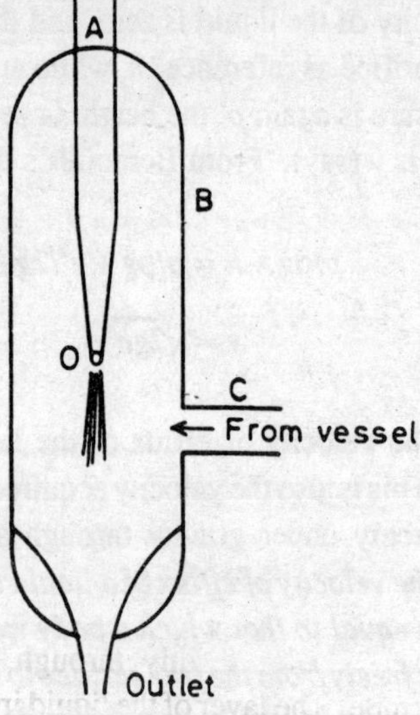

Fig. 8.8 *Filter pump*

(iii) Atomiser or Sprayer: This is also an illustration of the suction effect, produced on account of sudden increase in the velocity of air stream, and is used in spraying oils, scents, etc. When the bulb B (Fig. 8.9) is squeezed, the air blows through the central tube A with high speed causing a low pressure inside A. The atmospheric pressure at the liquid surface of the container pushes the liquid up the stem C, from where it is blown out with the air stream through nozzle O in the form of a fine spray. Motor-car carburettors also work on this principle.

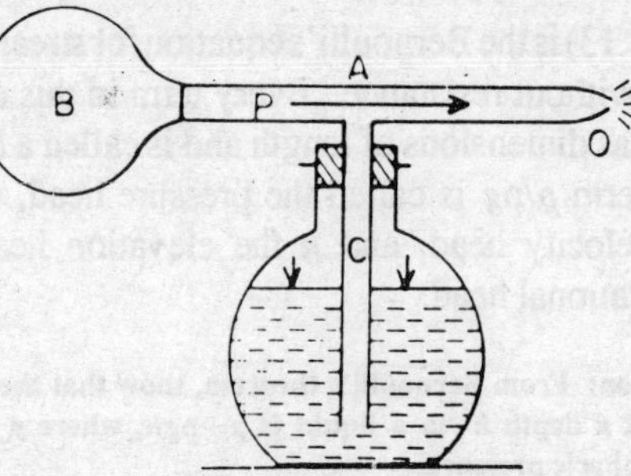

Fig. 8.9 *A sprayer*

(iv) Lifting a Sheet: A flat brass disc fitted with a central tube is placed over a cardboard sheet leaving a small space between the two. When air is blown downwards as shown in Fig. 8.10, the cardboard sheet is not blown away but is drawn towards the disc. When the air stream rushes through the narrow space between the brass disc and the cardboard sheet, its velocity is greatly enhanced, reducing the pressure considerably. The cardboard is thus pushed towards the brass disc due to atmospheric pressure.

Principle of Bunsen burner and domestic gas burner, blowing off of house-roofs during a wind storm, flight of an aeroplane (Section 8.6), the curved flight of a spinning ball, and many other phenomena can be explained on the basis of Bernoulli's theorem.

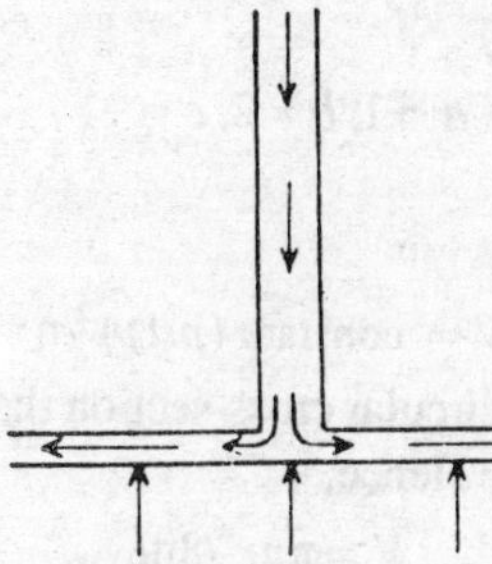

Fig. 8.10 *A demonstration of Bernoulli's theorem*

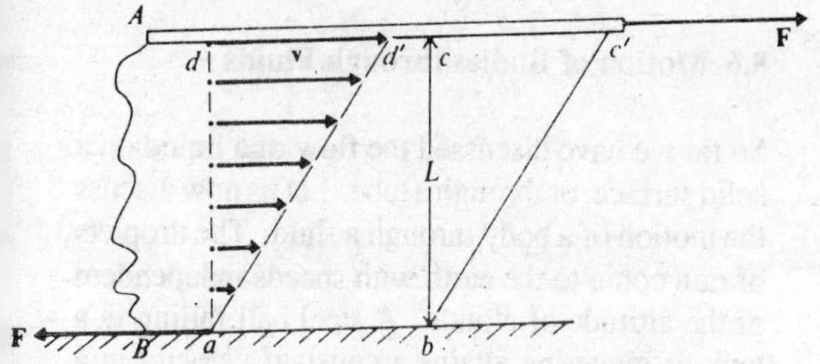

Fig. 8.11 *Laminar flow of a viscous fluid*

8.4 Viscosity

The internal friction offered by a fluid to its flow is called the viscosity of that fluid. If a liquid in a vessel is stirred and left to itself, the motion will disappear after some time. This should not happen unless a resisting force acts on the liquid. Therefore, in order to slide one layer of a liquid over another a force must be exerted to overcome a viscous force.

Consider a layer of liquid interposed between two plates *A* and *B* (Fig. 8.11). The plate *B* is stationary and *A* is made to slide with a velocity v. The liquid layer in contact with *A* is found to move with the velocity of *A* while the layer of liquid in contact with *B* remains stationary*. The velocity of intermediate layers decreases from *A* to *B* as shown. A portion of the liquid which at some instant had a shape *abcd* takes a distorted shape *abc'd'* at a later instant. The distortion will increase as the motion advances. The laminae of fluid slide over each other (like the cards in a pack placed on a table with the topmost card pushed in a certain direction), and the flow is called a *laminar flow.* Every lower layer tries to check the flow of the upper layer, which is equivalent to the statement that the flow of the liquid is opposed by a force offered by different layers along their surfaces. This force is found to be proportional to: (i) the area *A* of the layers in contact with each other, and (ii) the velocity gradient v/L. Or, the resisting force of any layer is given by

$$F \propto v/L$$

and $$F \propto A,$$

or $$F = \eta \frac{Av}{L} \qquad (8.14)$$

where η is the constant of proportionality and is called the *coefficient of viscosity* or simply *viscosity*. It is equal to a tangential force per unit area offered by a liquid layer when a unit velocity gradient is created in it. The SI unit of viscosity is pascal-second (Pa-s), and one Pa-s is equal to ten poise, where poise is the unit of viscosity in cgs system.

The coefficient of viscosity depends upon temperature. When temperature is increased η increases for gases, and decreases for liquids.

In order to maintain the motion of the liquid (Fig. 8.11), we will have to apply a force to overcome the viscous force given by Eq. (8.14). If the force applied at *A* is **F**, this force will drag the liquid and the lower plate as well to the right. Hence, an equal force – **F** must be exerted on the lower plate *B* to hold it stationary.

Questions: (i) Is a laminar flow irrotational ? (ii) What are the dimensions of η ?

8.5 Flow of a Liquid through a Tube

A liquid flows steadily through a horizontal narrow tube. The layer of the liquid in contact with the walls of the tube is stationary, and the speed of flow increases gradually towards the centre of the tube. In steady flow the work done by the agent

*Thin layer of dust on the blades remains intact even after a fan has been running at full speed !

causing motion is dissipated chiefly in overcoming the viscous drag exerted between different layers of the liquid. If the applied force is large, the liquid gets accelerated, and the flow changes from orderly to turbulent motion at a certain critical velocity. The factors which determine the average velocity v are the coefficient of viscosity (η) of the liquid, the density (ρ) of the liquid, and the radius (r) of the tube through which the liquid flows. Reynold showed by experiment that:

$$v = \eta R/\rho r \tag{8.15}$$

The constant R is called Reynold's number and lies between 0 and 1000 for a laminar flow. For values of $R > 1500$, the flow is turbulent. For R between 1000 and 1500 the flow is unstable and switches over from one type to another.

Example 8.3 Deduce Eq. (8.15) from dimensional analysis.

As pointed out above, the flow velocity v depends on η, ρ and r. Let

$$v = \text{constant}\ \eta^x \rho^y r^z$$

Writing dimensions of v, η, ρ and r, we get:

$$[LT^{-1}] = [ML^{-1}T^{-1}]^x [ML^{-3}]^y [L]^z$$

Equating the powers of M, L and T separately on both sides,

$$x = 1, y = z = -1$$

Thus,

$$v = R\eta/\rho r$$

where the constant R is the Reynold's number.

Let us now consider a liquid flowing through a horizontal tube of uniform cross-section A, and length l, under a constant pressure difference p applied between the ends of the tube. If the flow is streamline, the volume V of the liquid which flows through the tube per second will depend upon the pressure gradient p/l along the tube, the area of cross-section A, and the viscosity η of the liquid. Thus we can write the dimensional equation,

$$V = \text{constant}\ (p/l)^a A^b \eta^c \tag{8.16}$$

or

$$[L^3T^{-1}] = [ML^{-1}T^{-2}L^{-1}]^a [L^2]^b [ML^{-1}T^{-1}]^c$$

Equating the powers of M, L and T separately on both the sides,

$$a = 1, b = 2, c = -1$$

Thus

$$V = \text{constant}\ (p/l)A^2/\eta$$

For a tube of circular cross-section the constant is equal to $1/8\pi$. Hence,

$$V = \pi p r^4/8\eta l \tag{8.17}$$

Equation (8.17) is the Poiseulle's formula for the flow of a liquid through a capillary tube.

8.6 Motion of Bodies through Fluids

So far we have discussed the flow of a liquid on a solid surface, or through a tube. Let us now discuss the motion of a body through a fluid. The droplets of rain come to the earth with speeds independent of the altitude of clouds. A steel ball falling in a tank of glycerine attains a constant velocity in a short distance from the start of the motion. These observations and other experiments show that bodies moving in fluids are subject to resistive forces which depend on the size and shape of the body, its velocity and nature of the fluid. The layer of a fluid adjacent to the body is almost stationary with respect to the body. There is a region, called boundary layer, in which the velocity of the fluid grows very rapidly from almost zero (with respect to the body) to the velocity of undisturbed flow (if any) of the fluid. If the body moves very fast, the boundary is separated from the body and eddy currents are formed behind the body. The phenomena in the boundary layer and eddy currents are mainly responsible for the resistance offered by the fluid. The actual calculations for the resistance of the fluid are very complicated. However, one can find the relations for the resistive forces from the dimensional analysis.

The resistive force consists of two components: (i) the pressure drag D determined from the difference in pressure at the front and rear edges of the body, and (ii) the viscous (or frictional) force F_v due to large velocity gradient in the boundary layer.

The pressure drag D depends upon the density ρ of the fluid, its speed of flow v relative to the body, and the maximum area of cross-section A of the body perpendicular to the flow. The dimensional equation for D can be written as:

$$D = \text{constant } \rho^x v^y A^z \qquad (8.18)$$

Equating the two sides dimensionally,

$$[MLT^{-2}] = [ML^{-3}]^x [LT^{-1}]^y [L^2]^z,$$

or $\quad x = 1, y = 2, z = 1$

Thus the equation for drag can be written as:

$$D = C\frac{\rho v^2}{2}A \qquad (8.19)$$

where the constant ($C/2$) depends mainly upon the shape of the body passing through the fluid. For a sphere C lies between 0.2 and 0.4.

The viscous force, F_v, on a spherical ball of radius r will depend on the size of the ball, the relative speed through the fluid, and the viscosity η of the fluid. Thus,

$$F_v = \text{constant } r^a v^b \eta^c \qquad (8.20)$$

We determine the unknown exponents by equating the dimensions on the left hand side and the right hand side of Eq. (8.20). Thus,

$$[MLT^{-2}] = [L]^a [LT^{-1}]^b [ML^{-1}T^{-1}]^c$$

giving $a = b = c = 1$.

The frictional force F_v is, therefore, given by:

$$F_v = \text{constant } \eta r v$$

For a sphere the constant is equal to 6π. Thus,

$$F_v = 6\pi\eta r v \qquad (8.21)$$

Eq. (8.21) is called the Stokes' law.

The total force of resistance is a complicated combination of the pressure drag D and the frictional force F_v. Since F_v is proportional to v and D is proportional to v^2, the frictional force will be considerably greater than pressure drag at low speeds. But at higher speeds the pressure drag will be much more prominent than the frictional force. Now, for a spherical body:

$$D/F_v = C\rho v^2 A/12\pi\eta r v$$
$$= C\rho v r/12\eta \qquad (8.22)$$

Thus the relative importance of D and F_v depends on the Reynold's number $R = \rho v r/\eta$. When $R < 1$, the pressure drag can be neglected, and only the frictional force may be considered. For high values of R only the pressure drag is important, and the frictional force may be neglected.

Let us now consider a sphere of density ρ falling under gravity through a fluid of density ρ_0 (Fig. 8.12). The forces acting on the sphere are:

(i) Weight of the sphere (acting downward)

$$Mg = \frac{4\pi}{3}r^3\rho g$$

(ii) Force of buoyancy (acting upward)

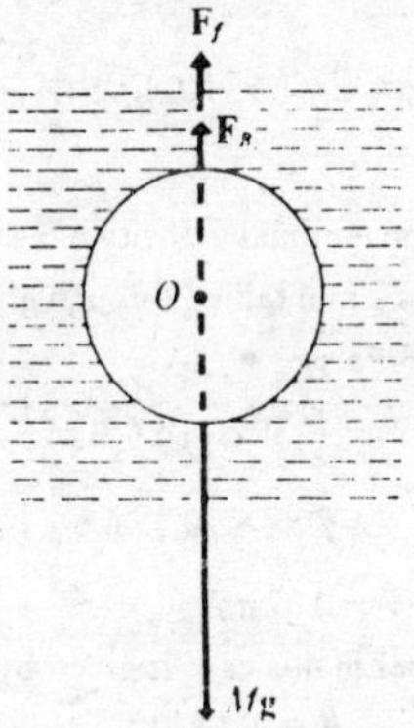

Fig. 8.12 *A sphere falling under gravity through a fluid*

$$F_B = \frac{4\pi}{3}r^3\rho_0 g$$

(iii) Resistive force of the fluid (acting upward)

$$F_f = 6\pi\eta r v + CA\rho_0 v^2/2$$

Therefore, the resultant acceleration a with which the sphere moves downwards is given by:

$$Ma = \frac{4\pi}{3}r^3\rho g - \frac{4\pi}{3}r^3\rho_0 g - 6\pi\eta r v - CA\rho_0 v^2/2 \qquad (8.23)$$

When the sphere is released from rest the velocity is zero, and the resistive force is also zero. The initial acceleration is therefore:

$$a_i = g(\rho - \rho_0)/\rho \qquad (8.24)$$

Owing to this acceleration the velocity of the sphere will continue to increase. With increase in velocity the fluid resistance will also increase, thus reducing the downwards acceleration. An ultimate velocity will be attained by the sphere when the downwards acceleration is reduced to zero. This velocity is known as terminal velocity, v_t. Putting $a = 0$ in Eq. (8.23), we get:

$$\frac{4\pi}{3} r^3(\rho - \rho_0)g = 6\pi\eta r v_t + CA\rho_0 v_t^2/2 \qquad (8.25)$$

For low velocities the expression containing v_t^2 can be neglected, and Eq. (8.25) reduces to:

$$\frac{4\pi}{3} r^3(\rho - \rho_0)g = 6\pi\eta r v_t,$$

or

$$v_t = 2r^2(\rho - \rho_0)g/9\eta \qquad (8.26)$$

Example 8.4 Find the terminal velocity of a steel ball ($\rho = 7.8 \times 10^3$ kg/m^3) of radius 2 mm falling through glycerine ($\rho_0 = 1.2 \times 10^3$ kg/m^3, $\eta = 0.83$ Pa-s).

$$v_t = 2r^2(\rho - \rho_0)g/9\eta$$

$$= 2 \times 4 \times 10^{-6} \times 6.6 \times 10^3 \times 9.8/9 \times 0.83$$

$$= 0.07 \text{ m/s}$$

The Reynold's number in this case is given by:

$$R = \rho_0 r v_t/\eta$$

$$= \frac{1.2 \times 10^3 \times 2 \times 10^{-3} \times 0.07}{0.83}$$

$$= 0.2$$

Thus, we are justified in using Stokes' formula (frictional force F_v only is considered) because $R < 1$.

For large Reynold's numbers the force of resistance is mainly due to the pressure drag D. The terminal velocity is then given by:

$$v_t = [8\pi r^3(\rho - \rho_0)g/3CA\rho_0]^{1/2} \qquad (8.27)$$

For motion of bodies through air ($\rho_0 \cong 1$ kg/m^3), the buoyant force may be neglected, and Eq. (8.27) can be written as:

$$v_t = [8\pi r^3 \rho g/3CA\rho_0]^{1/2}$$

$$= (2Mg/CA\rho_0)^{(1/2)} \qquad (8.28)$$

Here M is the mass of the sphere. For droplets of spherical shape ($C = 0.4$) the terminal velocity comes out to be about $250\sqrt{r}$ ms^{-1}. Thus larger drops will have higher terminal velocities, and the terminal velocity will be independent of the altitude of the cloud. One can also understand the principle of parachute from Eq. (8.28).

When a body moves with a high relative velocity through a fluid, the eddies are formed [Fig. 8.13 (a)]. The motion becomes turbulent causing extremely large drag. By shaping a body to the streamlines of the fluid through which it moves, the drag may be greatly reduced [Fig. 8.13 (b)]. The cross-sections of bodies of aeroplanes and ships are so designed that the fluid resistance is minimum for the speeds at which they are required to move.

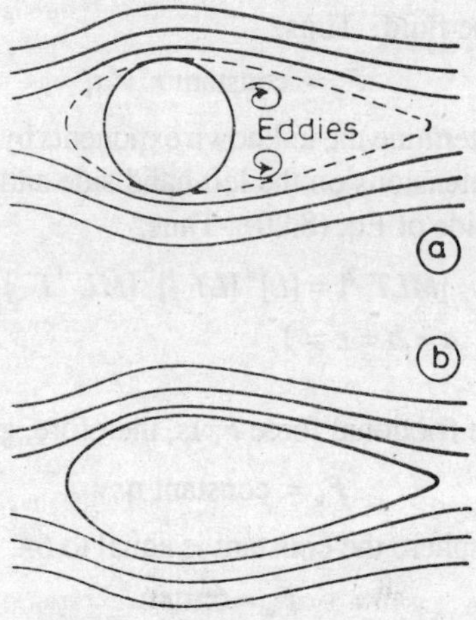

Fig. 8.13 (a) *Formation of eddies,* (b) *By properly shaping the moving bodies the eddies can be eliminated*

Besides reducing the drag D, the wings of an aeroplane are designed to provide sufficient lift to balance the weight of the plane. Figure 8.14 shows the cross-section (side view) of the wing of an aircraft. The cross-section is of such a shape that the total distance travelled by air flowing over the wing is longer than that of the air flowing under it. Thus air flows faster above the wing as compared to its flow under the wing. From Bernoulli's principle we deduce that the pressure above the wing will be lower than the pressure below the wing. This difference of pressure gives rise to the lift L required to keep the plane air-borne.

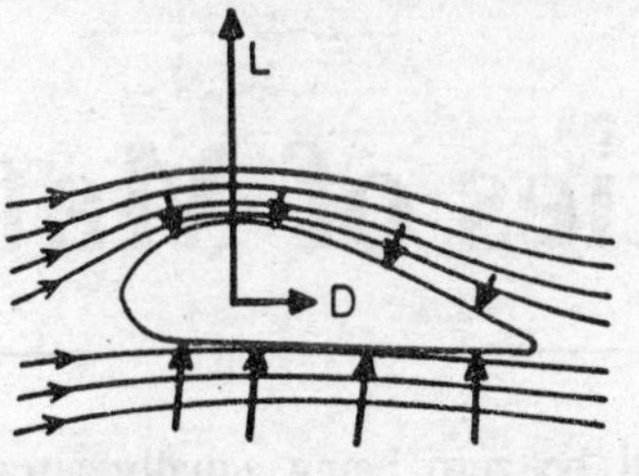

Fig. 8.14 *Principle of lift to an aeroplane*

Discussion and Problems

8.1 A glass plate of 1 sq.m is placed on a plane surface. If a 1 mm layer of glycerine is interposed between the plate and the surface, find the force required to keep the glass plate moving with a speed of 1 m/s. (η for glycerine is 8.3 poise.)

8.2 A sphere is dropped under gravity through a fluid of viscosity η. Taking the average acceleration as half of the initial acceleration, show that the time to attain the terminal velocity is independent of the fluid density.

8.3 The diameter of a cylindrical container is 15 cm and its height is 30 cm. Water is poured into this container from above continuously at the rate of 150 cm^3/s. But there is a circular hole of cross-section 1.0 cm^2 at its bottom. Upto what height will water be filled in the container?

8.4 In a horizontal water pipe of uniform cross-section the pressure decreases between two points 1 km apart by 10 N/m^2. What is the energy required for 1m^3 of water to move through a distance of 1m?

8.5 Water falls from a height of 40 m at the rate of 20 m^3/min, and drives a water turbine. What is the maximum power that can be developed by this turbine?

8.6 A pitot-tube is mounted on an aeroplane wing to measure the speed of the plane (Fig. 8.15). The tube contains alcohol and shows a level difference of 30 cm. What is the speed of the plane relative to air (density of alcohol = 0.8 g/cc, and density of air = 10^{-3} g/cc).

8.7 A solid is immersed in water. The water container falls freely. Will the Archimedes' principle still hold? What will happen if the container falls with acceleration less than g ?

8.8 A bucket of water of total mass 10 kg is kept on a spring-balance. An iron piece of 1.0 kg mass is suspended in water with the help of a thread tied to another spring-balance. The iron piece is completely immersed in water and does not touch any side of the bucket. What will be the reading of each balance ? Density of iron is 7.86 g cm^{-3}.

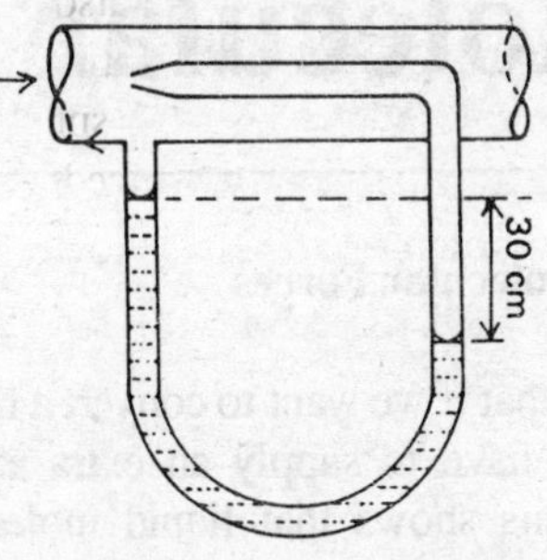

Fig. 8.15 *For Problem 8.6*

8.9 A container is kept on a table and is filled with water up to a height H. A hole is made in the side of the container at a height h. What will be the range of the outcoming water on the table ? For what value of h will this range be maximum ?

8.10 A horizontal water pipe of 30 cm^2 cross-section has been tapered to a thinner pipe of area of cross-section 10 cm^2. If the velocity of water in the wider region is 100 cm/s, what is its velocity in the narrower region ? If the gauge pressure in the former region is 30 cm-Hg, how much is it in the latter? The atmospheric pressure is 75 cm-Hg. (NOTE: The gauge pressure is the excess pressure over the atmospheric pressure.)

8.11 What should be the maximum average velocity of water at 20°C in a tube of diameter 2 cm so that the flow is laminar? The viscosity of water is 0.010 dyn/cm^2. [Hint: Use the criterion of Raynold's number.]

8.12 In an aeroplane, the profile of a wing is so designed that the velocity of the air above the wing surface is greater than that below it. If these velocities are v_1 and v_2 respectively and A is the area of the wing, show that the upward force (or lift) is: $\frac{1}{2}A\rho(v_1^2 - v_2^2)$. Here ρ is the density of the air.

8.13 Show that the path of a spinning ball is curved.

Chapter 9

Molecular Properties of Matter

9.1 Intermolecular Forces

It is known that if we want to convert a liquid into vapour, we have to supply an extra amount of energy. This shows that liquid molecules are bound with one another. To separate two molecules the energy supplied to them should be at least equal to intermolecular binding energy. The attractive forces between similar molecules are called *cohesive forces.* At the same time, it takes a lot of effort to compress something. It indicates that at closer separations molecules repel one another strongly. Finite size of matter is an evidence of both attraction and repulsion—molecules attract one another at larger separations, and repel one another at closer separations.

The attractive forces between the molecules is primarily an electrostatic phenomenon due to the dipole moment and polarizability of molecules. Three types of dipole interactions can be recognized. They are as follows:

(i) Dipole-Dipole Interaction : In many molecules the charges are so arranged that the centre of mass of negative charge does not fall on the centre of mass of the positive charge. The separation between the positive and negative charges gives rise to a dipole moment. Orientation forces between molecular electric dipoles operate in a system of such polar molecules. These forces tend to bring the interacting system into a position of minimum potential energy. If the position of one polar molecule is considered to be fixed, a neighbouring molecule cannot take all possible orientations. If it were so, the force between the pair will be zero being equally attractive and repulsive. The attractive low potential orientations are favoured over the repulsive high potential orientations. But for the electrostatic forces, the thermal distribution of molecules would have resulted in every orientation with equal probability. H_2O is a polar molecule.

(ii) Dipole-Induced-Dipole Interaction: The presence of permanent dipole in a molecule results in an additional contribution which is not included above. The dipole of one molecule can interact and polarize the electron distribution of a neighbouring molecule. The charge distribution of the second molecule is distorted in such a manner as to interact with the dipole of the first molecule attractively.

(iii) Dispersion Forces: All molecules including those which do not have permanent dipole moments attract one another. The example of this is the attraction between molecules of noble gases which can be liquefied and solidified. To understand the basis of this attraction, consider a pair of neighbouring molecules. The electron distribution in a non-polar molecule is symmetrical on the average, but at any instant it is asymmetrical. Thus it possesses an electric dipole moment at every instant. This instantaneous moment induces a dipole moment in the neighbouring molecule, and then the two molecules attract each other. The forces arising from such instantaneous dipole moments are present not only between all molecules, but also between all atoms including those of rare gases which do not otherwise attract. This force is, therefore, the primary intermolecular

force and acts between all molecules whether they are polar or non-polar. Quantum mechanical treatment of dispersion forces was given by London.

All the above three types of dipole forces, also called van der Waals forces, decrease rapidly with increase in separation between the molecules—as a whole these forces are proportional to $1/r^7$, where r is the distance between the molecules. The interaction potential energy varies as $1/r^6$. Thus these attractive forces are significant upto distances of a few angstroms only. Forces between polar molecules are relatively more strong. Molecules being polar, water is liquid at room temperature. Dispersion force is the weakest of all —it is very difficult to liquefy rare gases. Apart from dipole interactions multipole interactions are also possible.

When two molecules are brought too much close, their electron clouds start overlapping. But Pauli's principle prevents two electrons from occupying the same quantum state. As a result there is a strong repulsion between the molecules. The interaction potential varies exponentially with the separation. Approximate variation of the force between two molecules as a function of their separation is shown in Fig. 9.1. Variation of potential is also shown. Distance OC is of the order of a few angstroms; and AB, the depth of the potential well, is of the order of 10^{-21} J. OA is the equilibrium distance between two molecules at which the force is zero and the potential is minimum.

9.2 Atomic and Molecular Bindings

When two (or more) atoms of similar or different elements come close together they may bind themselves to make molecules. Following are the common types of bindings.

(i) Ionic Binding: Ionic binding is due to Coulomb's law. There is an attractive force between positive and negative ions. If the electrostatic energy of attraction exceeds the amount of energy needed to form two ions, a stable

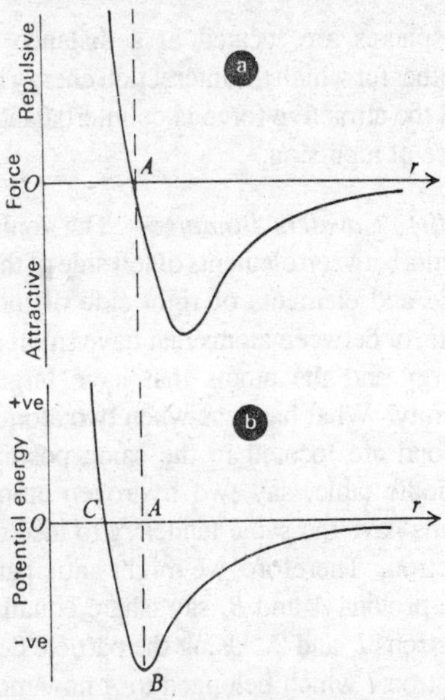

Fig. 9.1 (a) *Intermolecular force as a function of separation between two molecules* (b) *Intermolecular potential energy as a function of distance between the molecules*

molecule is possible. Consider a molecule of sodium chloride. Ionization potential of sodium is 5.14 eV. Electron affinity of chlorine is 3.61 eV.* Therefore, the net energy required to form the two ions is 5.14 eV–3.61 eV = 1.53 eV. Now, as the positive sodium ion comes closer to negative chlorine ion a decrease in energy results. The decrease in energy can be calculated to be 5.18 eV. If we add this to the ion formation energy requirement we find that 3.65 eV of energy is released on the formation of sodium chloride molecule. Since energy is released in this process the molecule is stable. The binding of most inorganic compounds is due to ionic binding.

The ions formed in ionic crystals have a spherically symmetrical electric field so that the Coulomb interaction is of same value in all directions. This enables the description of ionic crystals as a system of spheres of a definite radius closely packed into a certain lattice. The centres

*It means that 3.61 eV of energy is released when chlorine accepts an electron to complete its octet, that is, eight electrons in its outermost orbit. The outermost orbit then gets saturated.

of spheres are located at a distance from one another for which the interaction energy is minimal and the attractive force is counterbalanced by the force of repulsion.

(ii) Covalent Binding: The ionic binding occurs between elements of left side of the periodic table and elements of right side of the periodic table, or between atoms that have small ionization energy and the atoms that have large electron affinity. What happens when two atoms forming a bond are located in the same position of the periodic table, say two hydrogen atoms. Both atoms have the same tendency to lose or gain an electron. Therefore, we might anticipate that the two protons A and B, say, share equally the two electrons *1* and *2*. Using the particle description, electron *1* which belonged to A now moves in an orbit circumscribing the two protons A and B. Same is the case with electron *2*. True, electrons *1* and *2* repel each other, but forces of attraction on them due to protons are greater than the forces of repulsion. Thus in a covalent bond the electrons are shared between the two atoms.

A correct description of the covalent bond requires quantum mechanics. In essence, the bond is the result of sharing of electrons now pictured as electron clouds. The sharing results in the concentration of the charge cloud in the region between the two protons. The Coulombic attraction of the charge cloud to protons binds the atoms together. The electrons in this description lose their identities—it is no more possible to divide the charge cloud between the two protons. The total electron cloud belongs to the whole molecule. The electron spins being opposite, there is no violation of Pauli's principle.

The covalent binding has spatial directivity —the interaction between the atoms depends on their orientation with respect to one another. This is why the geometry of these crystals cannot be described by packing of spheres. Typical examples of covalent bond are the crystals of diamond, graphite, zinc sulfide, berryllium oxide and many organic compounds.

(iii) Metallic Binding: This occurs in elements with unsaturated outer orbit of electrons. When the atoms are pushed together so that electron clouds may overlap, then due to the exclusion principle the outer electrons move to higher energy. The net energy is still less than when the atoms were far apart. Consider the crowding of lithium atoms in the lithium metal around a central lithium atom. The atoms are pushed till their inner closed shells touch each other. Then the neighbouring nuclei will be inside the outer electron cloud of the central atom. This outer electron cloud is, therefore, attracted by neighbouring nuclei resulting in both increase in its binding energy and spreading of the electron cloud. This permits it to be near even more remote neighbours which in turn pull out the electron cloud even more. The electron wave functions finally uniformly spread out over the entire crystal. The crystal can be described as a large number of positive ions suspended in a cloud of electron gas. The metallic bond is accomplished by the attraction between ionic lattice formed by positively charged ions and the negatively charged electron cloud which is called electron gas. The availability of free electrons lends the metals a good conductivity.

(iv) Hydrogen Bond: If hydrogen is covalently bonded to a highly electronegative* element such as fluorine, oxygen, chlorine, etc., its single electron will be confined largely to the zone of electron cloud overlap. As a consequence, the part of the hydrogen atom farthest from the overlap consists of an exposed proton because, unlike other atoms, once hydrogen shares its single electron no inner electron shells remain to shield its exposed nucleus. As a result the exposed hydrogen nucleus attracts negative ions and the force of this attraction is known as hydrogen bond.

(v) Molecular Binding: This binding is found in many inorganic and most organic compounds in which the molecules retain their individuality in the process of crystallization. Here molecules are located at lattice points. Examples are crystals of

*The ability of an atom in a crystal or a molecule to attract electrons into its outer shell is called its electronegativity.

bromine, iodine, dry ice, solid carbon dioxide, naphthalene, paraffins, fatty acids, etc. A typical feature of crystals of this kind is that distances between atoms of the molecule are considerably less than the intermolecular distance. As the force (the van der Waals force) between the molecules is very weak, the melting points of these crystals are low. They evaporate easily, and they are soft. Crystals of inert gases also come in this category.

9.3 Three States of Matter

Combination of molecular forces and thermal motion gives rise to three states: solids, liquids and gases. In gases, because of weak attractive forces, molecules are almost free to move anywhere they like. There is a complete disorder in spatial distribution of particles. The behaviour of gases is discussed in Chapters 15 and 16. Solids have their own shape (unless distorted by some external agency) and attractive forces are so strong that thermal motion cannot shake the molecules away. The molecules are very closely packed, and there is no room for them to move about. They more or less stay at one place, and vibrate about their positions of equilibrium. In liquids the molecules are not fixed but the forces are strong enough to hold them together—they adopt the shape of the vessel containing them.

Solids: Mostly solids are in the form of crystals —molecules are arranged in a particular geometry. The geometrical structure is that in which the energy of the system is minimum, and the attractive forces are counter-balanced by the repulsive forces. The unit of minimum size, which represents the molecular arrangement in a particular crystal, is called the unit cell of that crystal. By stacking up these unit cells one can get as large a crystal as one desires. Thus crystalline substances have periodicity of atomic structure. Some simple crystal structures are given below.

Spherically symmetric molecules can be treated like spheres. When spheres of equal size are packed into a container, the closest packing may be achieved by constructing a bottom layer in which each sphere is surrounded by 6 neighbouring ones (Fig. 9.2). Here, there is a straight row of spheres with adjacent rows staggered so that the spheres in the second row can fit comfortably against those in the first row. Exactly similar layer on the top of this one is then put using the triangular interstices between the spheres in the lower layer to hold the spheres that are added to the upper one. The manner in which the third layer is added to the above two gives rise to two types of structures to this close packing.

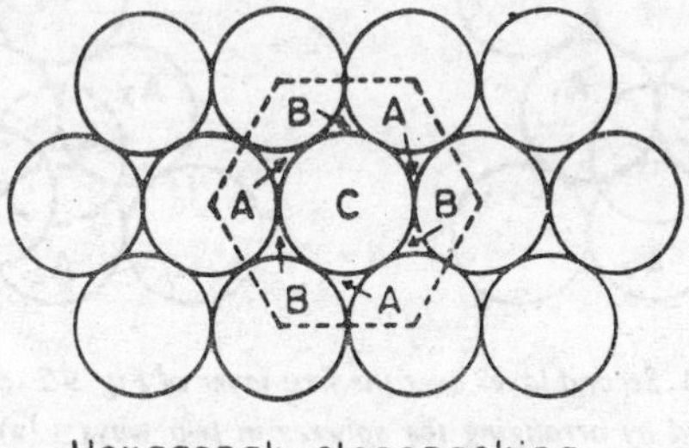

Fig. 9.2 *Close packing of spheres of equal size in a plane*

Consider sphere C in Fig. 9.2. We see that between it and its nearest six neighbours there are six depressions marked $ABABAB$. The spheres of the second layer either occupy the depressions A_1, A_2, A_3, or B_1, B_2, B_3 [Figs. 9.3 (a) and (b)]. If we place a sphere at A_1, we cannot place a second sphere at B_1 because A_1 and B_1 are so close to each other that the spheres will have to penetrate each other. Let us, therefore, fill A spaces by the second layer spheres. What are the possibilities of the third layer? We can place the spheres in the third layer over the spheres of the first layer [Fig. 9.3 (c)]. This excludes the use of the depressions over B. If we follow the AC choice consistently, we get a series of layers $ACACAC$, and so on, throughout the entire crystal. This structure has hexagonal symmetry and is referred to as the *hexagonal close-packed (hcp) structure*. This structure is found in the crystals of beryllium, magnesium, zinc, cadmium and many other metals, and also in many inorganic compounds.

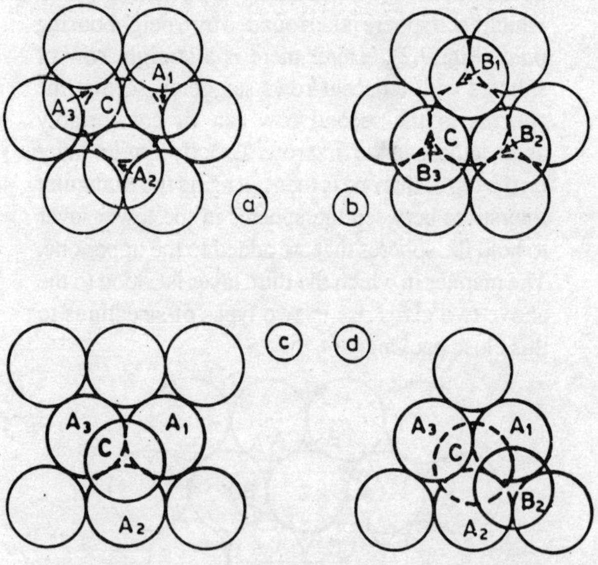

Fig. 9.3 *Second layer over the first layer of Fig. 9.2 can be obtained by arranging the spheres in two ways:* (a) *over depressions A,* (b) *over depressions B. Third layer can be had in two ways:* (c) *by repeating the first layer.* (d) *by putting the spheres over depressions B (or A) in (a) [or (b)] above*

Alternatively, the third layer can be formed by putting spheres in depressions over B_1, B_2, B_3 [B_2 shown in Fig. 9.3 (d)], instead of C spheres in the first layer. This three layer pattern repeated again and again yields the scheme *CABCAB*.... This type of packing is called *cubic close packing*. Metals such as nickel, copper, gold, silver show cubic close packing structure. This structure is also called *face-centred cubic (fcc) structure* because the unit cell is cubic and the atoms are at the corners of the cube and the centres of the faces of the cube.

In both the above structures a particular sphere has 12 nearest neighbours—6 in its own plane, 3 in the plane below it and 3 in the plane above it. Both the structures affect maximum economy of space —74 per cent of the space is filled.

Another important structure is the sodium chloride structure shown in Fig. 9.4. It has alternating Na and Cl atoms at the lattice points of a simple cubic lattice. It can be seen from Fig. 9.4 that Na atoms lie on the lattice points of a fcc lattice, as do the Cl atoms. This type of structure is exhibited by LiH, KCl and PbS also.

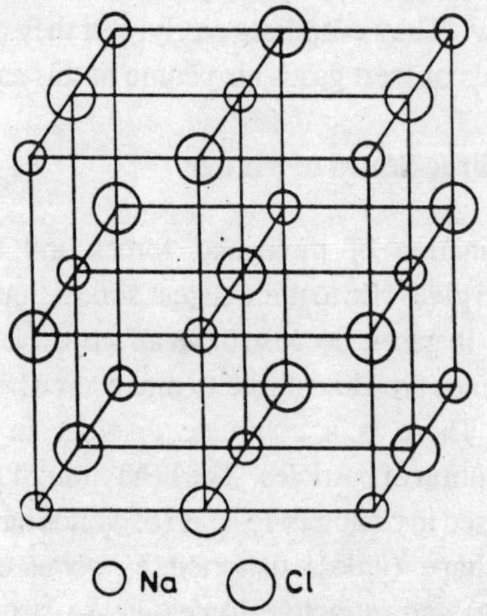

Fig. 9.4 *NaCl structure*

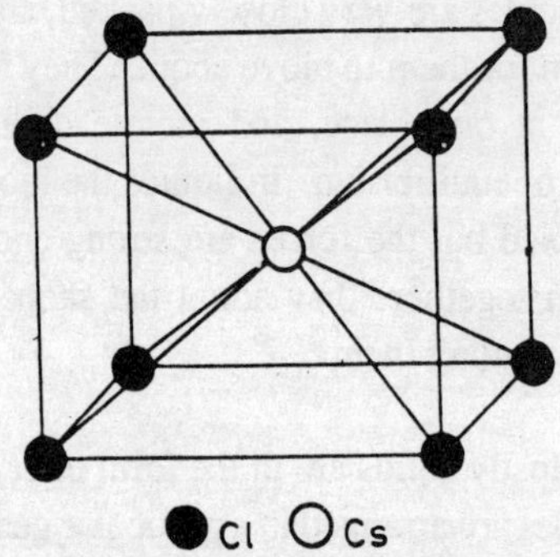

Fig. 9.5 *CsCl structure*

Cesium chloride structure is shown in Fig. 9.5. It is basically body-centred, with Cs atoms at the body-centre positions and Cl at the cube corners, or *vice versa*. Materials like CuZn (β-brass), AgMg also crystallize in this structure.

Diamond structure can be taken as composed of two fcc lattices displaced from each other by one-quarter of a body diagonal. An atom has four nearest neighbours, each neighbouring atom being located at the corners of a tetrahedron surrounding that atom (Fig. 9.6). To visualize this, think of an atom at the centre of a cubical room, two atoms at opposite corners of the floor, and two atoms at the opposite corners of the ceiling other than the two above the atoms at the floor. This structure is remarkably 'open'—only 34 per cent of the available volume will be filled if the atoms were like hard spheres. Carbon, silicon, and germanium crystallize in diamond structure. This structure is the result of covalent bonding.

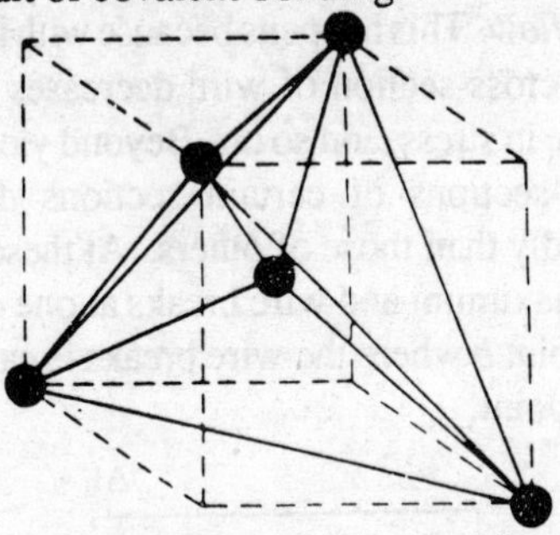

Fig. 9.6 *Tetrahedral structure of diamond*

The cubic zinc sulfide or zinc blende structure results from the diamond structure when Zn atoms are placed on one fcc lattice and *S* atoms on the other fcc lattice. Other examples of cubic zinc sulfide structure are the III-V semiconductor compounds InSb, GaAs, etc.

Depending upon the type of interaction responsible for holding the atoms (or molecules) of the crystal together, solids generally fall into one of the four categories—ionic, covalent, metallic, or molecular. Ionic crystals, for example, NaCl, have relatively high binding energies, and as a result, quite high melting and boiling points. Covalent crystals (like silicon) are also similar. The metallic elements in the free state form metallic crystals in which free electrons are present. Molecular crystals, for example, He, Ne, A, etc., in the solid state, have small binding energies. As a consequence of this, they have low melting and boiling points.

Fig. 9.7 *Structure of liquids and amorphous solids*

Liquids: X-rays analysis has shown that molecules have a definite order in liquids, especially when their temperatures are near the freezing points. At freezing point the density of a liquid is slightly different from that of solid. The molecular arrangement in a plane of a liquid consisting of molecules like uniform spheres is shown in Fig. 9.7. Here each particle has five nearest neighbours in a plane and about eleven neighbours spatially. The corresponding numbers for a close-packed solid are 6 and 12, respectively. Thus there is empty space available for molecules to slide around. The average distance, d, between the molecules of a liquid is approximately the same as in the corresponding solid, because density is nearly the same. But actually the length, l, of the sides of pentagons are greater than d in certain directions and less than d in other directions. In crystals, all particles are located at the distance d which corresponds to the separation for minimal energy. Both for $l > d$ and $l < d$ the interaction energy is larger. Thus the internal energy of matter in crystalline state is less than that in liquid state. Hence, an additional energy has to be supplied to a solid to convert it into liquid at its melting point. This additional energy per unit mass is called latent heat of fusion.

Amorphous substances: Substances like glass and resins have no definite melting points. As their temperature is raised, they become soft and start behaving like ordinary viscous liquids. These amorphous substances have molecular packing like that of liquids, and we may say that they are supercooled liquids. Obviously, amorphous bodies have surplus energy as compared to crystalline substances. As time passes amorphous bodies start crystallizing. This explains why sugar crystals are formed on sugar candy after certain time, and glass loses its transparency with age.

Transparent silicate glasses contain mostly silica (SiO_2) and oxides of some metals. Glasses having metallic properties are prepared from a melt containing metallic elements instead of their oxides. These glasses have useful electrical, superconducting and magnetic properties.

9.4 Elasticity

This property of solids is a manifestation of intermolecular forces. Let a block of material (or a wire of radius r) of length l and cross-section A (Fig. 9.8) be pulled at its ends with a force F. The length increases by an amount Δl. Fractional increase in length, $\Delta l/l$, is called tensile strain. For small value of strain, $\Delta l/l$, experiments show that F is proportional to Δl. *This is called Hooke's law.*

Owing to the stretching force, intermolecular distances increase above their equilibrium values, and the attractive forces tend to bring the molecules back to their equilibrium positions. Since the block is in equilibrium, any portion of it will be pulled at its ends with equal and opposite forces, each of magnitude F. Thus at any section of the block the restoring force per unit area, called *tensile stress*, S, is given by

$$F/A = Y(\Delta l/l) \tag{9.1}$$

where the constant Y is called *Young's modulus.*

Question: If the block is accelerated, will it be under uniform stress throughout?

When a wire is subjected to increasing stress the strain is found to change as in Fig. 9.9. The straight line portion Oe shows the region of perfect elasticity and Hooke's law is applicable. In this region the wire will return to its original length as soon as the stress is removed. If the stress is increased beyond the *elastic limit e*, to any point upto y and then removed, the material does not return to its original length. A permanent strain is retained by the wire. Beyond the *yield point y*, the wire is elongated without further increase in stress, that is, the material of wire flows. This is known as *plastic flow*. This happens because with increase in length cross-section of wire decreases leading to increase in stress, and so on. Beyond yield point the cross-sections of certain sections decrease more rapidly than those of others. At these points stress is maximum and wire breaks at one of these points. Point b where the wire breaks is called the breaking point.

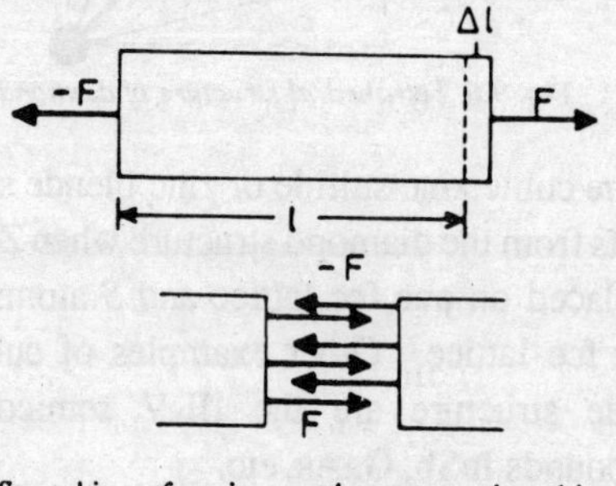

Fig. 9.8 *Stretching of a wire, and stress produced inside the wire*

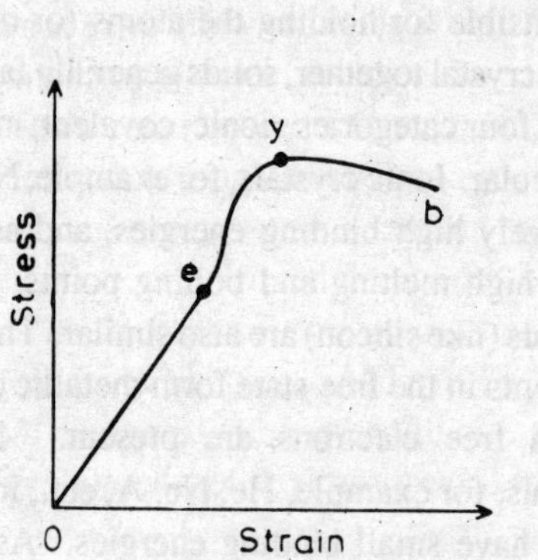

Fig. 9.9 *Stress-strain relationship for a particular wire*

There is yet another part to Hooke's law: when we stretch the block in one direction it contracts in width, say, r. The contraction in r is proportional to r and also to the longitudinal strain $\Delta l/l$.

$$\Delta r/r = -\sigma \Delta l/l \qquad (9.2)$$

where constant σ is called Poisson ratio. It is always positive in sign and is a number less than 0.5.

Bulk Modulus: Consider a block of square cross-section as in Fig. 9.10(a). It is compressed normally on all its faces with equal force F. The stress on the block is F/A, where A is the area of each face. Under this type of stress there is no change in shape, though there is a change in volume. If V is the original volume of the block, and it changes by an amount ΔV on the application of the stress, then $\Delta V/V$ is called the *volume strain*. *Bulk modulus*, B, of the material of the block is defined as,

$$B = -\frac{F/A}{\Delta V/V} \qquad (9.3)$$

The minus sign is included since an increase in stress causes a decrease in volume, and we want B to be a positive quantity. Reciprocal of the bulk modulus is called the *compressibility*, k, of the material.

Shear Modulus: Let us now apply a tangential force F to the upper face of the block, while holding the lower face rigidly, say, by fixing it on a table [Fig. 9.10(b)]. In this case while there is a change in the shape of the block, there is no change in its volume. The change in shape takes place by the movement of contiguous layers of the block relative to one another. The force F is called the shearing force, and the front face (facing the reader) is said to be sheared through an angle θ. This angle θ (in radians), through which a plane of the solid originally perpendicular to both the fixed face and the force turns, is called the angle of shear, or the shear strain.

The *shear modulus*, η, of the material of the block is defined as,

$$\eta = \frac{\text{Shear stress}}{\text{Shear strain}} = \frac{F/A}{\theta} \qquad (9.4)$$

The shear modulus is also called the *modulus of rigidity* or the *torsion modulus*. For most materials it is one-half to one-third as great as Young's modulus.

Since fluids can permanently withstand only a hydrostatic pressure (normal compressional stress), the only elasticity they possess is the bulk modulus. Various elastic constants of some of the common substances are given in Table 9.1.

Table 9.1 Elastic Constants of Some Common Substances

Substance	Y 10^{11} Pa	B 10^{11} Pa	η 10^{11} Pa	σ
Tungsten	3.6	2.0	1.5	0.20
Nickel	2.1	2.6	0.77	0.36
Steel	2.1	1.6	0.82	0.28
Iron	1.9	1.0	0.70	0.27
Copper	1.1	1.4	0.42	0.32
Brass	0.91	0.61	0.36	0.26
Aluminium	0.72	0.70	0.27	0.33
Glass	0.55	0.37	0.23	0.19
Lead	0.16	0.077	0.056	0.43
Mercury	—	0.27	—	—
Glycerine	—	0.048	—	—
Water	—	0.022	—	—
Ethyl alcohol	—	0.0091	—	—

Although four types of elastic constants have been discussed above separately, they are not completely independent. For homogeneous isotropic solids only two of these are really independent. For example, the bulk and shear moduli may be expressed in terms of Young's modulus and Poisson's ratio as,

$$B = Y/3(1-2\sigma) \qquad (9.5)$$

and

$$\eta = Y/2(1+\sigma) \qquad (9.6)$$

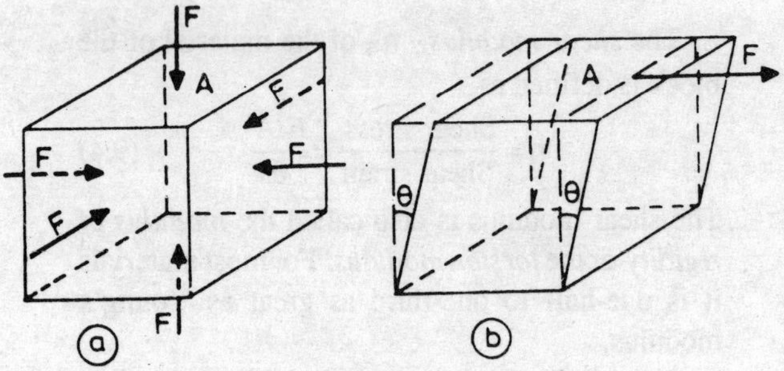

Fig. 9.10 (a) *Bulk strain,* (b) *shear strain*

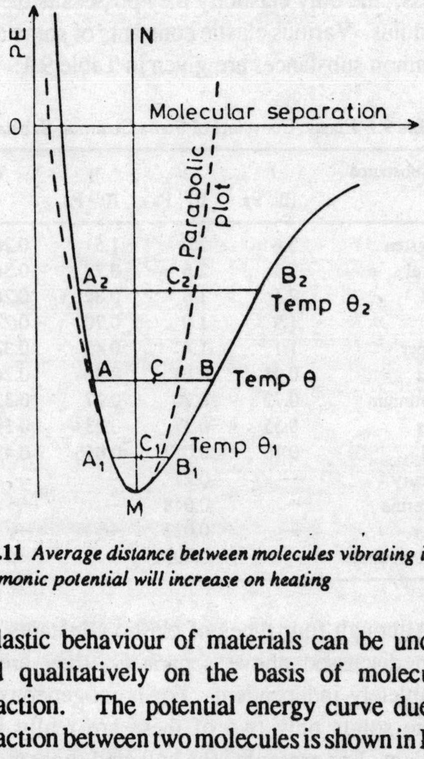

Fig. 9.11 *Average distance between molecules vibrating in an anharmonic potential will increase on heating*

Elastic behaviour of materials can be understood qualitatively on the basis of molecular interaction. The potential energy curve due to interaction between two molecules is shown in Fig. 9.11. For small displacements from equilibrium positions the curve is very nearly a parabola. It has been seen in Chapter 7 that for a parabolic curve the force needed to displace a particle from its equilibrium position is proportional to the displacement. See also the straight portion of the force curve around A in Fig. 9.1(a).

Take a wire of length l. Assume that the distance between two molecular planes perpendicular to l is r_0. Therefore,

Total number of molecular planes, $n = l/r_0$

If on the application of force F parallel to l the increase in length is Δl and the distance between the planes becomes $(r_0+\Delta r)$,

$$\Delta l = n\Delta r,$$

and

$$\frac{\Delta l}{l} = \frac{n\Delta r}{nr_0} = \frac{\Delta r}{r_0}$$

If there are N molecules in a plane, then at steady state,

$$F = Nk\Delta r$$

where k is the force constant between two opposite molecules. Interaction between other molecules has been neglected. From above equations we get,

$$\frac{\Delta l}{l} = \frac{F}{Nkr_0} = \left(\frac{A}{Nkr_0}\right)\left(\frac{F}{A}\right) = \frac{1}{Y}\left(\frac{F}{A}\right)$$

Here A is the transverse cross-section of the wire. Hooke's law thus gets verified, and for Young's modulus we get,

$$Y = Nkr_0/A$$

For larger values of Δr (or Δl) the potential energy curve will not be parabolic, restoring force will not be proportional to Δl, and, hence, Hooke's law will not hold good.

At very large values of tension, intermolecular distances will increase too much and molecular rearrangement will take place, thereby permanently deforming the body. When the stress is beyond the yield point, attractive forces can no more hold the molecules and plastic flow starts, till at some point (site) where stress becomes so large that molecules separate from one another. By similar reasoning one can understand the compressibility and rigidity also.

9.5 Thermal Expansion

With a few exceptions, the dimensions of all substances increase with increase in temperature. This can be understood qualitatively on the basis of the molecular theory.

At a given temperature θ the molecules of a solid or liquid are located at definite distances from one another and vibrate about the equilibrium position C, between A and B, (Fig. 9.11). At temperature θ_1 less than θ the molecules vibrate between points A_1 and B_1, while at temperature θ_2 ($> \theta$) corresponding to higher energy they vibrate between the points A_2 and B_2. When we go from θ to θ_2 it is seen that the point A is displaced to the left and B is displaced to the right. As the potential energy curve is not symmetrical, the displacement of B is more than that of A. Hence, with increase in temperature the equilibrium position C is also displaced to the right. In other words we may say that the distance between molecules increases with increase in temperature. However, it is not simple to obtain a quantitative relation from this analysis.

An important thing may be noted here. If the potential in Fig. 9.11 were parabolic, shown by dotted curve, there would be no thermal expansion. The mean position of a particle will remain at the same place (line MN in Fig. 9.11) at all temperatures. Only the amplitude of vibrations will change. As pointed out earlier (Chapter 7), we get parabolic curve for harmonic vibrations only. Hence, the thermal expansion is due to anharmonicity of vibrations.

Experimentally, it is known that the relative change in volume of a substance $\Delta V/V$, is proportional to change in temperature $\Delta\theta$. Thus,

$$\Delta V/V = \gamma\Delta\theta \qquad (9.7)$$

where the constant γ is known as the *coefficient of volume expansion*. If the volume at 0°C is V_0, then the volume V at temperature θ is given by

$$(V - V_0)/V_0 = \gamma(\theta - 0),$$

or

$$V = V_0(1 + \gamma\theta) \qquad (9.8)$$

Similarly, by taking a solid in the form of a rod of uniform cross-section and length l_0 at 0°C, the length l at θ°C is given by

$$l = l_0(1 + \alpha\theta) \qquad (9.9)$$

The constant α is called the *coefficient of linear expansion*.

Example 9.1 Show that $\gamma = 3\alpha$.

Consider a cube of side l_0 at 0°C. The volume at 0°C is $V_0 = l_0^3$. At θ°C the length of each side becomes $l = l_0(1 + \alpha\theta)$ and the volume V is given by

$$V = l^3 = l_0^3(1 + \alpha\theta)^3,$$

or

$$V_0(1 + \gamma\theta) = l_0^3(1 + \alpha\theta)^3$$

Hence,

$$1 + \gamma\theta = (1 + \alpha\theta)^3$$

$$= 1 + 3\alpha\theta,$$

or

$$\gamma = 3\alpha$$

As α is a small quantity (~10^{-5} to 10^{-6}) the higher powers in the expansion of $(1 + \alpha\theta)^3$ are neglected.

9.6 Surface Tension

This phenomenon is also a manifestation of intermolecular forces. Place a sewing needle on a blotting paper floating on the surface of water. After sometime the paper sinks and we observe that the needle keeps on floating. The weight of the needle is not sufficient to pull apart the water molecules. The surface becomes depressed till the upward force F_B is equal in magnitude to the weight of the needle (Fig. 9.12). If the needle is heavy, the weight of the needle will pull the water molecules apart and the needle will sink. Compare it with a weight lying on a stretched sheet of cloth or paper. Further, we observe that rain drops, fog drops and soap bubbles, etc., assume a spherical shape as they fall through air. These phenomena show that there exists a boundary surface between a liquid and some other substance. In the following paragraphs we will study the properties of the liquid surface.

Consider a liquid contained in a vessel. Some molecules are at the surface of the liquid while the others are inside it. A molecule inside is surrounded by similar molecules from all directions. The cohesive forces are, therefore, acting equally in all the directions on this molecule. Hence, the net cohesive force on a molecule inside the liquid is zero (Fig. 9.13). A molecule nearer the surface has liquid molecules below it and on its sides, while above it are the vapour molecules. We neglect the forces due to the molecules of the liquid vapour above the surface because their number density is very small as compared with that of the liquid molecules. We can resolve the attractive forces on the molecule along the surface of the liquid and along a direction perpendicular to it. The net force acting on the molecule is in the downward direction, because the forces along the surface cancel one another. Same is the fate of all other molecules at the surface of a liquid.

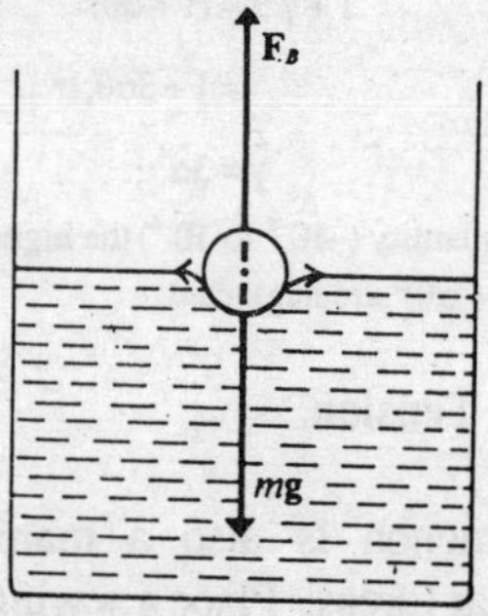

Fig. 9.12 *A needle floating on the surface of a liquid*

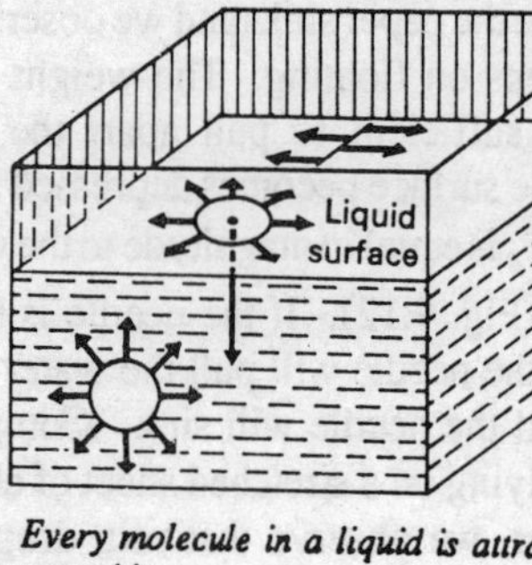

Fig. 9.13 *Every molecule in a liquid is attracted by all the molecules around it*

It has been seen above that the resultant force on a molecule at the surface of a liquid is into the liquid. Thus, if a molecule is brought to the surface from within the liquid, a work will have to be done on it. In other words, an external work will have to be done to increase the surface area of the liquid. This work gets stored as surface energy of the liquid. This energy per unit surface area is called free surface energy of the liquid, and is denoted by σ.

Let a molecule within the liquid be surrounded on an average by Z molecules ($Z \cong 11$). Let the energy of interaction between two nearest molecules be $-\varepsilon$, the negative sign appears because of the attractive nature of forces. The energy of the molecule inside the liquid is $E_i = -Z\varepsilon$. A molecule at the surface is surrounded by molecules from underneath only. The concentration of vapour molecules being very small, their effect can be neglected. The energy of molecule at the surface is, therefore, $E_s = -\frac{1}{2}Z\varepsilon$. Thus the work done on a molecule in bringing it from within the liquid to its surface is given by:

$$\Delta W = \Delta E = E_s - E_i \cong -\frac{1}{2}Z\varepsilon + Z\varepsilon$$

$$= \frac{1}{2}Z\varepsilon \tag{9.10}$$

If area of the liquid is increased by an amount ΔA and there are n molecules in it, the energy needed for this purpose will be:

$$E_{excess} = n\Delta E = nZ\varepsilon/2 \tag{9.11}$$

$$\therefore \quad \sigma = E_{excess}/\Delta A = nZ\varepsilon/2\Delta A \tag{9.12}$$

We know that physical systems come to a state of least energy if they are able to. To reduce the number of its high energy molecules and, therefore, its total energy, a piece of liquid will do its best to make its surface as small as possible. The geometrical shape that has least surface for a given volume is a sphere. It is for this reason that fog drops, rain drops and soap bubbles assume a spherical shape.

In other words, we can compare a liquid surface with a stretched membrane. If an imaginary line is drawn in the liquid surface, there will be a stretching force on both sides of this line. Force per unit length acting perpendicular to an imaginary line drawn on the surface of a liquid and tangential to the surface is defined as the surface tension of the liquid. Solids also have surface tension, but due to lack of mobility of their molecules they do not exhibit surface tension phenomenon.

We can show that free surface energy is same as surface tension. Take a piece of wire that can slide on two vertical wires dipped in a liquid (Fig. 9.14). Attach the sliding wire through a thread passing over two pulleys to a weighing pan. Put the weights on the pan till a thin liquid film is formed and the film is at the breaking point. If at this stage $(F+w)$ is the upward force (w is the weight of the wire) acting on the sliding wire, the surface tension will be $F/2l$. Here l is the length of the wire. The factor '2' appears because there are two surfaces of a thin film. Now, when the sliding wire is pulled up by a distance d, the work done is given by

$$\Delta W = F \times d = \text{ surface tension } \times 2l \times d$$

$$= \text{surface tension} \times \text{increase in area of the film,}$$

or

$$\text{surface tension} = \text{work done in creating a unit area of surface}$$
$$= \text{free surface energy}$$
$$= \sigma \qquad (9.13)$$

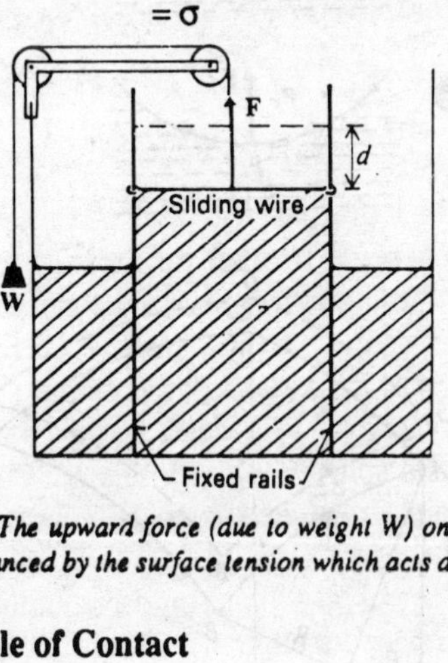

Fig. 9.14 *The upward force (due to weight W) on the liquid film is balanced by the surface tension which acts downwards*

9.7 Angle of Contact

We have so far considered surface tension of a free liquid surface. The only thing in contact with the liquid surface was its vapour. We denote this surface tension by σ_{LV}. If there is another medium above the liquid surface (liquid, solid, or compressed gas), then the surface tension at the interface will be different. The molecules at the surface of the liquid will interact with the molecules of the medium above it, apart from their interaction with the molecules of the parent liquid. These two forces will be in opposite directions. Due to change in the resultant force on the surface molecules, the surface tension will also change. Adhesive force being opposite to the cohesive force the surface tension will decrease.

Take a liquid in a vessel. On the liquid-water boundary three surfaces, viz. liquid-vapour, solid-vapour, and solid-liquid, are formed [Fig. 9.15(a)]. The shape of the liquid meniscus in a tube, or the shape of a drop placed on a surface, will depend upon the tension of the various surfaces. If $\sigma_{SV} = \sigma_{SL}$, the liquid drop will be hemispherical [Fig. 9.15(b)]. σ_{LV} will be normal to the solid surface and equal to the force of adhesion per

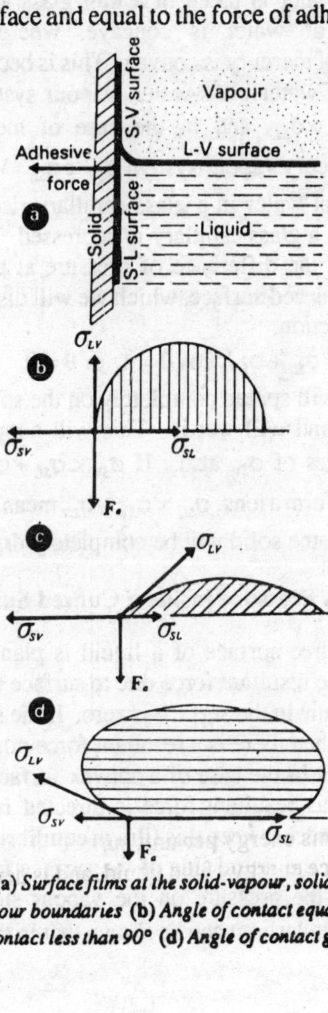

Fig. 9.15 **(a)** *Surface films at the solid-vapour, solid-liquid and liquid-vapour boundaries* **(b)** *Angle of contact equal to 90°* **(c)** *Angle of contact less than 90°* **(d)** *Angle of contact greater than 90°*

unit length, F_a, which always acts perpendicular to the walls. If $\sigma_{SV} > \sigma_{SL}$, the liquid drop will be flattened [Fig. 9.15(c)] and σ_{LV} will make an acute angle with σ_{SL}. On the other hand, if $\sigma_{SL} > \sigma_{SV}$, the drop will be contracted [Fig. 9.15(d)] and σ_{LV} will make an obtuse angle with σ_{SL}. The angle between σ_{LV} and σ_{SL} is called the angle of contact. We can define the angle of contact as the angle between the tangent to the liquid surface, at the point of contact, and the solid surface taken inside the liquid. If this angle is θ, then:

$$F_a = \sigma_{LV} \sin \theta,$$

and $$\sigma_{SV} = \sigma_{SL} + \sigma_{LV} \cos \theta$$

When water is taken in a thin glass tube the meniscus of water is concave, whereas the meniscus of mercury is convex. This is because in the case of water-glass-water vapour system we have $\sigma_{SV} > \sigma_{SL}$, and in the case of mercury–glass –mercury vapour system $\sigma_{SV} < \sigma_{SL}$. We also observe that water in a glass capillary rises, and mercury in a glass capillary is depressed. This is because of the difference of pressure at the two sides of a curved surface, which we will discuss in the next section.

If $\sigma_{SV} = \sigma_{SL} + \sigma_{LV}$, $\cos \theta = 1$, or $\theta = 0$ Hence, the liquid will spread completely on the solid as a thin film, and will wet it. This will happen for larger values of σ_{SV} also. If $\sigma_{SV} < \sigma_{SL} + \sigma_{LV}$, the liquid will form a lens. $\sigma_{SL} = \sigma_{SV} + \sigma_{LV}$ means $\theta = \pi$. In this case the solid will be completely dry.

9.8 Excess Pressure Inside a Curved Surface

When the free surface of a liquid is plane [Fig. 9.16(a)], the resultant force due to surface tension on a molecule in the surface is zero. If the surface is curved, then there is a resultant force normal to the surface. In the case of a convex surface [Fig. 9.16(b)], this resultant force is directed into the liquid. In order to keep this film in equilibrium the pressure on the liquid side of the surface must be more than the pressure on the vapour side, i.e. $p_L > p_V$. Similarly, it can be shown that in the case of a concave surface [Fig. 9.16(c)] $p_V > p_L$. Hence, we conclude that the pressure on the concave side of a film is always greater than the pressure on the convex side.

Let us consider a curved surface *ABCD* [Fig. 9.16(d)] having radii of curvature R_1 (of *AB* and *CD*) and R_2 (of *BC* and *AD*). From Fig. 9.16(d), we can see that:

$$AB = CD = R_1\theta_1,$$

and $$BC = AD = R_2\theta_2$$

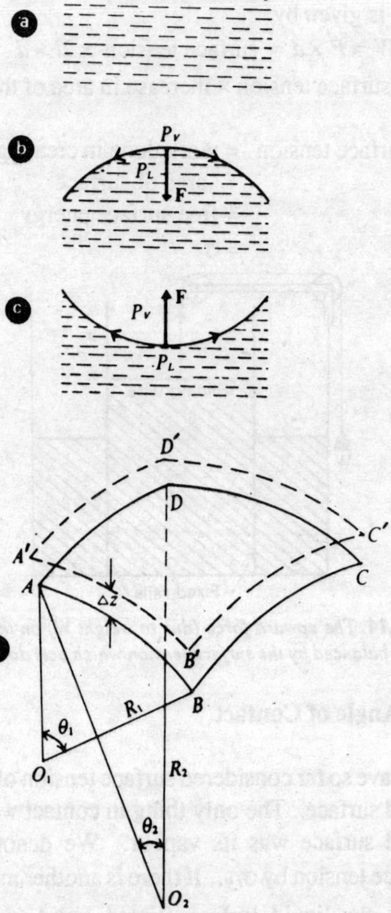

Fig. 9.16 (a) *Pressure on both the sides of a flat surface is the same.* (b) *Pressure inside a convex surface is more than outside.* (c) *Pressure inside a concave surface is less than outside.* (d) *Excess pressure inside a surface of radii of curvature R_1 and R_2*

The area of the film,

$$\Delta S \cong R_1R_2\theta_1\theta_2 \quad (9.14)$$

If the surface $ABCD$ is allowed to expand by a small distance Δz, the new surface $A'B'C'D'$ has an area:

$$\Delta S' = (R_1 + \Delta z)\theta_1(R_2 + \Delta z)\theta_2$$

$$= [R_1R_2 + (R_1 + R_2)\Delta z]\theta_1\theta_2 \quad (9.15)$$

neglecting the quantity $(\Delta z)^2\ \theta_1\theta_2$ which is very small.

The increase in area of the film

$$= (R_1 + R_2)\Delta z\,\theta_1\theta_2$$

$\therefore$ Increase in free surface energy =

$$(R_1 + R_2)\Delta z\ \theta_1\theta_2\sigma$$

The work done in the expansion of the film is given by:

ΔW = excess pressure $\times$ area of the film $\times$ distance moved

This work should be equal to the increase in free surface energy. Hence

$\therefore$ excess pressure $\times R_1R_2\theta_1\theta_2 \times \Delta z$

$$= \sigma(R_1 + R_2)\Delta z\,\theta_1\theta_2$$

$\therefore$ the excess pressure, $\Delta p = \sigma\left(\frac{1}{R_1} + \frac{1}{R_2}\right)$ (9.16)

For a cylindrical surface, $R_1 = R$, the radius of the cylinder, and $R_2 = \infty$,

$\therefore$ the excess pressure, $\Delta p = \sigma/R$ (9.17)

For a spherical drop of a liquid, $R_1 = R_2 = R$ (the radius of the sphere). Hence, the excess of pressure inside a spherical drop is:

$$\Delta p = 2\sigma/R \quad (9.18)$$

Question: What is the excess of pressure inside a soap bubble?

We see from Eq. (9.18) that smaller the drop greater is the internal pressure. A good illustration is to be found in the ease with which ice-skates slide over a surface of smooth ice. As the sharp metal edges of the skates exert enormous pressure on ice, it melts forming tiny droplets. Due to extremely large internal pressure these droplets have rigidity property like solids, and the skates run along as if these droplets were ballbearings.

It can be shown that due to surface tension (and, hence, excess pressure) the saturated vapour pressure above a convex surface of a liquid is greater than that at a flat surface of that liquid. It is clear from Fig. 9.17(b). (Description of this figure is given in next section). The smaller the radius of curvature the more is this excess vapour pressure. Therefore, smaller drops have more tendency to evaporate. Some nuclei of finite size are required for the vapour to condense on and be stable for some time. Smaller the particle the higher is the degree of supersaturation required for the vapour (around it) to condense on it. Vapour cannot condense on nothing. If the particles on which the vapours condense are charged, the condensation becomes still easier. This is because the surface tension effect becomes less due to electrostatic repulsion between the charges on it.

9.9 Rise of a Liquid in a Capillary Tube

If we dip a capillary of glass in water, the water level rises in the capillary [Fig. 9.17(a)]. But if we dip this capillary in mercury the column of mercury in the capillary is depressed [Fig. 9.17 (b)]. If the pressure on the concave side of the meniscus is p and that on the convex side is p', then we can very easily see that:

$p - p'$ = the pressure due to liquid column AB

In case of water in a glass capillary : p is the atmospheric pressure, and as $p' < p$, the water must rise in the capillary till the weight of the water column is such that it exerts a pressure $(p - p')$ at its base. If the angle of contact of the meniscus is θ and the radius of the capillary is R, the radius of the meniscus is given by $R \sec\theta$ [Fig. 9.17(c)]. The excess of pressure is given by (Eq. 9.18):

$$p - p' = 2\sigma/R \sec\theta \quad (9.19)$$

Therefore

$$\frac{2\sigma\cos\theta}{R} = h\rho g \quad (9.20)$$

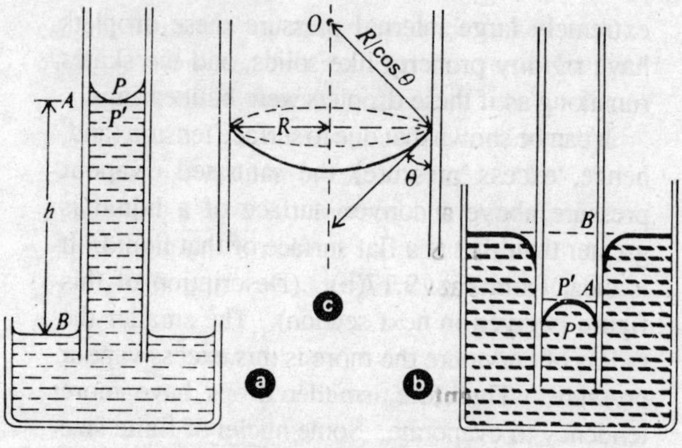

Fig. 9.17 (a) *Rise of water in a capillary tube* (b) *Depression of mercury in a capillary tube* (c) *If angle of contact is less than 90°, liquid rises in a capillary*

Here, we have neglected the small volume of the meniscus. For water-glass-vapour surface the angle of contact can be taken as zero, giving $\cos\theta = 1$. The volume of the meniscus will then be $\pi R^3/3$. Hence

$$\frac{2\sigma}{R} = h\rho g + \frac{1}{3}R\rho g,$$

or

$$\sigma = \frac{1}{2}R\rho g\left(h + \frac{1}{3}R\right) \qquad (9.21)$$

When we insert a capillary in a trough of mercury, the meniscus is convex and p' = atmospheric pressure. As $p' < p$, mercury must be depressed in the capillary till $p - p' = h\rho g$, where h is the depression of mercury.

For a considerable rise of a liquid in a tube the radius of the tube must be very small. The phenomenon of the rise of liquid in a capillary is called *capillarity action*. Because of this capillarity action oil reaches the top of the wick in a lamp, and water rises up in trees.

We can find out the surface tension of a liquid if we know the angle of contact, which is zero in the case of water. The rise of water level in a capillary and the radius of the capillary are measured with the help of a travelling microscope. It should be noted that before inserting the tube in the liquid, the tube must be cleaned. If there is any impurity at the inside surface of the capillary, the angle of contact will change. This will amount to change in the height, and the measurements will not be correct.

Question: Water can rise to a certain height h in a capillary tube. If the height of the tube is less than h above the water level in a vessel, will we get a fountain of water?

Example 9.2 A U-tube (Fig. 9.18) is such that the diameter of one limb is 0.5 mm and that of the other is 1 mm. Find the difference in the level of water in the limbs (σ = 72 dyn/cm).

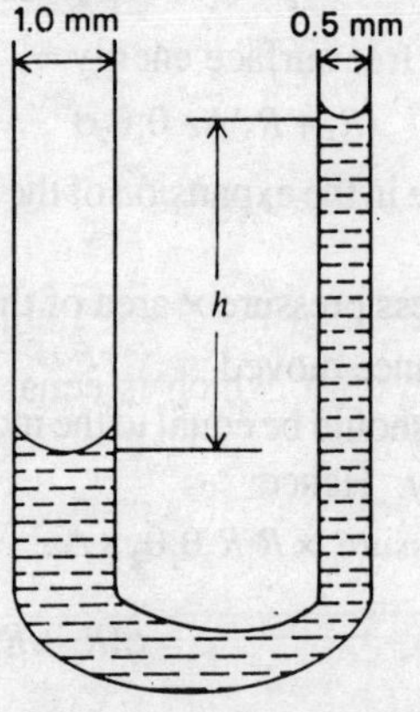

Fig. 9.18 *For Example 9.2*

As the angle of contact for glass-water-vapour surface is nearly zero, we can take the radius of the tube as the radius of the meniscus. As the pressure just beneath the meniscus of the tube of smaller bore is less than that beneath the meniscus of the wider tube, the level of water in the thinner tube will be higher than the level in the wider tube. Let this difference in heights be h. Then:

$$h\rho g = \text{difference of pressure}$$

$$= \frac{2\sigma}{r} - \frac{2\sigma}{r'}$$

or

$$h \times 1 \times 980 = 2 \times 72\left(\frac{2}{0.05} - \frac{2}{0.1}\right)$$

$$= \frac{4 \times 72 \times 0.05}{0.1 \times 0.05},$$

or

$$h = \frac{4 \times 72 \times 10}{980} = 2.94 \text{ cm}$$

9.10 Jaeger's Method

A tube with a small orifice is dipped into the liquid whose surface tension is to be determined (Fig. 9.19). The tube is connected to a Woulf's bottle through a manometer containing a liquid of density d. Air is forced into the glass tube by dropping water from a funnel into the Woulf's bottle. The surface of the liquid in the capillary tube is thus pressed down. When the air pressure inside the capillary is quite large, it protrudes a hemispherical bubble into the liquid. The pressure inside the bubble over and above the atmospheric pressure is given by:

$$h\rho g + \frac{2\sigma}{r} \qquad (9.22)$$

where ρ is the density of the liquid and r is the radius of the drop, which is equal to that of the orifice of the tube.

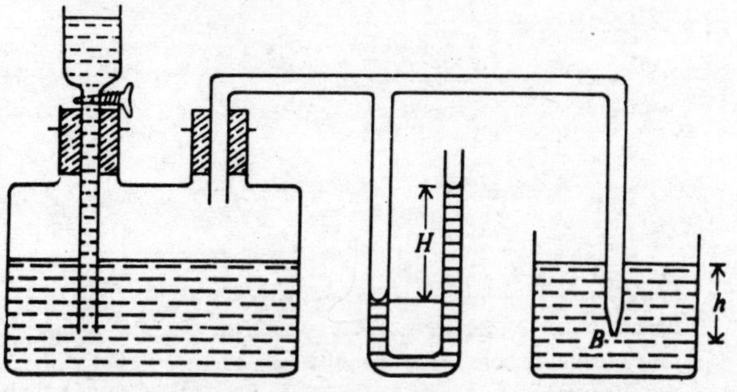

Fig. 9.19 *Jaeger's apparatus for measuring surface tension of a liquid*

If the difference of height in the liquid columns of the manometer is H, then:

$$Hdg = h\rho g + \frac{2\sigma}{r},$$

or

$$\sigma = \frac{1}{2} rg(Hd - h\rho) \qquad (9.23)$$

Hence, knowing H and h at the time of formation of the bubble, we can find the surface tension.

In this method the bubble is formed inside the liquid where temperature can be controlled very easily. This method is, therefore, very useful in determining the surface tension of a liquid as a function of temperature. The errors due to non-uniform bore and contamination in capillary are also avoided. Surface tension of molten metals, and solutions at various concentrations can also be determined.

9.11 Detergent Action

When one liquid is dispersed in another, what we get is called an emulsion. The dispersed particles are about 0.1 to 1.0 micron (10^{-6} m) in size. The formation of an emulsion leads to a large increase in interfacial area between the components forming the emulsion, and hence a large increase in surface energy. As a result such emulsions are not very stable. A third substance, known as emulsifying agent, is added to the emulsion to make it stable. Such agents, such as soaps and detergents, reduce the interfacial tension between the two components forming the emulsion. These compounds have long hydrocarbon chains attached to a polar group. At an oil-water interface the hydrocarbon chain is adsorbed on the oil side of the interface, and the polar group on the aqueous side.

If the action of a detergent solution on a grease layer on fibre is considered, it acts in three stages. First, the detergent is more strongly adsorbed at the fibre-water interface than the grease, wets the fibre and so displaces the grease layer. Second, as the grease is displaced it is rolled into a globule. Third, the globule is detached from the fibre into the water and is then emulsified with the detergent acting as an emulsifying agent.

Discussion and Problems

9.1 Two parallel plates are dipped in a liquid. Find the rise of liquid level between the two plates in terms of the distance between them.

9.2 A liquid drop of radius R breaks into 64 tiny drops. Find the resulting change in energy. Surface tension of the liquid is σ.

9.3 A capillary tube is dipped in water with its lower end 15 cm below the surface. If water rises to a height 5 cm above that of the surrounding surface, with what gauge pressure should air be blown inside the tube to blow a hemispherical bubble?

9.4 When a liquid leaves a nozzle it breaks away only after reaching a quite definite size. Why?

9.5 You might have observed that when two mercury drops are brought into contact, they form one drop. Explain.

9.6 Show that the pressure inside a spherical bubble of radius R exceeds the outside pressure by an amount $4\sigma/R$.

9.7 What would be the gauge pressure inside an air bubble of 0.10 mm radius situated just below the surface of water? Surface tension of water is 70 dyn/cm.

9.8 The thickness of a water film enclosed between two parallel plane glass plates of surface area 10 cm^2 each is one micron. How much force will have to be applied perpendicular to the surface of the plates to separate them? Take surface tension of water to be 70 dyn/cm.

9.9 A drop of an immiscibe liquid-2 is put on the surface of another liquid-1 (as oil on water). The drop takes the shape of a convex lens. At the circular interface surrounding the lens the surface tensions are σ_{12}, σ_{13} and σ_{23}. Here air is taken as medium-3. Draw the diagram and show that,

$$\sigma_{13}^2 = \sigma_{23}^2 + \sigma_{12}^2 + 2\sigma_{23}\sigma_{12}\cos\theta$$

where θ is the angle between the directions of σ_{12} and σ_{23}.
Hint: At equilibrium of liquid-2 the vector sum of the components of the three forces parallel to the free surface of liquid-1 and perpendicular to it will be zero.

When will liquid-2 completely spread on liquid-1 ?

Chapter 10

Waves in Elastic Media

When a particle is fired from one place and it strikes a target at some other place, there is a transfer of energy from the source to the receiver. Energy can be transmitted from one place to the other through one more agency, the wave. We will deal here with various properties of waves. The difference between the wave-behaviour and the particle-behaviour will also be brought out.

10.1 Types of Waves

Suppose a person A standing on the bank of a lake wishes to draw the attention of his friend B sitting in a boat in the lake. There are a number of ways of doing this. He can row down to B in another boat. He can swim down to B—not a very convenient proposition. He can throw a stone at B. This is dangerous. But suppose he drops a big boulder in the lake. The falling boulder will send out disturbances all around in the lake. The water does not move out from the site where the boulder fell. It is only the disturbance caused at the site that will spread out. This disturbance will rock the boat, and if the rocking of the boat is violent enough it can put B into alarm. Here A has made use of a phenomenon called wave motion. Anyone standing on the banks of lakes and rivers must have noticed wave motion. It is different from the current in which mass of water moves from one place to another. In wave motion, the disturbance created at a point is carried to different parts through the medium. There are some more ways in which A could contact B. He can, for example, shout at him. That would be making use of a sound wave. A could throw a beam of light on B by reflecting sunlight from a mirror. We will learn in Chapter 28 that light is also a wave motion. If A and B had a walkie talkie set, they could communicate by what in common language is called radio waves. In all such types of modes of communication a disturbance has to be caused at a location. The spreading out of this disturbance is called the wave motion.

It is apparent from the above discussion that there can be different types of waves. Can we have a system to classify them? Yes, broadly wave motions can be divided into the following groups:

(1) A wave motion which occurs only in a material medium. The disturbance produced at any point of the medium deforms the medium calling forth the elastic forces into action. Examples of such waves are water waves, sound waves, seismic waves, etc.

(2) Electromagnetic waves (Ch. 26) do not need any material medium. They can exist in vacuum also. Light waves, radio waves, X-rays, etc., fall in this category.

(3) Wave-like property is also found to be associated with every material particle (Ch. 33). These waves are called matter waves. It may be pointed out that the above types of waves are not similar to one another in all respects.

It is worthwhile to make clear that wave motion is not a mechanical phenomenon because a wave is not a material object but a form or a pattern that moves. This form or pattern is different for different waves—crests or troughs in water waves, compressions or rarefactions in sound waves, field configurations in electromagnetic waves, etc. A wave cannot be assigned a mass, and the concept of acceleration has no meaning. The motion of a wave is very different from that undergone by the medium in which it travels. In fact, we can have a wave without any movement of matter at all, for example waves of category (2) above.

The elastic waves described in category (1) can be sub-divided into two types : transverse waves and longitudinal waves.

Transverse waves : The disturbance of the medium, in transverse wave motion, occurs at right-angles to the direction of motion of the wave. Consider an infinitely long stretched cord [Fig. 10.1(a)]. The cord is shaken, up and down, at a point *A* in a periodic manner continuously. Initially the cord will take up a complex shape. But after some time the movement of the cord will settle down to synchronise with the periodic shaking at *A*. If the periodic disturbance at *A* is a simple harmonic motion the cord will assume the shape of a sine curve [Fig. 10.1(b)]. The disturbance from *A* moves out along the length of the cord to the left of *A* and to the right of *A*. But the cord is moving up and down, i.e. at right-angles to the wave motion. This is called a *transverse progressive wave*. An infinitely long cord is an idealization. In practice the cord will end somewhere, say at a rigid wall on the right [Fig. 10.1(c)]. In this case the wave motion starting from *A* meets an obstacle at the wall, gets reflected, and starts moving back. In Fig. 10.1(c), the final shape of the cord is shown as a combination of the direct and the reflected progressive transverse waves (shown by solid and dotted curves). This is called a *stationary wave*.

The shaking of the cord at *A* need not be continuous. Suppose the cord is flicked occasionally at *A*. Each flick sends out a pulse along the cord. After some time there will be a series of pulses travelling along the cord giving it a complex shape. In Fig. 10.1(d) the direct and the reflected pulses are shown as solid and broken lines.

Let us now consider what happens when point *A* on the cord is given a vertical displacement. As *A* moves up, the part of the cord adjacent to *A* also receives a pull. Due to inertia the neighbouring parts of *A* do not respond to the pull immediately. But after some delay they follow the motion of *A*. Once this adjacent part moves, it drags up the portion next to it and so on. The quickness with which the up-and-down motion spreads out depends on the twin properties: the elasticity and the inertia of the medium. Inertia of the medium is decided by the density of the medium. Thus, the speed with which the wave spreads is directly related to the modulus of elasticity of the medium and inversely related to its density.

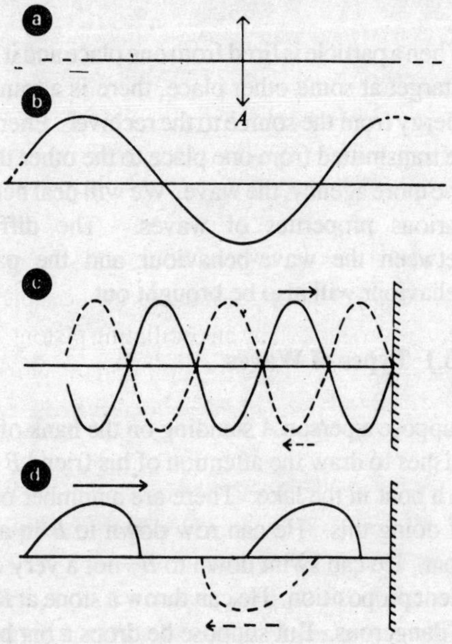

Fig. 10.1 *Transverse waves on a stretched cord:* **(a)** *An infinitely long undisturbed cord.* **(b)** *When the cord is moved up and down at A, transverse waves go in both the directions.* **(c)** *When the cord is fixed at one end the waves get reflected here.* **(d)** *Propagation of pulses along a cord fixed at one end*

Longitudinal waves: In this type of wave motion the disturbance of the medium is in the direction of wave motion. Consider a spring (Fig. 10.2). Suppose it is pushed to right at the end *A*. The part of the spring nearest to *A* is compressed. The compressed portion, however, has a tendency

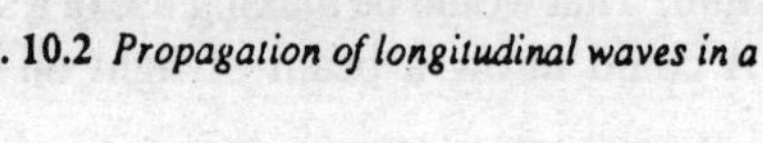

Fig. 10.2 *Propagation of longitudinal waves in a spring.*

to recover its shape. As a free end exists next to it, the compressed part gets relaxed by transmitting the compression ahead. Like this a wave of compression travels along the spring. The speed with which the compression travels down the spring depends upon the quickness with which the consecutive parts of the spring respond to the push. This has been found to depend on the stiffness and the inertia of the spring. The compression of portion near A needed some work to be done which was transformed into the potential energy of its compression. During expansion, the potential energy is converted into Kinetic energy. In compressing the next portion, the expanding portion near A transfers its kinetic energy to the potential energy of the compression of this portion. The disturbance and the energy go through the spring in this way. Let the end A of the spring be subjected to a to and fro motion by an oscillating piston. This releases waves of *compression* and *rarefaction*. The left-ward motion of the piston gives rise to the expansion of the spring nearest to A. This expansion is called a rarefaction. Thus the oscillating piston releases a wave of compression and rarefaction into the spring. As an alternative the spring can be pushed occasionally. Then a series of pulses of compression and rarefaction travel down the spring. Such pulses in this case, as well as in the case of transverse waves, can be Fourier analysed into simple harmonic wave motions (Sec. 10.8).

In the case of transverse waves the similarity of the shape assumed by the medium at once reveals some connection to simple harmonic motion which is shown by a sine curve. The relation is not so obvious in the case of longitudinal waves. It is, however, possible to represent a longitudinal wave by a sine curve. Consider Fig. 10.3(a). Mass points $A, B, C,...,$ etc., are connected by small springs, and the system is in equilibrium. In Fig. 10.3(b) the shape of the system is shown at some instant after a longitudinal wave is set into the system. At

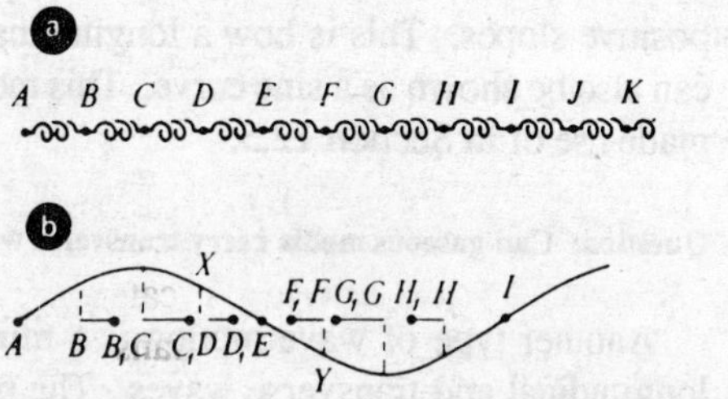

Fig. 10.3 *Graphical representation of longitudinal waves:* **(a)** *Mass points A, B, ..., etc., coupled by small springs and in equilibrium* **(b)** *Positions of masses A, B, ..., etc., after a longitudinal wave has been set up in the system. The displacement of a mass from its equilibrium position is represented along a direction perpendicular to the direction of displacement*

this instant point A has just returned to its equilibrium position and is to move to the left. Point B follows the motion of A after some delay. This is because of the inertial property of the spring. The displacement of B, therefore, lags behind A, and its new position is B_1. The delay between the motion of successive points gives them different displacements. Fractions of the oscillations completed by successive points are also different. For example, if point A has completed three oscillations the successive points upto I have completed oscillations given by $2\frac{7}{8}, 2\frac{3}{4}, 2\frac{5}{8}, 2\frac{1}{2},$ $2\frac{3}{8}, 2\frac{1}{4}, 2\frac{1}{8}$ and 2, respectively. Thus successive points lag behind those ahead of them by $\frac{1}{8}$th of the periodic time. Since the periodic time is equivalent to an angle of 2π, the phase difference between successive points in $\pi/4$. The displacement of the different points at the instant chosen is shown in Fig. 10.3 (b). Draw ordinates at B, C, D, etc., equal to their displacements, i.e. BB_1, CC_1, etc. The displacements to the right are shown by positive ordinates. The displacements of point F, G, H are to the left and they are shown by negative ordinates. A curve passing through the ordinates results into the familiar sine curve. At this instant the compression is about E. Therefore, in the sine curve the portions having negative slopes (e.g. XEY) represent compression. Similarly, the rarefactions (as about I) are shown by the portions having positive slopes. This is how a longitudinal wave can also be shown as a sine curve. This method is made use of in Section 12.5.

Question: Can gaseous media carry transverse waves?

Another type of wave motion is a mixture of longitudinal and transverse waves. The particles of the medium have simultaneous movements along the wave motion and at right-angles to it. The motion of a particle is, therefore, similar to the motion of a point acted upon by two orthogonal SHMs (Section 7.7). The particles move along a circle or an ellipse. An example of such a wave motion is water waves (Fig. 10.4).

These waves are produced on the surfaces of ponds by the force of wind, or by dropping a stone in the pond. The undisturbed surface of water is plane and horizontal. If from the impact of an object, force of wind, etc., a portion of water is either heaped above or depressed below the surface level, restoring forces of gravity and surface tension come into play. If the waves are large as in sea waves, i.e. wavelength is much larger than the curvature of the surface, the force of gravity is the principal restoring force. If, on the other hand, the curvature of the waves is larger as in ripples or capillary waves, surface tension becomes the predominant force. This happens when the wavelength of the wave is small.

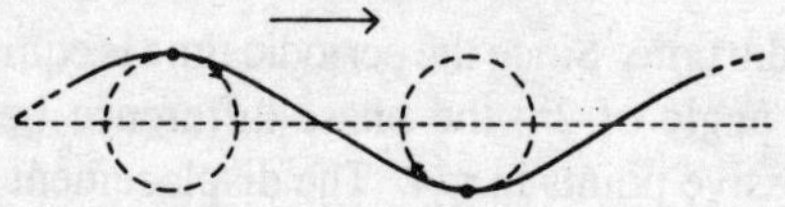

Fig. 10.4 *Combination of transverse and longitude waves as found on the surface of a liquid*

Waves can also be classified in terms of dimensions involved in their description. The longitudinal wave moving along a spring, and the transverse wave moving on a cord are one-dimensional waves. The wave motion on a water surface or a shaking sheet of cloth is a two-dimensional wave. A sound wave spreads out in all directions and is, therefore, called a three-dimensional wave.

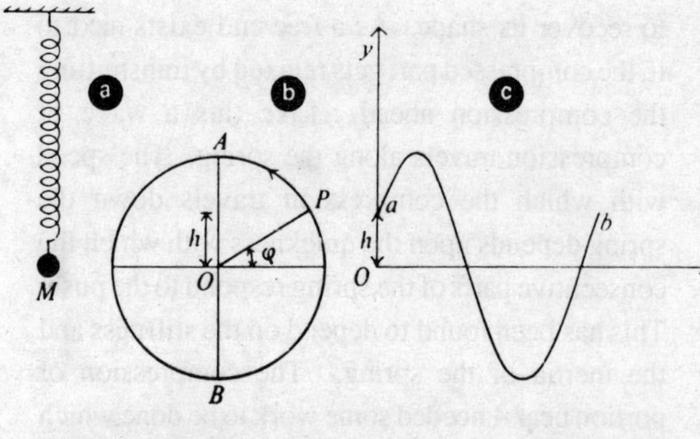

Fig. 10.5 *Oscillation of a bob tied to a spring:* **(a)** *Loaded spring* **(b)** *Reference circle for oscillation of M* **(c)** *Displacement of M as a function of time*

At an instant a large number of particles may be in the same phase of wave motion. It means that if an arbitrarily chosen particle has executed $50\frac{1}{4}$ oscillations at an instant, a large number of particles can be found to have completed the same number of oscillations, namely $50\frac{1}{4}$. By joining all these points, we obtain a *wavefront*. While the wavefront for a three-dimensional wave is a surface, for a two-dimensional wave it is a line. For a one-dimensional wave there is only a point at a particular phase. But the direction of the wave motion is always normal to the wavefront in every case.

10.2 Progressive Waves

Progressive waves have been discussed qualitatively, along with the types of waves, in the preceding section. In this section we will try to represent them mathematically. It was pointed out that there is an intimate connection between wave motion and harmonic oscillations. For simplicity we will first discuss wave motion caused by simple harmonic motion. This type of wave can be represented by a sine curve as stated earlier. If any point of an elastic medium is given simple harmonic displacement, then the transference of the displacement to the other points of the medium gives rise to a wave motion. For the wave motion of type 2 (Section 10.1) simple harmonic variation of electric and magnetic fields takes place.

Consider an oscillating loaded spring [Fig. 10.5(a)]. The mass M executes a simple harmonic motion along the vertical diameter AOB of a reference circle carrying a point P moving with an angular speed ω [Fig. 10.5(b)]. From Eq. (7.19):

$$\omega = \sqrt{k/M}$$

Suppose the point P has already moved through an angle ϕ at time $t = 0$. This corresponds to an initial upward displacement of M through a distance h. During time t the point P will move through an angle ωt. Adding the initial angular displacement ϕ, the total angle moved is $(\omega t + \phi)$. The displacement of M is, therefore,

$$y = A \sin(\omega t + \phi) \tag{10.1}$$

If t is plotted along x-axis and the displacement y is plotted along y-axis, then the resultant curve is shown in Fig. 10.5(c). At $t = 0$, the position of M in Fig. 10.5(c) is at a. At time $t = T$ the displacement is again h shown by point b. Point P has, therefore, returned back to the position where it made an angle ϕ at $t = 0$, and has thus gone through a complete circle. Or:

$$\omega T = 2\pi$$

Let us now consider a series of vertically loaded similar springs (Fig. 10.6). The springs are coupled by some hypothetical medium. As the first spring starts oscillating it will impel (through the coupling) the other springs also to execute a similar simple harmonic motion after some delay in time. Suppose the speed with which the disturbance moves out is c. This speed depends on the property of the couplings. The state* of oscillation of the first spring reaches a spring at a distance x from it after time x/c. At an instant t the spring at a distance x will thus be in a state of oscillation of the first spring x/c time earlier, i.e. at a time t_1 given by:

$$t_1 = \left(t - \frac{x}{c}\right) \tag{10.2}$$

Substituting t_1 in Eq. (10.1) in place of t, the displacement of the spring which is at a distance x is given as:

$$y = A \sin\left[\omega\left(t - \frac{x}{c}\right) + \phi\right], \tag{10.3a}$$

or

$$y = A \sin\left(\omega t - \frac{\omega x}{c} + \phi\right) \tag{10.3b}$$

This is the equation of a wave motion, and is true only for a simple harmonic wave motion. In general, the equation can be written as

$$y = A f[\omega(t - x/c)] \tag{10.4}$$

where $f[\omega(t - x/c)]$ is some function of t and x.

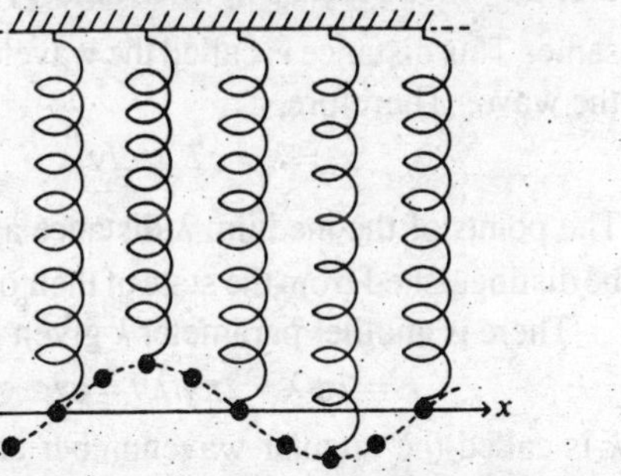

Fig. 10.6 *A progressive wave on a series of loaded springs (only a few of them are shown here) coupled together*

Note that the sine function associated with the wave [Eq. (10.3)] is a function of both time and space coordinates. The up-and-down motion of any arbitrarily chosen point is a sine function of time (x is fixed). But at a particular instant, the displacements of various masses is a sine function of x.

10.3 Wave Parameters

The parameters occurring in Eq. (10.3) are A, ω, ω/c and ϕ. A is the amplitude of the wave, and it decides the energy of the wave (Section 10.4). Also:

---

*The word 'state' should be noted very carefully. The velocity c thus defined is also called the phase velocity. The energy may or may not propagate with this velocity (Section 10.7).

$$T = 2\pi/\omega,$$

or

$$\omega = 2\pi/T = 2\pi\nu$$

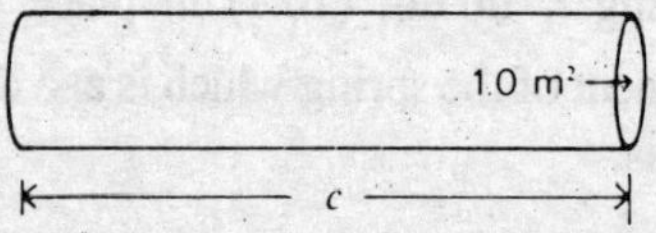

Fig. 10.7 *Intensity of a wave: Energy passing per second through an area is equal to the energy contained in a cylinder with its axis along the direction of propagation of the wave, the area as its base, and the speed of the wave as its length*

Now, in Fig. 10.5, the displacement of M becomes the same after every multiple of T. In other words, the displacement is indistinguishable after time T. In a row of springs (Fig. 10.6) the displacement of the first spring and that of another spring at a distance x_1 from the first spring such that the wave reaches x_1 after time T, will be the same. This distance is called the wavelength, λ, of the wave. Therefore,

$$x_1 = \lambda = cT = c/\nu \qquad (10.5a)$$

The points of the medium λ distance apart cannot be distinguished from the state of their oscillations.

There is another parameter k given by*

$$k = 2\pi/\lambda = 2\pi\nu/\lambda\nu = \omega/c \qquad (10.5b)$$

k is called the angular wavenumber and its units are radian per unit length. It is also called the propagation constant of the wave motion. Eq. (10.3) can now be written as:

$$y = A \sin(\omega t - kx + \phi) \qquad (10.6)$$

10.4 Intensity of a Wave

Wave motion involves transmission of energy from one point to another. Intensity of a wave at any point is defined as the energy carried by the wave per second through unit area around that point and normal to the direction of propagation. It is denoted by I and its units are joule/m^2 sec or watt/m^2. The value of I can be determined if the energy carried by the wave is known. Since the wave moves through a distance c in one second, the energy contained in a cylinder of unit area and length c will be equal to the intensity of the wave in magnitude (Fig. 10.7). While discussing simple harmonic motion it was pointed out that for any displacement x the sum of potential and kinetic energies of the oscillating body is constant. At the position of equilibrium the total energy is in the kinetic form. The velocity which is maximum here is given by $A\omega$ (Eq. 7.22). Therefore, the maximum kinetic energy which is also the total energy of the simple harmonic motion is given by:

$$E = \frac{1}{2} M \omega^2 A^2 \qquad (10.7)$$

Suppose there are n mass points in the cylinder of unit area and length c (Fig. 10.7). Therefore total energy in the cylinder is:

$$I = \frac{1}{2} n M \omega^2 A^2 \qquad (10.8)$$

But $nM = \rho c$, where ρ is the density of the medium. Hence, Eq. (10.8) becomes:

$$I = \frac{1}{2} \rho c \omega^2 A^2 \qquad (10.9)$$

10.5 Superposition Principle

If two moving balls cross each other they will get deflected from their original paths. But sound waves coming from different directions glide through each other without change in the content. In fact, this is the characteristic of all waves which distinguishes them from the particles. The presence of one wave does not alter the ability of the medium to transmit another wave.

It was pointed out in Section 7.7 that the displacement of the simple harmonic motion can be represented as a vector. It was also pointed out (see foot-note of Example 7.2) that it is the linear character of simple harmonic motion which makes the vectorial addition of the displacements possible. We drew Lissajous diagrams on the same principle. This is called the superposition principle. It applies to waves also unless the intensity of the waves is so large as to give rise to nonlinear effects. Examples of such waves are

*Do not confuse it with spring constant.

intense laser waves, and the shock waves produced during an explosion or a sonic bang. The following types of the superposition can occur among waves.

1. There can be superposition between two similar* waves travelling in opposite directions. This gives rise to a standing wave, and has been discussed in Chapter 12.

2. Two waves of different frequencies superpose to give the phenomenon of beats. This is discussed in Section 10.6.

3. There can be superposition between two similar waves travelling in the same direction. This is called interference, and has been discussed separately for sound and light waves.

It is interesting to note that the measurement of wavelength and velocity of a wave makes use of the superposition of waves. For example, the wavelength of sound is measured by a resonance column, Kundt's tube, sonometer, etc. In all these cases standing waves are formed. The measurements on the standing waves give us the value of λ and c. As the wavelength becomes small, the standing wave method fails. Interference between the waves is then exploited for the determination of λ and c. Light waves are of small wavelength. The wavelength of light is, therefore, measured by interference. But even for the electromagnetic waves standing wave method is workable at longer wavelengths, e.g. in the microwave region of the electromagnetic spectrum. For microwaves standing wave patterns are set up in waveguides. The measurement of the maximum and minimum voltages along the length of the waveguide permits measurement of the wavelength.

10.6 Beats

Consider the progressive waves of angular frequencies $(\omega+\Delta\omega)$ and $(\omega-\Delta\omega)$ travelling with the same velocity along the x-axis. Let us concentrate on their superposition at one particular value of x. For convenient handling of the equations, consider that both x and ϕ are zero. The equations of the waves then reduce to:

$$y_1 = A_1 \sin(\omega+\Delta\omega)t, \tag{10.10}$$

and $$y_2 = A_2 \sin(\omega-\Delta\omega)t \tag{10.11}$$

Since the waves are of different frequencies, even if they are in the same phase at an instant they are bound to fall out of step with each other at a later instant. This is evident from Fig. 10.8. The waves are in the same phase at instants represented by A and C. The displacements being in the same direction, the amplitude of the resultant wave is a maximum. At instants like the one represented by point B, the displacements of the two waves are opposite and, therefore, they tend to cancel each other. Thus the resultant at these times is a minimum. The two waves, by beating against each other in this manner, give rise to maximum and minimum amplitudes. As the intensity is proportional to the square of the amplitude [Eq. (10.9)], there will be waxing and waning in it as time passes.

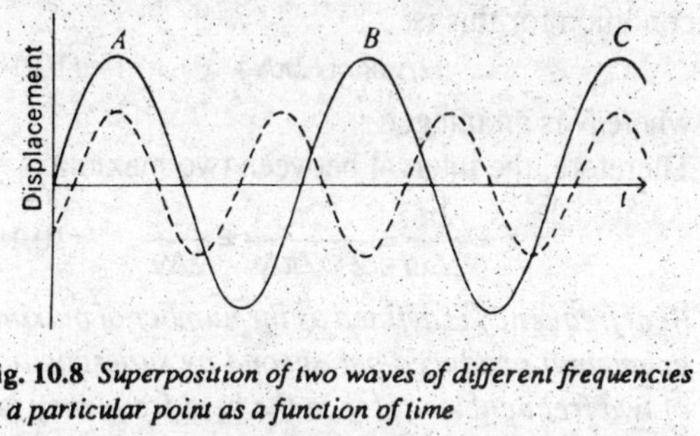

Fig. 10.8 *Superposition of two waves of different frequencies at a particular point as a function of time*

The resultant of the two displacements given by Eqs. (10.10) and (10.11) is:

$$y = y_1 + y_2 = A_1 \sin(\omega+\Delta\omega)t + A_2 \sin(\omega-\Delta\omega)t \tag{10.12}$$

$$= A_1 \sin\omega t \cos\Delta\omega t + A_1 \cos\omega t \sin\Delta\omega t + A_2 \sin\omega t \cos\Delta\omega t - A_2 \cos\omega t \sin\Delta\omega t$$

$$= (A_1+A_2) \sin\omega t \cos\Delta\omega t + (A_1-A_2)\cos\omega t \sin\Delta\omega t \tag{10.13}$$

Writing

$$(A_1-A_2)\sin\Delta\omega t = A\sin\theta, \tag{10.14}$$

and $$(A_1+A_2)\cos\Delta\omega t = A\cos\theta \tag{10.15}$$

Eq. (10.13) becomes:

$$y = A\sin\omega t\cos\theta + A\cos\omega t\sin\theta$$

$$= A\sin(\omega t+\theta) \tag{10.16}$$

*The word 'similar' means the waves have same frequency and same amplitude.

The resultant wave is, therefore, of a frequency equal to the mean of the frequencies of the two superposing waves, i.e. $(\omega + \Delta\omega + \omega - \Delta\omega)/2$. The amplitude A of the resultant wave is found by squaring and adding Eqs. (10.14) and (10.15). Thus,

$$A^2 = [A_1^2 + A_2^2 + 2A_1A_2 \cos 2\Delta\omega t] \qquad (10.17)$$

The resultant A^2 has a maximum value $(A_1 + A_2)^2$ and a minimum value $(A_1 - A_2)^2$. If the amplitudes A_1 and A_2 are equal, say A_1, the resultant amplitude varies between $2A_1$ and zero. Now, the resultant intensity I is proportional to A^2 [Eq. (10.9)]. It is maximum whenever $\cos 2\Delta\omega t$ is equal to +1. The condition for this is:

$$2\Delta\omega t = (2\pi N) \qquad (10.18)$$

where N is an integer.
Therefore, the interval between two maxima is

$$\tau = \frac{2\pi}{2\Delta\omega} = \frac{2\pi}{2 \times 2\pi\Delta\nu} = \frac{1}{2\Delta\nu} \qquad (10.19)$$

Beat frequency is defined as the number of maxima or minima produced per second by superposition of two frequencies. Hence the beat frequency is:

$$f = 1/\tau = 2\Delta\nu \qquad (10.20)$$

Now, $2\Delta\nu$ is the difference of the frequencies of the two waves. The number of beats produced is, therefore, equal to the difference of the frequencies of the beating waves.

In the case of sound waves, beating of two waves causes a pulsation in the intensity of sound. The ear, fortunately, responds to the square of the amplitude, i.e. the intensity. The ear is, therefore, called a square-law detector. Expert musicians tune their stringed instruments by making use of the phenomenon of beats. Since the persistence of sound is about 1/12th of a second, the time difference between two maxima should be more than this.

Example 10.1 A sound wave of unknown frequency gives 10 beats with a wave of frequency 320 Hz, and 5 beats with a wave of frequency 315 Hz. What is the frequency of the unknown wave?

Let the unknown frequency be n. Then

$$(n - 320) \text{ or } (320 - n) = 10, \qquad (10.21)$$

and $$(n - 315) \text{ or } (315 - n) = 5, \qquad (10.22)$$

or $$n = 310 \text{ Hz}$$

You should note that the number of beats give the difference of the two frequencies, but they do not reveal whether the unknown frequency is higher or lower. For example, from Eq. (10.21) n can be 330 or 310. But if it is 330, it will give 15 beats with a wave of frequency of 315. So it must be 310.

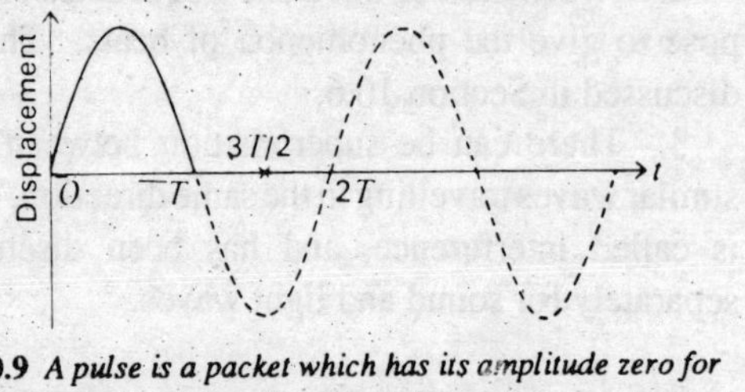

Fig. 10.9 *A pulse is a packet which has its amplitude zero for all times except in a given time interval*

Consider a pulse in Fig. 10.9. With the extended broken portion the pulse is found to be of a width in time scale equal to the half-period of the completed sinusoid. This is called the pulse width T. Therefore:

$$T = 2\pi/2\omega_0 = \pi/\omega_0 \qquad (10.23)$$

where ω_0 is the angular frequency of the sinusoid shown by the broken part of the wave. To get the pulse from the sinusoidal curve, the sine wave has to be destroyed everywhere except between O and T. For convenience we do not consider the portion of the time axis between O and $-\infty$.

Now suppose a wave of angular frequency ω_1 is mixed with the sinusoid ω_0 to give beats with minimum at time O, T, $2T$, etc. Using Eq. (10.19) we get

$$T = 2\pi/(\omega_1 - \omega_0),$$

or $$2\pi/(\omega_1 - \omega_0) = \pi/\omega_0,$$

or $$\omega_1 = 3\omega_0 \qquad (10.24)$$

Similarly, if a wave of angular frequency ω_2 produces a beat minimum at O and $3T/2$, then :

$$3T/2 = 2\pi/(\omega_2 - \omega_0),$$

or $$\omega_2 = 7\omega_0/3 \qquad (10.25)$$

Go on mixing waves like this till a beat minimum is produced at all the points from T to ∞. To produce a beat at infinity the frequency needed is:

$$\infty = 2\pi/(\omega_{\infty} - \omega_0),$$

or $$\omega_{\infty} = \omega_0 \qquad (10.26)$$

Therefore, if infinite number of waves of the same amplitude and with angular frequencies lying between ω_0 and $3\omega_0$ are superposed, a pulse of the width T (Fig. 10.9) is produced. The bandwidth (BW) of such a combination is the spread of the frequency. In terms of angular frequency:

$$BW = 3\omega_0 - \omega_0 = 2\omega_0$$

$$= 2\pi/T,$$

or $$BW \times T = 2\pi \qquad (10.27)$$

This is a very important wave equation and imposes a limitation on the purity of the waveform. For the purest waveform the bandwidth is zero. The time T for such a waveform must then be infinite, i.e. the wave should have existed for all the time. Since all waves are to be produced at some time and are of finite duration, any attempt to produce pure single-frequency wave is bound to be futile.

If matter has a wave-like character, the above limitation must extend into the domain of wave mechanics also. A principle similar to this was enunciated by Heisenberg and it goes by the name of uncertainty principle.

In wave mechanics the energy of a particle is given by (Ch. 33):

$$E = h\nu = h\omega/2\pi \qquad (10.28)$$

A small change ΔE in E will be accompanied by a similar change in ω, say $\Delta\omega$. or:

$$\Delta E = \frac{h}{2\pi}\Delta\omega \qquad (10.29)$$

But $\Delta\omega$ is the bandwidth. Using Eq. (10.27):

$$\Delta E \times T = h \qquad (10.30)$$

Thus, the spread in the energy of a particle is related to its lifetime. For example, in the structure of atom the ground state is very stable. An electron in the ground state will remain there for infinite time. ΔE for the ground state is thus zero. The excited states in an atom have lifetimes of the order of 10^{-8} second. The excited energy levels are also diffused. This can be interpreted as a direct consequence of the wave-like nature of matter particularly at atomic scale. The uncertainty principle is described in detail in Chapter 33.

10.7 Group Velocity and Wave Velocity

Consider the superposition of a large number of waves with wavelengths lying between λ and $(\lambda + \Delta\lambda)$. It has been shown in the preceding section how an infinite number of waves lying between two frequencies or between two wavelengths combine to give a pulse. Now the sum of the energies of the individual waves is concentrated in this pulse. Therefore, the velocity with which the energy travels is equal to the velocity of the pulse, and is called the group velocity. In certain media the velocity of wave changes with frequency. It is this variation of velocity with wavelength that causes dispersion through a prism. Such media are called dispersive media. In Fig. 10.10(a), only two extreme waves of wavelengths $(\lambda + \Delta\lambda)$ and λ of a group of waves travelling towards right are shown at a particular instant. If all the intermediate waves are filled in, the result at this instant will be a pulse at (A, C), where the maximum due to λ and $(\lambda + \Delta\lambda)$ occurs. Suppose the velocity of the wave increases with wavelength. The point D on the wave $(\lambda + \Delta\lambda)$ will, say after time t, catch up with point B on the wave of wavelength λ causing a shift in the position of the pulse [Fig. 10.10(b)]. It is clear from Fig. 10.10 that:

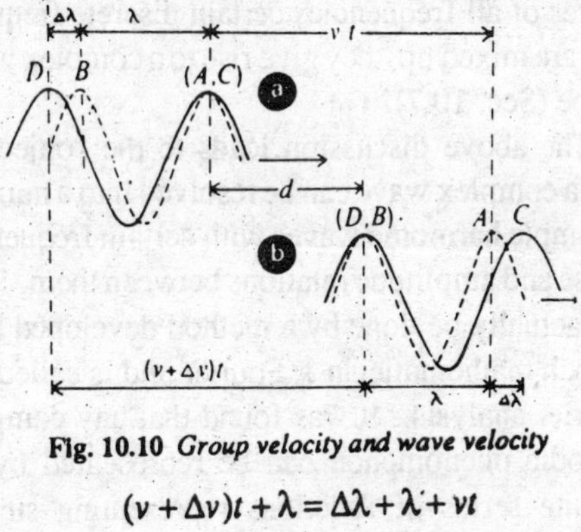

Fig. 10.10 *Group velocity and wave velocity*

$$(v + \Delta v)t + \lambda = \Delta\lambda + \lambda + vt$$

$$\therefore \quad t = \Delta\lambda/\Delta v \qquad (10.31)$$

Here, v is the velocity of the wave of wavelength λ, and $(v + \Delta v)$ is the velocity of the wave of wavelength $(\lambda + \Delta\lambda)$. Again from Fig. 10.10, the distance travelled by the pulse is given by:

$$d = (v + \Delta v)t - \lambda - \Delta\lambda$$

Substituting the value of t from Eq. (10.31),

$$d = v\frac{\Delta\lambda}{\Delta v} - \lambda \qquad (10.32)$$

Therefore, the group velocity of the wave combination is given by

$$v_g = d/t$$

Making use of Eqs. (10.31) and (10.32),

$$v_g = v - \lambda\frac{\Delta v}{\Delta\lambda} \qquad (10.33)$$

If $(\Delta v/\Delta\lambda) = 0$, i.e. if the medium is non-dispersive:

$$v_g = v, \text{ or}$$

The group velocity = The wave velocity

10.8 Fourier Analysis

Sinusoids and cosinusoids are the simplest functions to be handled mathematically. But all audio and visual signals are much more complex functions. It has been seen in the preceding sections that the superposition of a large number of simple harmonic waves gives rise to a complex wave. It was shown in Section 10.6 that a pulse is the resultant of the sine waves with continuously varying frequencies. If, instead of combining waves of all frequencies certain discrete frequencies are mixed up, they give rise to a complex wave shape (Sec. 10.7).

The above discussion leads to the conjecture that a complex wave can be resolved into a number of simple harmonic waves with certain frequency, phase and amplitude relations between them. This can actually be done by a method developed by a French mathematician J. Fourier and is called the Fourier analysis. It was found that any complex periodic phenomenon can be represented by an infinite series of functions representing simple harmonics.

A periodic physical phenomenon can be represented by a periodic function, i.e. a function which has the same value after certain interval of the variable. If the variable is time, the function repeats itself after a certain time interval, say T. Thus, if a function $f(t)$ is periodic and well defined in an interval from $t = 0$ to $t = T$, it can be written as:

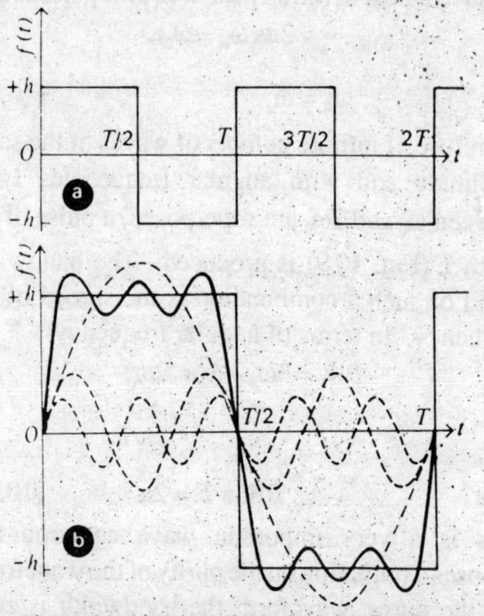

Fig. 10.11 **(a)** *A rectangular wave* **(b)** *Resultant of three sinusoidal waves of appropriate amplitude and phase. The resultant of an infinite number of properly selected sinusoidal waves will generate the original rectangular wave*

$$f(t) = f(t+T) = f(t+2T)\ldots., \text{ etc.}$$

If the variable is a space coordinate, say x, and the function is well defined in the interval $x = 0$ to $x = L$:

$$f(x) = f(x+L) = f(x+2L)\ldots., \text{etc.}$$

Fourier showed that functions like $f(t)$ and $f(x)$ above can be represented as:

$$f(t) = \sum_{n=0}^{\infty} [A_n \cos n\omega t + B_n \sin n\omega t] \qquad (10.34)$$

$$\text{and } f(x) = \sum_{n=0}^{\infty} [C_n \cos nkx + D_n \sin nkx] \qquad (10.35)$$

where A_n, B_n, C_n and D_n are constants. All the terms corresponding to $n = 0$ represent unvarying components.

The frequency of the fundamental component ($n = 1$) in Eq. (10.34) is equal to the rate at which the complex wave repeats itself in one second, i.e. $1/T$. The higher values of n give harmonics, the frequencies of which are integral multiples of the fundamental frequency. The amplitudes go on falling with the rise in frequency. Therefore only a first few members of the series describe the wave in a sufficiently detailed manner.

As an example, let us consider a square wave of time-period T and amplitude h [Fig. 10.11(a)]. The Fourier expansion of this function is given by*:

$$f(t) = \frac{4h}{\pi}\left(\sin \omega t + \frac{1}{3} \sin 3\omega t + \frac{1}{5} \sin 5\omega t + \ldots\right) \tag{10.36}$$

The first three terms of this series, along with their resultant, have been plotted in Fig. 10.11(b) for one period of the wave. The original wave has been amplified for the sake of clarity. One can note that the resultant of these terms is roughly a square wave with a few overshoots and undershoots. The inclusion of more and more terms takes the resultant more close to the original wave. You can take five terms and verify this conclusion.

Two unidirectional waves commonly used in electronic instruments are: (1) a square wave [Fig. 10.12(a)], and (2) a saw-tooth wave [Fig. 10.12(b)]. Their Fourier analysed series respectively are:

$$f(t) = \frac{h}{2} + \frac{2h}{\pi}\left(\sin \omega t + \frac{1}{3} \sin 3\omega t + \frac{1}{5} \sin 5\omega t \ldots\right) \tag{10.37}$$

*Multiplying both sides of Eq. (10.34) by dt and integrating:

$$A_0 = \frac{1}{T}\int_0^T f(t)dt$$

Multiplying both sides of Eq. (10.34) by $\cos n\omega t dt$ and integrating:

$$A_n = \frac{2}{T}\int_0^T f(t) \cos n\omega t dt.$$ Similarly:

$$B_n = \frac{2}{T}\int_0^T f(t) \sin n\omega t dt.$$

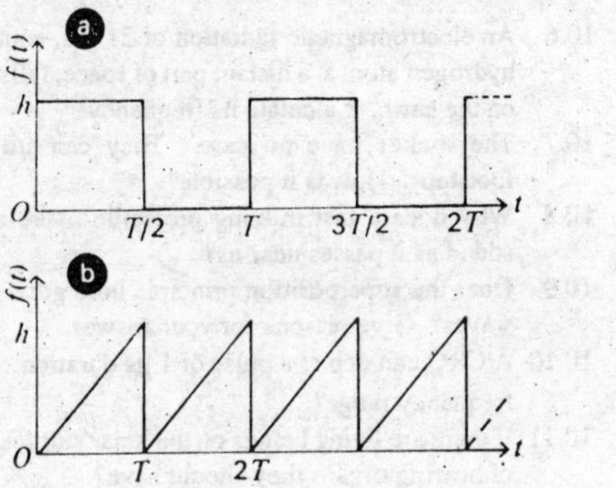

Fig. 10.12 (a) *A unidirectional rectangular wave* (b) *A saw-tooth wave*

and

$$f(t) = \frac{h}{2} - \frac{h}{\pi}\left(\sin \omega t + \frac{1}{2} \sin 2\omega t + \frac{1}{3} \sin 3\omega t + \ldots\right) \tag{10.38}$$

Exercise: Plot the first four terms in each of the above series. Find the resultants and compare them with the original wave shapes.

To study the response of the passive and active circuits to such waves as given above we can think of the input to consist of a large number of harmonic waves. After passing through the circuit these harmonic waves recombine to give the resultant wave. If the attenuation or the amplification is not the same for all the components, the output wave will get distorted. Thus the sinusoidal frequency response of a circuit gives a lot of information about its pulse response, or vice versa.

Discussion and Problems

10.1 There are always ripples on the surface of lakes. Why?

10.2 Cite some examples of phase velocities being greater than the velocity of light.

10.3 In the transverse representation of the longitudinal waves (Fig. 10.3) which parts of the curve represent compression and rarefaction for a wave travelling in negative direction?

10.4 There are compressions and rarefactions in a longitudinal wave. What are the corresponding quantities for a transverse wave?

10.5 Is the velocity of vibration of the medium same as the velocity of the wave motion?

10.6 An electromagnetic radiation of 21 cm, emitted from a hydrogen atom at a distant part of space, falls constantly on the earth. Calculate its frequency.

10.7 The snakes have no ears. They can still hear our footsteps. How is it possible?

10.8 Why does a fast moving projectile make a whistling sound as it passes near us?

10.9 Does the superposition principle hold good for intense waves? Give reasons for your answer.

10.10 A CRO can depict a pulse of 1 μs duration. What is its frequency range?

10.11 If there are living beings on the lunar surface, what type of hearing organs they should have?

10.12 A progressive wave moves in a wire of linear density 4×10^{-3} kg/m. x and t are chosen to be zero at a point on the wire. Radius of the wire is 3.7×10^{-4} m, and the equation of progressive wave is:

$$y = 0.01 \cos (512\pi t - 30x) \text{ m}$$

Find: (i) the velocity of the wave, and (ii) the intensity of the wave.

10.13 A wave travelling along a wire is watched through slits A, B and C. The distances AB and AC are 0.02 m and 0.03 m, respectively. From each slit the wire is found to oscillate with a frequency of 30 Hz. If the phase difference of the oscillations as observed from A and B is $\pi/3$, and between A and C is $\pi/2$, find the velocity of the wave on the wire.

10.14 What is the velocity of a wave along a perfectly rigid rod?

10.15 What is the qualitative difference between the transverse waves occurring in solid and liquid media?

Chapter 11

Sound Waves

11.1 Velocity of Longitudinal Waves

It has been pointed out in Section 10.1 that sound in air moves as longitudinal waves. In as extended medium a vibrating body gives rise to three-dimensional waves. Consider the simplest vibration of a sphere (Fig. 11.1), which expands and contracts periodically. A train of compressions and rarefactions will radiate out in all directions. In an isotropic medium the sound waves moving out radially give rise to spherical wavefronts. The energy needed for the displacement of the layers of the medium and the potential energy of compression is supplied by the vibrating source.

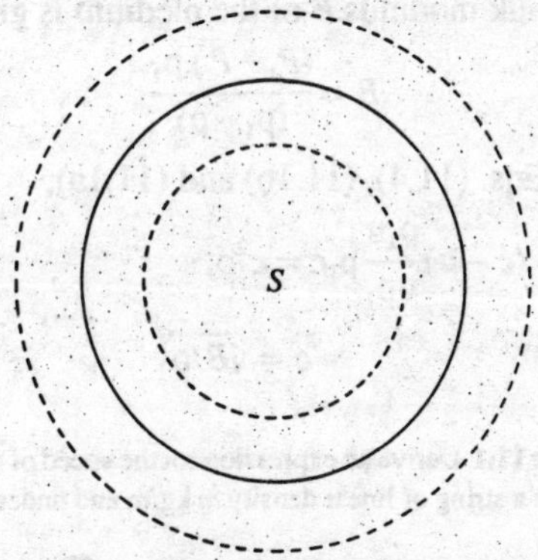

Fig. 11.1 *A sphere vibrating in a medium releases waves of compression and rarefaction in all directions*

For a closer inspection of the motion of longitudinal waves let us consider a very long tube of area of cross-section A, and filled with a homogeneous medium of density ρ and pressure P [Fig. 11.2(a)]. The medium is isotropic, and its bulk modulus of elasticity is B. A piston M at one end is given a simple harmonic motion by some arrangement so that it moves between M_1 and M_2.

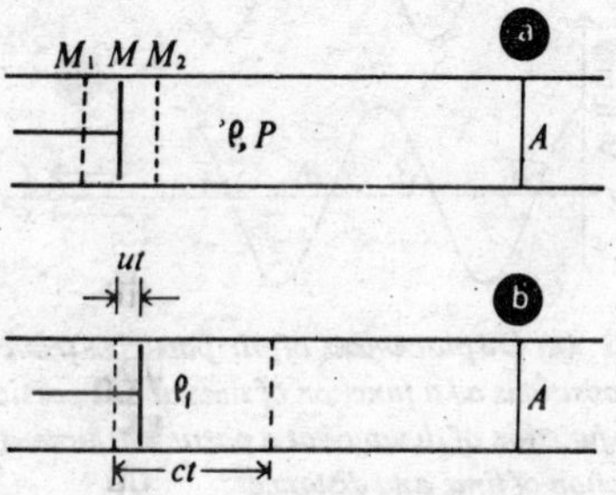

Fig. 11.2 **(a)** *Generation of longitudinal waves in air enclosed in a pipe. When piston M moves to and fro between positions M_1 and M_2, waves of compression and rarefaction move along the pipe.* **(b)** *When the piston moves a distance ut the compression wave moves a distance ct*

As the piston oscillates it sets the layers of the medium parallel to the piston into vibrations. As the tube is very long, the reflected waves need not be considered. When the piston moves from M_1 to M_2 the layers of the medium next to it are displaced. There is some phase lag between the successive layers of the medium. As a result a bunch of layers come closer and form a compression. As the piston moves back from M_2 to M_1 the layers in the compressed part of the medium relax and, as is the characteristic of the elastic bodies, move to the other side of their positions of equilibrium. The layers which were close together now fall apart giving rise to a rarefaction. At the same time a bunch of adjacent layers get compressed. If the layers could be made visible they will be found executing a simple harmonic motion like the piston. Figs. 11.3(a) and (b) show the displacement and the velocity curves of the layers. The displacement and the velocity of a simple harmonic motion are at quadrature (Section 7.4). Points A_1, A_2 and A_3 correspond to zero displacement and maximum velocity. Points B_1, B_2 and B_3 correspond to maximum displacement and zero velocity. Remember that the velocity of vibration of the layers is not equal to the speed of sound.

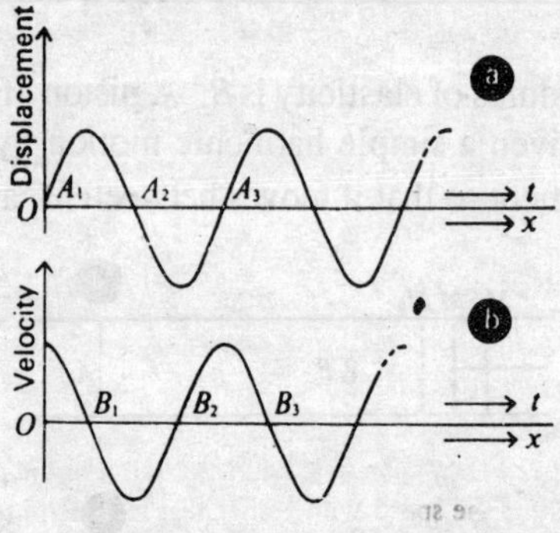

Fig. 11.3 (a) *Displacement of air particles from their equilibrium positions as a function of time at one particular place, and as a function of distance at a particular time* (b) *Velocity as a function of time and distance*

A medium is made of atoms and molecules. But while considering the propagation of sound, layers of medium are being considered. The reason for this is that the particles of the medium have random thermal motion. The displacement due to wave motion is in addition to this chaotic thermal motion. It is necessary that the region of compression be larger than the mean free path of the particles. Otherwise the colliding particles will, by the process of transfer of momentum, equalise the pressures of compression and rarefaction. No wave propagation can then occur.

If layers of small thickness, say of the order of 0.001 mm, are considered, they enclose a large number of molecules moving at random. As a result the centre of mass of the layer can be taken to be at rest when the medium as a whole is not disturbed. Yet the layer is so thin that there is no appreciable change in the properties of the gas in it. Moreover, the difference in the displacements of the molecules at the two opposite faces of the layer is also negligible. The propagation of sound can then be represented by the displacement of the centre of mass of the layer.

To calculate the speed of sound in the medium consider an impulsive motion of the piston M to the right [Fig. 11.2(b)]. A pulse of compression wave travels down the tube. Let the velocities of the piston and sound be respectively u and c. The distances moved by the piston and the compression wave in time t will be ut and ct respectively. Therefore, the length occupied by compression is $(ct - ut)$. Initially the particles in this compression had a length ct. Thus the density of the medium increases due to compression. Let it be ρ_1. By conservation of mass:

$$(ct - ut)\,A\rho_1 = ct\rho A, \qquad (11.1a)$$

or

$$c(\rho_1 - \rho) = u\rho_1 \qquad (11.1b)$$

Now, the momentum gained by the medium in time t is equal to the impulse given by the piston to the medium. Or :

$$\text{Momentum gained} = ct \times \rho A u \qquad (11.2a)$$

Using Eq. (11.1a), Eq. (11.2a) becomes:

$$\text{Momentum gained} = (ct - ut)\,A\rho_1 u \qquad (11.2b)$$

But,

$$\text{gain in momentum} = \text{applied force} \times \text{time}$$

$$= (P_1 - P)\,At \qquad (11.3)$$

where $(P_1 - P)$ is the excess of pressure due to compression. Combining Eqs. (11.3) and 11.2(b), we get,

$$(ct - ut)\,A\rho_1 u = (P_1 - P)\,At \qquad (11.4)$$

Also, bulk modulus B of the medium is given by:

$$B = \frac{(P_1 - P)\,\rho_1}{(\rho_1 - \rho)} \qquad (11.5)$$

Using Eqs. (11.4), (11.1b) and (11.1a),

$$B = (c - u)\frac{\rho_1 u}{u\rho_1}\rho_1 c = c^2\rho, \qquad (11.6)$$

or

$$c = \sqrt{B/\rho} \qquad (11.7)$$

Example 11.1 Derive an expression for the speed of transverse waves on a string of linear density m kg/m and under a tension of T N.

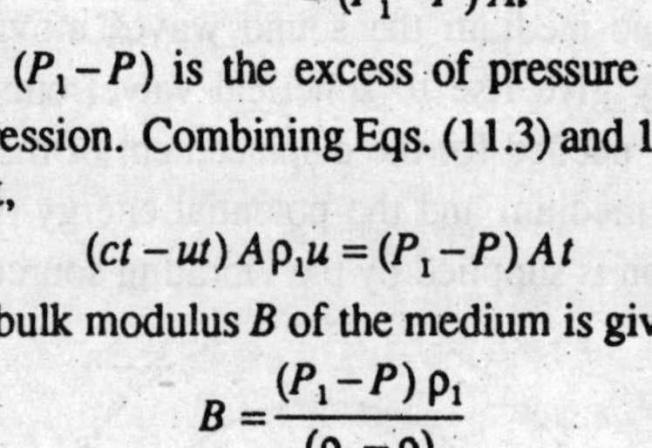

Fig. 11.4 *For Example 11.1*

Consider the wire in Fig. 11.4(a). At time $t = 0$, the end A of the string is given an impulse in the y-direction and, therefore, the string gets a displacement in this direction [Fig. 11.4(b)] with a speed u. This impulse will give rise to a transverse wave on the string in the x-direction. The string is considered to be sufficiently long not to have the reflected wave. If the speed of transverse wave on the string is c, it reaches the distance ct in time t. From Fig. 11.4(b), the transverse component of the tension T is $T \sin \theta$. For small θ

$$T \sin \theta = T \tan \theta = Tu/c \qquad (11.8)$$

$$\text{Transverse impulse in time } t = Tut/c \qquad (11.9)$$

Now,

Transverse impulse in time t = momentum gained in time t

$$\therefore \quad Tut/c = mcut, \qquad (11.10)$$

$$\text{or} \quad c = \sqrt{T/m} \qquad (11.11)$$

Newton derived an expression for the bulk modulus of elasticity of a gas for the isothermal case, and found B to be equal to the pressure P of the gas.* Eq. (11.7) then becomes:

$$c = \sqrt{P/\rho} \qquad (11.12)$$

At STP

$$c = \sqrt{\frac{1.013 \times 10^5 \text{ N/m}^2}{1.293 \text{ kg/m}^3}} = 280 \text{ m s}^{-1}$$

Experimental value of the speed of sound in air is 331 m s^{-1}. This difference in theoretical and experimental values (about 15%) remained an enigma to Newton. It was explained a century later by Laplace who reasoned that the compression and rarefaction for sound waves cannot occur isothermally. He, therefore, proposed that sound is propagated in adiabatic conditions. It means that the compressed layers which warm up do not dissipate their heat. Similarly, the rarefied layers cool down. This leads to the breakdown of Boyle's law. The appropriate perfect gas law is :

$$PV^{\gamma} = \text{constant} \qquad (11.13)$$

where, $\gamma = C_P/C_V$

---

*For isothermal changes in a perfect gas:

$$PV = \text{constant}$$

$$\therefore \quad PdV + VdP = 0. \text{ Now,}$$

$$B = -V\frac{dP}{dV} = P$$

The value of B for an adiabatic change is γP.* Eq. (11.7), therefore, becomes:

$$c = \sqrt{\gamma P/\rho} \qquad (11.14)$$

Now, air is a mixture of diatomic gases. By using γ value for diatomic molecules in Eq. (11.14) the correct value of the speed of sound is obtained. For a perfect gas of molecular weight M, since :

$$P/\rho = R_0T/M = kT/m$$

$$\therefore \quad c = \sqrt{\gamma kT/m} \qquad (11.15)$$

Here m is the mass of one molecule and k is the Boltzmann constant.

Do the compressions and rarefactions really occur adiabatically ? Let us find out. To be adiabatic, the compressions and rarefactions should occur fast enough to prevent transmission of heat. A compression or a rarefaction occupies a length $\lambda/2$. They change in half a period. For the process to be isothermal the speed with which heat is transmitted should be $(\lambda/2)/(T/2)$, i.e. equal to c. Since heat is carried by randomly-moving particles, the speed of heat transmission is equal to the thermal velocity of the particles. This in a given x-direction is (Example 15.2):

$$v_{rms} = \sqrt{kT/m} \qquad (11.16)$$

From relations (11.15) and (11.16) we note that c and v_{rms} are of the same order of magnitude. But since the thermal motions are random, the particles go on suffering collisions. The mean free path is of the order of 10^{-7} m. This is very much smaller than the wavelength of sound which lies in the range of 16.5 m to 16.5×10^{-3} m. Therefore, heat equalisation cannot take place.

Effect of pressure, temperature and humidity: Since B is equal to γP, an increase of P makes the medium more elastic. At the same time the density of the medium also increases leading to greater inertia. As at constant temperature, (P/ρ) is constant, the speed of sound is independent of pressure. But if the temperature changes, then from Eq. (11.15)

---

*Differentiating Eq. (11.13):

$$P\gamma V^{\gamma-1}dV + V^{\gamma}dP = 0,$$

$$\text{or} \quad V\frac{dP}{dV} = -\gamma P,$$

$$\therefore \quad B = \gamma P$$

$$c_t \propto \sqrt{T} \qquad (11.17)$$

$$\propto \sqrt{t+273},$$

and $$c_0 \propto \sqrt{273}$$

$$\therefore \quad c_t/c_0 = \sqrt{1+t/273},$$

or $$c_t = c_0\sqrt{1+t/273} \qquad (11.18)$$

Thus the speed of sound at t°C can be calculated from Eq. (11.18), if its value at 0°C is known, and vice versa.

The presence of humidity in air changes the speed of sound slightly. The values of γ for moist and dry air are almost equal. The change in velocity occurs due to change in ρ. The speed of sound in moist air is higher than that in dry air.

Example 11.2 The speed of sound in dry air at STP is 331 m s^{-1}. Find its speed in air saturated with water vapour at 27°C and at pressure 75 cm-Hg. It is given that the vapour pressure at 27°C is 2.5 cm-Hg, and the ratio of the densities of dry air and aqueous vapour is 2.5 : 1.

We first find out the effect of temperature. Using Eq. (11.18):

$$c_{27} = 331 \text{ m s}^{-1}\sqrt{\frac{300 \text{ K}}{273 \text{ K}}} = 347 \text{ m s}^{-1}$$

Let the densities of dry and moist air be ρ and ρ_1. Suppose the volume occupied by water vapour in 1 cm^3 of moist air is x cm^3. This is the volume occupied by water vapour at 75 cm-Hg. Or:

$$x \times 75 = 1 \times 2.5$$

assuming water vapour to be a perfect gas. By Dalton's law of partial pressures:

75 = pressure of dry air + 2.5

$\therefore$ Pressure of dry air = 72.5 cm-Hg

Now, the mass of moist air is equal to the sum of the mass of dry air and that of the vapour in mixture. Therefore :

$$1 \times \rho_1 = (1-x)\rho + \frac{x\rho}{2.5},$$

or $$\rho_1 = 0.98\rho$$

$\therefore$ speed of sound in moist air

$$= \text{speed in dry air} \times \sqrt{\frac{\rho}{\rho_1}}$$

$$= 347\sqrt{\frac{1}{0.98}}$$

$$= 351 \text{ m s}^{-1}$$

11.2 Intensity of Sound

As already stated, a longitudinal wave gives rise to a wave of compressions and rarefactions in a medium. A longitudinal wave can, therefore, be represented by a pressure wave analogous to a displacement wave given by Eq. (10.6). It turns out to be a cosinusoidal wave, very much like the velocity wave [Fig. 11.3(b)]. We will now derive an expression for the pressure wave. Consider a medium enclosed in a cylindrical volume. The layers of the medium are represented by vertical lines a, b, c, d, etc. (Fig. 11.5). Fig. 11.5(a) represents their positions at equilibrium. The flow of a longitudinal wave displaces these layers. As stated earlier the successive layers lag behind each other in phase. Thus, if the layer a has a maximum

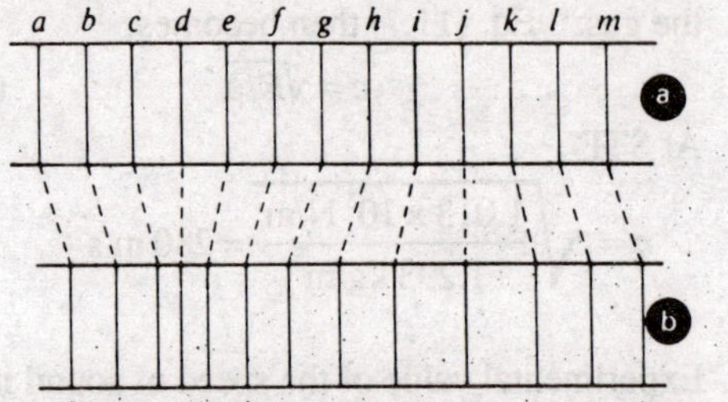

Fig. 11.5 *Air enclosed in a cylindrical volume:* **(a)** *Various layers of air at equilibrium,* **(b)** *Displaced positions of these layers at a particular time when a longitudinal wave passes through this region.* **(c)** *Inverse square law of intensity*

displacement to the right, b will have a smaller displacement, c still smaller till the layer d has zero displacement. The layer e has a displacement to the left. The displacement goes on increasing from layer to layer up to g. The layer j has a zero displacement [Fig. 11.5(b)]. The layers in the portion AB occupy smaller volume as they have come closer together. The maximum pressure exists at d because the layers c and e on left and right of d have both closed upon it. Thus, the region of maximum pressure coexists with the region of zero displacement. The same thing can be said about BC—the region of rarefaction. Now, the bulk modulus of elasticity B is given by:

$$B \equiv -\frac{\Delta P}{\Delta V/V} = \frac{\text{excessive pressure}}{\text{volume strain}} \qquad (11.19)$$

Let $\quad V = A\Delta x \qquad (11.20)$

where Δx is a small region of the medium and A is the area of cross-section of the cylinder. The layers are executing simple harmonic motion. Let the medium originally occupying length Δx have a length $(\Delta x + \Delta y)$ at a particular time. Thus,

$$\Delta V = A\Delta y \qquad (11.21)$$

$$B = -\frac{\Delta P \Delta x}{\Delta y},$$

or

$$\Delta P = -\frac{B\Delta y}{\Delta x} \qquad (11.22)$$

Let Δx be the distance between two layers at x_1 and x_2. Using Eq. (10.6) and taking $\phi = 0$:

$$\left.\begin{array}{l} y_2 = A\sin(\omega t - kx_2), \\ \text{and } y_1 = A\sin(\omega t - kx_1) \end{array}\right\} \qquad (11.23)$$

$$\therefore\ y_2 - y_1 = A\sin(\omega t - kx_2) - A\sin(\omega t - kx_1)$$

$$= -2A\cos\left[\omega t - \frac{k(x_1+x_2)}{2}\right]$$

$$\times \sin k\frac{\Delta x}{2} \qquad (11.24)$$

where $\Delta x = x_2 - x_1$. We write $(x_1+x_2)/2 = x$, and since $(k\Delta x)/2$ is small, we write $\frac{k\Delta x}{2}$ for $\sin\frac{k\Delta x}{2}$. Eq. (11.24) then becomes

$$\Delta y = -\frac{2Ak\Delta x}{2}\cos(\omega t - kx),$$

or

$$\frac{\Delta y}{\Delta x} = -Ak\cos(\omega t - kx)$$

Using Eq. (11.22) and writing p for ΔP

$$p = AkB\cos(\omega t - kx)$$

$$= p_0\cos(\omega t - kx) \qquad (11.25)$$

Here, p_0 is the amplitude of the pressure variation of the wave. The intensity of a wave is given by [Eq. (10.9)]:

$$\left.\begin{array}{l} I = \frac{1}{2}\rho c A^2\omega^2 \\ \text{Also } \quad p_0 = BkA \end{array}\right\} \qquad (11.26)$$

From Eq. (11.26), writing γP for B (where P is the atmospheric pressure),

$$I = \frac{1}{2}\frac{c^3\rho p_0^2}{\gamma^2 P^2} \qquad (11.27)$$

$$= \frac{1}{2}\frac{c^4\rho p_0^2}{\gamma^2 P^2 c}$$

But $\qquad c^2 = \gamma P/\rho$

$$\therefore \qquad I = \frac{1}{2}p_0^2/c\rho \qquad (11.28)$$

Substituting the value of c in Eq. (11.28):

$$I = \frac{1}{2}\frac{p_0^2}{\rho\sqrt{\gamma P/\rho}},$$

or

$$I = \frac{1}{2}\frac{p_0^2}{\sqrt{P\rho\gamma}} \qquad (11.29)$$

The term $\sqrt{P\rho\gamma}$ is written as Z. It is called the *acoustical impedance of the medium.* Or :

$$I = \frac{1}{2}\frac{p_0^2}{Z} \qquad (11.30)$$

Now, the power dissipated in an ac electric circuit is given by:

$$\text{Power} = \frac{1}{2}\frac{V^2}{R} \qquad (11.31)$$

The term 'acoustical impedance' becomes meaningful when we compare Eqs. (11.31) and (11.30). The acoustical impedance absorbs power from the source of sound just as the resistive part of a circuit absorbs power from a source of electricity. The potential difference across an impedance in the electric circuit opposes the source of emf. The excess pressure also acts as a reaction against the source of sound.

Example 11.3 The loudest painless sound produces a pressure amplitude of 28 N/m^2. Calculate the intensity of this sound at STP.

To find I we use Eq. (11.30).

$$Z = \sqrt{P\rho\gamma} = c\rho \qquad (11.32)$$

$$= 331 \times 1.293 \text{ m/s} \times \text{kg/m}^3$$

$$= 428 \frac{\text{N/m}^2}{\text{m/s}} \qquad (11.33)$$

$$\therefore \quad I = \frac{1}{2}\frac{28 \times 28}{428}\frac{\text{N}^2/\text{m}^4}{\text{N/m}^2}\text{m s}^{-1}$$

$$= \frac{1}{2}\frac{28 \times 28}{428}\frac{\text{N m}}{\text{s}}/\text{m}^2$$

$$= 0.916 \text{ W/m}^2$$

$$= 91.6 \times 10^{-6} \text{ W/cm}^2 \qquad (11.34)$$

The unit used to describe intensity levels is W/cm^2.

The maximum amplitude of displacement corresponding to the loudest sound comes out to be of the order of 10^{-5} m. For the faintest sound at the frequency of 1000 Hz the pressure amplitude is of the order of 10^{-5} N/m^2. The corresponding displacement amplitude is of the order of 10^{-11} m. This is smaller than the molecular size which is of the order of 10^{-10} m. This justifies the earlier statement that displacement of the centre of mass of the layers in a medium is usually smaller than the mean free path of the particles of the medium.

The response of an ear to sound is logarithmic. It means that the response of the ear to a sound is dependent on the state of the ear conditioned by the immediate previous stimulii. Denoting the intensity of sound in logarithmic scale as I_L, it can be expressed as

$$I_L = K \log_{10} (I/I_S) \qquad (11.35)$$

For $K = 1$ the logarithmic intensity is expressed in 'bel' after the name of Grahm Bell. If $K = 10$, I_L is expressed in decibel, and is abbreviated as dB. A difference of 1 dB in loudness produces approximately the smallest change which a normal ear can detect.

In Eq. (11.35) I_S is a standard level of sound. All the other sound levels are expressed in comparison with I_S. It is chosen as 10^{-16} watt/cm^2, which is the minimum audible intensity for a frequency of 1000 Hz. Audibility of human beings, which depends upon the frequency of sound, lies roughly between I_S and $10^{12}\, I_S$. For intensities of 10^{12} to 10^{13} times I_S or more the sounds become painful.

The loudest painless sound can be expressed in dB by substituting Eq. (11.34) in Eq. (11.35). Thus,

$$I_L = 10 \log \frac{91.6 \times 10^{-6}}{10^{-16}}$$

$$\cong 120 \text{ dB} \qquad (11.36a)$$

The intensity of normally spoken sound at the listener's ear is about 10^{-10} watt/cm^2. In dB it is

$$I_L = 10 \log \frac{10^{-10}}{10^{-16}}$$

$$= 60 \text{ dB} \qquad (11.36b)$$

Thus, though the power in the loudest sound is 10^6 times the normally spoken sound, the ear records it as double only. Table 11.1 lists the intensities of familiar sounds expressed in dB.

Table 11.1 *Approximate Intensity Levels of Various Common Sounds*

Source of sound	*Intensity*, dB
Airplane (near by)	120
Noises of a busy street	75
Ordinary conversation	70
Whisper	20
Rustle of leaves	10

Consider a source of sound at A [Fig. 11.5(c)]. If there are no obstacles, sound spreads out equally in all directions. The energy contained in a spherical shell of radius R_1 and thickness ΔR is

E = volume of the shell × energy per unit volume

$= 4\pi R_1^2 \Delta R I_1/c$

where I_1 is the intensity at a distance R_1 due to the source at A. After a time t this energy occupies a shell of radius R_2 such that the energy contained in a shell of thickness ΔR at R_2 is given by

$$E = 4\pi R_2^2 \Delta R I_2/c$$

$$\therefore \quad 4\pi R_2^2 I_2 = 4\pi R_1^2 I_1,$$

or $$\frac{I_2}{I_1} = \frac{R_1^2}{R_2^2}$$

This is the *inverse square law of the intensity of sound. It states that intensity of sound varies inversely as square of the distance from the source.*

11.3 Reflection, Refraction, Interference and Diffraction of Sound

The velocity of sound is different in different media. This difference in velocity gives rise to the phenomena of reflection and refraction at the discontinuity as sound travels from one medium to another. The motion of sound can be depicted by the motion of wavefronts (Section 10.1). For simplicity consider a train of plane wavefronts in medium (1) entering medium (2) normally (Fig. 11.6). It is found that a part of sound energy is reflected at the discontinuity giving rise to a reflected wave. The rest of it gets refracted to the other medium. The amount of energy reflected is decided by a reflection coefficient r given by:

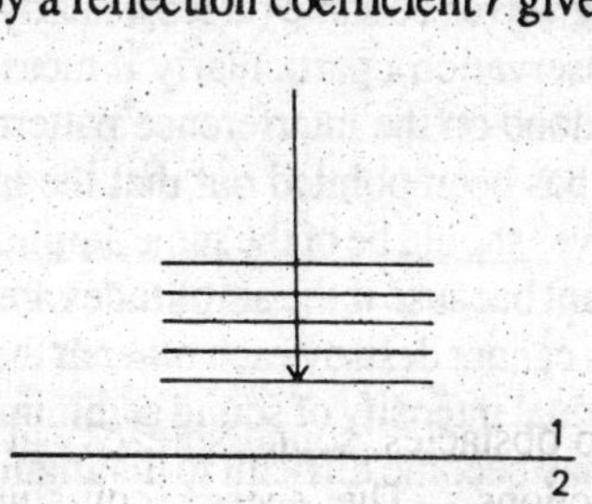

Fig. 11.6 *A plane wave meeting normally the boundary between media (1) and (2)*

$$r = \frac{c_1 - c_2}{c_1 + c_2} \tag{11.37}$$

where c_1 and c_2 are the velocities of sound in media (1) and (2) respectively. The following conclusions can be drawn about r.

(1) The value of r lies between −1 and +1. The negative values of r mean that sound is reflected with a phase change of π in the displacement wave.

(2) If either $c_1 \gg c_2$ or $c_2 \gg c_1$, the magnitude of r is unity. Its sign is decided by whether sound goes from a denser to a rarer medium ($r = 1$) or a rarer to a denser medium ($r = -1$). The velocity of sound in water is ten times the velocity of sound in air. The noise made by the sea animals under water cannot, therefore, be heard out of water.

Eq. (11.37) is true for normal incidence only. For an angular incidence the equation is more complicated. For oblique incidence of the wavefront into another medium, refraction also occurs. Consider that $c_2 > c_1$. A plane wavefront meets the second medium in an inclined way (Fig. 11.7). Disturbance at point A enters the medium with a higher velocity. In the time A travels to A_1, B which is moving at a slower velocity travels only upto B_1. The final wavefront is thus inclined. This is shown by the inclination of the two arrows drawn at right-angles to the wavefronts AB and A_1B_1.

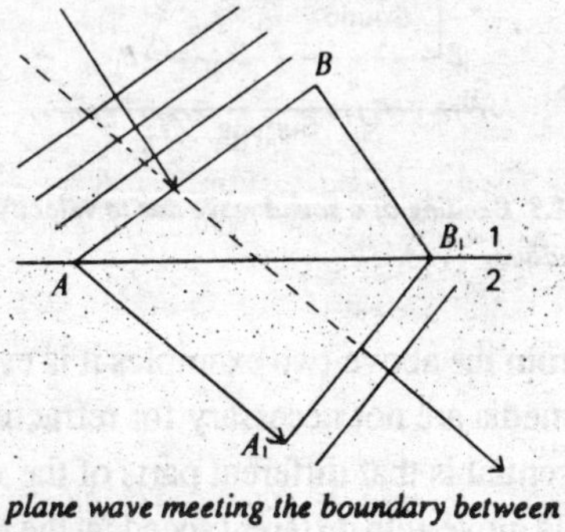

Fig. 11.7 *A plane wave meeting the boundary between media (1) and (2) obliquely*

We observe the refraction of sound on a hot day. It is observed that sound dies off quickly over a hot surface. What happens is shown in Fig. 11.8. The temperature over the surface falls with height. Sound travels faster near the ground surface than at higher altitudes. Therefore, by the time point A of the wavefront has reached A_1, B has travelled upon to B_1 only. The

direction of motion of the new wavefront is shown by the arrow. Thus, sound is curling up. For an observer on the ground sound seems to die off quickly over the hot surface.

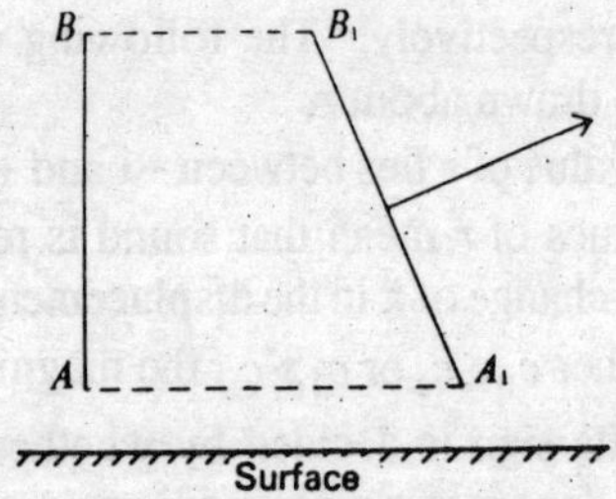

Fig. 11.8 *Bending of a sound wave due to temperature gradient in the medium*

In Fig. 11.9, a plane wavefront AB moves against the wind. The velocity of sound at different points of the wavefront is determined with respect to the wind by subtracting the wind velocity w from the velocity of sound c. The velocity of the wind is lowest near the surface of the earth and increases with the altitude. Therefore, the lower parts of the wavefront move faster than the upper parts. This results in the curling up of the wavefront. The range of transmission of the sound is, therefore, smaller against the wind.

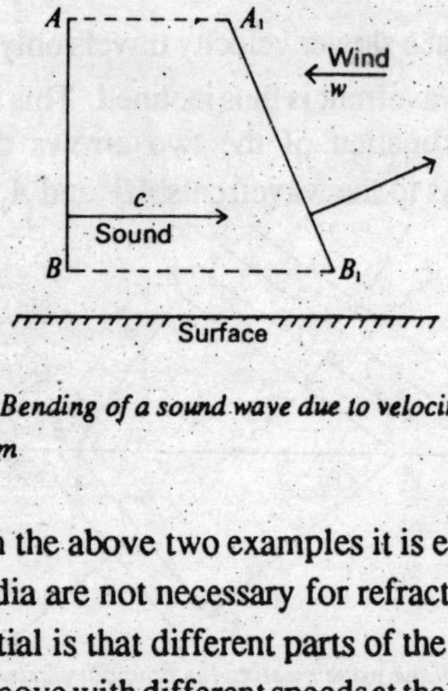

Fig. 11.9 *Bending of a sound wave due to velocity gradient in the medium*

From the above two examples it is evident that two media are not necessary for refraction. What is essential is that different parts of the wavefront should move with different speeds at the same time. If this happens even in one medium, sound will appear to 'bend'.

Question: Is frequency conserved (remains unchanged) when a wave goes from one medium to another?

The principle of superposition for waves was discussed in Chapter 10. It applies to sound waves also. Two sound waves of different frequencies can 'beat' with each other. Advantage of this phenomenon is taken in measuring the frequency of sound. Two sound waves moving in opposite directions give rise to standing waves. This is made use of for measuring the velocity of sound (Chapter 12).

Two sound waves of the same frequency and amplitude can also lead to interference under the following conditions.

(1) As pointed out earlier, the way waves superpose distinguishes them from particles. In the phenomenon of beats (Section 10.6) the result of combining the waves was observed at a fixed point in space. Since the frequencies of the mixing waves were different, the phase difference between them continued to change with time giving rise to maximum and minimum intensity. In interference, on the other hand, the frequencies of the waves are equal. The required phase difference between the waves is the result of the path difference between them. Since the path difference will go on varying at different points of observation, a series of maxima and minima can be obtained. But if the two sources of sound have different frequencies at a particular point, the resultant amplitude of the combined wave will also go on varying just as happens in beats. This will happen at every point and the maxima and minima will, therefore, be shifting their positions. If a maximum or a minimum at a point shifts before 1/12th of a second, which is the persistence of audibility, the maxima and minima will become indistinguishable inspite of the interference between the sound waves. It is, therefore, necessary that the frequencies of the interfering waves be same (at least nearly same for simple observation), particularly if measurements are to be done on the interference pattern.

(2) It has been pointed out that the interfering sound waves should be of the same amplitude. This is important because if the amplitudes are different the waves cannot destroy each other at the minima. If the residual intensity of sound at minima is fairly large, it may become difficult to distinguish it from the maxima.

(3) It is important that the initial phase difference between the two sources remains the same during the observation of the interference pattern. This condition is necessary because the resultant amplitude at the point of observation is decided by the sum of the initial phase difference and the additional phase difference resulting from the path difference. If the initial phase difference fluctuates, so will the resulting amplitudes at various points. The maxima and minima do not remain stationary, again leading to the difficulty pointed out in (1) above. Two sources of sound with constant initial phase difference are called *coherent sources.* It is much easier to get coherent sources for sound than for light waves. Two tuning forks of the same frequency will successfully demonstrate interference if their relative positions with respect to each other and with respect to the points of observation remain the same.

(4) The interfering waves should be of one frequency only. If the waves contain many frequencies the maxima and minima of the different frequencies are formed at different places. A maximum of one may lie at the minimum of the other. This leads to overlapping of the interference patterns. This condition is relaxable to an extent for sound waves because of their longer wavelengths.

Consider two pairs of coherent sources of sound S_1 and S_2, and S_3 and S_4 (Fig. 11.10). The initial phase difference between S_1 and S_2 is zero or a multiple of 2π [Fig. 11.10(a)], and between S_3 and S_4 it is an odd multiple of π [Fig. 11.10(b)]. In Fig. 11.10(a), the distances S_1O and S_2O are the same. The same path difference makes the phase difference from S_1 and S_2 at O equal to zero as their initial phase difference is also zero. Therefore, the compressions (and rarefactions) from S_1 and S_2 reach O at the same time leading to a maximum there.

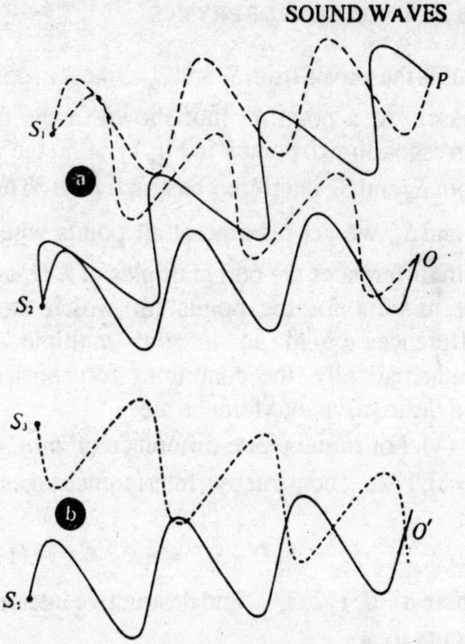

Fig. 11.10 *Superposition of two waves from coherent sources:* **(a)** *Initial phase difference zero* **(b)** *Initial phase difference π*

Question: Has a compression or a rarefaction reached *O* in Fig. 11.10(a) ? Since rarefaction is a diminished pressure, will the combined arrival of two rarefactions cause a minimum?

At a point P the difference of distances between S_2P and S_1P is $\lambda/2$. Therefore, compressions and rarefactions from S_1 reach P with rarefactions and compressions arriving there from S_2. The two waves cancel each other at all times and a minimum results. At another point Q the path difference between S_2Q and S_1Q will be λ. Again the waves arrive in phase and lead to a maximum. Thus at all points where path difference between the waves from S_1 and S_2 is an integral multiple of λ, a maximum exists. Similarly, at points where the path difference is an odd multiple of $\lambda/2$, a minimum exists. The interference causing a maximum is called constructive interference. The destructive interference leads to a minimum. The series of points of maximum and minimum energy is called the *interference pattern.*

Let us now seek the interference pattern for S_3 and S_4. They give a minimum at O' unlike S_1 and S_2, because though $(S_3O' - S_4O')$ is still zero the initial phase difference of π between them causes the waves from S_3 and S_4 to meet in opposite phase. At a point P' (not shown in the figure) corresponding to point P in Fig.11.10(a) the waves from S_3 and S_4 interfere constructively. Thus for S_3 and S_4, we get maxima at all points where the path differences are odd multiples of $\lambda/2$, and we get minima for the points for which the path differences equal an integral multiple of λ. Mathematically, the conditions for constructive and destructive interference are:

(1) For initial phase difference of $2m\pi$, where $m = 0, 1, 2, \ldots$, constructive interference takes place for

$$S_2P - S_1P = n\lambda \qquad (11.38a)$$

where $n = 0, 1, 2, 3, \ldots$, and destructive interference occurs if

$$S_2P - S_1P = \left(n + \frac{1}{2}\right)\lambda \qquad (11.38b)$$

where $n = 0, 1, 2, 3, \ldots$

(2) For initial phase difference of $(2m+1)\pi$, where $m = 0, 1, 2, 3, \ldots$, constructive interference occurs for

$$S_2P - S_1P = \left(n + \frac{1}{2}\right)\lambda \qquad (11.39a)$$

where $n = 0, 1, 2, 3, \ldots$, and destructive interference occurs for

$$S_2P - S_1P = n\lambda \qquad (11.39b)$$

where $n = 0, 1, 2, 3, \ldots$. Here, P is any point and not the particular point in Fig. 11.10(a).

The absence of sound energy from the minima does not lead to breakdown of the principle of conservation of energy. The energy is actually redistributed. The energy lost at the minima supplies the increased energy at the maxima. This can be shown as follows:

$$\text{Energy of a wave} \propto A^2$$

$$= KA^2$$

where K is a constant. If E_1 and E_2 are the energies of the first and the second waves and we consider their amplitudes to be the same, we get

$$E_1 = KA^2$$

$$E_2 = KA^2$$

At a maximum the combined amplitude is $2A$, and at a minimum the combined amplitude is zero.

$$\text{Energy at a maximum} = 4KA^2$$

$$\text{Energy at a minimum} = 0$$

$$\text{Hence, the average energy} = \frac{4KA^2 + 0}{2}$$

$$= 2KA^2$$

$$= E_1 + E_2$$

It is a common observation that the motion of particles is hindered by rigid obstacles. Waves, on the other hand, can bend around the edges of such an obstacle if the dimensions of the obstacle are small as compared with the wavelength of the wave. This phenomenon is called *diffraction* and is again a characteristic of the waves. As an example, if a pole is fixed in a water pool, the water waves move about unimpeded. But the same water waves will break against a boat. This happens because the λ of water waves is smaller than the size of the boat.

Sound waves show diffraction. The λ of sound waves lies between 16.5 m and 1.65 cm for the whole audible range. The obstacles encountered by them are comparatively smaller. They can, therefore, bend around these obstacles. They are however scattered, like any other wave, by large obstacles. For a complex sound an obstacle can sometimes scatter the smaller wavelengths. Then the quality of the diffracted sound is changed because of the absence of the higher frequencies. You will read more about diffraction of waves in Chapter 28.

11.4 Doppler Effect

The frequency of sound as heard by a listener is called the *pitch*. The pitch of the sound is decided by the frequency of sound emitted and the relative motion between the listener and the source of sound. Since the listener observes the phenomenon, it is a practice to call him observer. The dependence of the pitch of sound on the relative motion of the observer and the source of sound can be noted when a hooting engine rushes past us. The pitch of the hoot keeps on varying. An explanation for this phenomenon was given by Doppler in 1842 AD. He had actually explained the phenomenon as observed for light*. The same explanation holds in the case of sound also. It is convenient to explain the phenomenon with either the source or the observer moving. The effect due to their combined motion can then be found.

1. First we consider the motion of the source :
The observer is at rest. It is given that :

(i) Velocity of source S towards O is u
(ii) Velocity of sound is c
(iii) Distance between S and O is c
(iv) The frequency of source is ν
(v) The pitch recorded by O is ν_{SF}

If the source were stationary, ν waves would have been present between S and O. But in one second S moves towards O by a distance u [Fig. 11.11(a)]. Now, as ν waves are confined in length $(c-u)$ they seem to have been compressed. Let the reduced wavelength be λ'. Therefore,

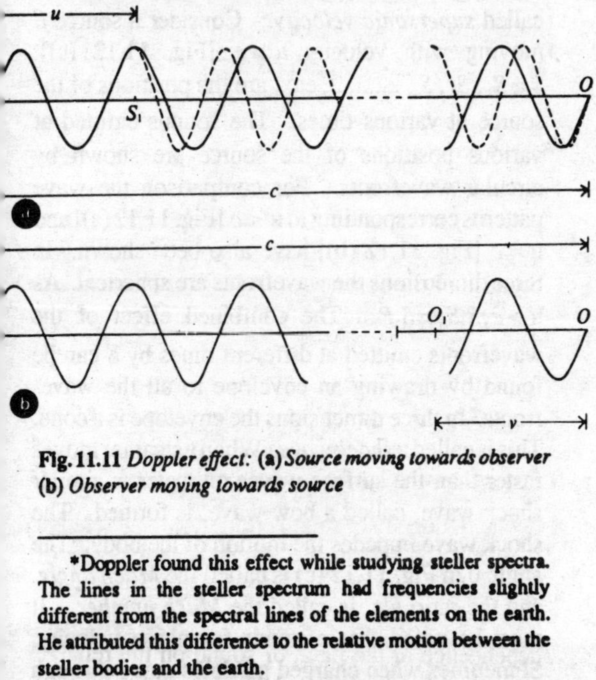

Fig. 11.11 ***Doppler effect: (a) Source moving towards observer (b) Observer moving towards source***

***Doppler found this effect while studying steller spectra. The lines in the steller spectrum had frequencies slightly different from the spectral lines of the elements on the earth. He attributed this difference to the relative motion between the steller bodies and the earth.**

$$\lambda' = \frac{c-u}{\nu} \tag{11.40}$$

Hence, $$\nu_{SF} = \frac{c}{\lambda'} = \frac{c\nu}{c-u} \tag{11.41}$$

Thus, $$\nu_{SF} > \nu$$

For S moving away from O the pitch observed, ν_{SR}, is given by:

$$\nu_{SR} = \frac{c\nu}{c+u} \tag{11.42}$$

Therefore, $$\nu_{SR} < \nu$$

2. *S is at rest and O moves towards S.* It is given that :

(i) Velocity of O towards S is v
(ii) Frequency of S is ν
(iii) Distance SO is c.
(iv) Wavelength of the sound emitted is λ
(v) The pitch recorded by O is ν_{OF}

If O were stationary, ν waves would have passed by him per second. Since O moves towards S, it receives a few more waves. This additional number of waves are those that lie between OO_1 [Fig. 11.11(b)]. Therefore :

$$\nu_{OF} = \nu + v/\lambda$$

$$= \frac{c}{\lambda} + \frac{v}{\lambda}$$

$$= \left(\frac{c+v}{c}\right)\nu \tag{11.43}$$

Thus, $$\nu_{OF} > \nu$$

If O moves away from S, the pitch observed will be :

$$\nu_{OR} = \left(\frac{c-v}{c}\right)\nu \tag{11.44}$$

Or, $$\nu_{OR} < \nu$$

Question: Derive the Equations (11.42) and (11.44)

From the above we see that the Doppler effect due to the motion of the source is essentially different from that due to the motion of the observer. Put $u = v = c$ in relations (11.41) and (11.43) and compare the results. The two are equivalent only to a first approximation, i.e. when higher powers of u/c can be neglected.

Question: Does the above difference due to the relative motion between the source and the observer violate the principle of relativity?

Example 11.4 Two locomotives moving with a speed of 36 km/hr recede from each other. One of the locomotives whistles with a frequency of 200 Hz. Find the pitch of the whistle as heard by an observer O in the other locomotive. Assume c = 340 m s^{-1}.

Here, S and O both are moving. First treat O as stationary. Using Eq. (11.42)

$$\nu_{SR} = \frac{200 \times 340}{(340+10)} \text{ Hz}$$

$$= \frac{40 \times 34}{7} \text{ Hz} \tag{11.45}$$

Take the source to be stationary and emitting a sound of the frequency given by Eq. (11.45). Now, consider the motion of O. Using Eq. (11.44):

$$\nu_{OR} = \frac{40 \times 34}{7} \times \frac{330}{340}$$

$$= 188 \text{ Hz}$$

Doppler effect is also observed for light waves. But now the change in frequency is due to two factors: (1) time period of the vibrations emitted by the source gets dilated for the observer because of relative velocity; (2) distance between two waves spread in space changes with the motion of the source.

If the proper time period of the source is T_0, the time-period for the observer will be

$$T = \gamma T_0, \tag{11.46}$$

$$\text{or} \quad \nu = \frac{1}{T} = \frac{1}{\gamma T_0} = \frac{\nu_0}{\gamma} = \nu_0 \left(1 - \frac{u^2}{c^2}\right)^{1/2} \tag{11.47}$$

Here, ν_0 is the frequency of the source in its own frame, and ν is its frequency as observed by the observer. The velocity of the source relative to the observer is u.

If the source is moving towards the observer, then it can be seen from the wavelength spread in space, Eq. (11.41), that the frequency observed by the observer would be

$$\nu_{SF} = \frac{c}{\lambda'} = \frac{c}{c-u}\nu$$

$$= \frac{c}{c-u} \times \nu_0 \times \frac{\sqrt{c^2 - u^2}}{c},$$

$$\text{or} \quad \nu_{SF} = \nu_0 \sqrt{\frac{c+u}{c-u}} = \nu_0 \sqrt{\frac{1+u/c}{1-u/c}} \tag{11.48}$$

The Doppler formulae (11.41) to (11.44) need modification if the velocities u and v become comparable to c. Consider Eq. (11.41). As u increases ν_{SF} also increases. Suppose the periods corresponding to ν and ν_{SF} are T and T_S. If the energy released by the source per period T is E, the same energy passes the observer in time T_S. As $T_S \ll T$, the energy per second, i.e., the power, as well as intensity for the sound wave, are increased by Doppler shift of the frequency. The sound appears to be louder. The intensity becomes very large as the value of u approaches c leading to a non-linear response of the medium.

When u exceeds c, the Doppler formula has no meaning. Velocity greater than that of sound is called *supersonic velocity*. Consider a source S moving with velocity $u > c$ [Fig. 11.12 (c)]. S_1, S_2, S_3, S_4, ..., etc., are the positions of the source at various times. The sounds emitted at various positions of the source are shown by circular wavefronts. For comparison the wave patterns corresponding to $u < c$ [Fig. 11.12 (a)] and $u = c$ [Fig. 11.12 (b)] have also been shown. In three dimensions the wavefronts are spherical. As $u > c, S_1S_6 > S_1P$. The combined effect of the wavefronts emitted at different times by S can be found by drawing an envelope to all the wavefronts. In three dimensions the envelope is a cone. This is called a *shock wave*. When a steamer moves faster than the surface waves on water, a similar shock wave, called a bow wave, is formed. The shock wave impedes the motion of the body. The angle θ in Fig. 11.12 (c) is called the *Mach angle*, and the ratio u/c is called the *Mach number*. It must be noted that c stands for phase velocity. Sometimes when charged particles move through a medium with a velocity greater than the velocity of light in that medium, a shock wave in the form of electromagnetic radiation is emitted. This is called *Cherenkov radiation*.

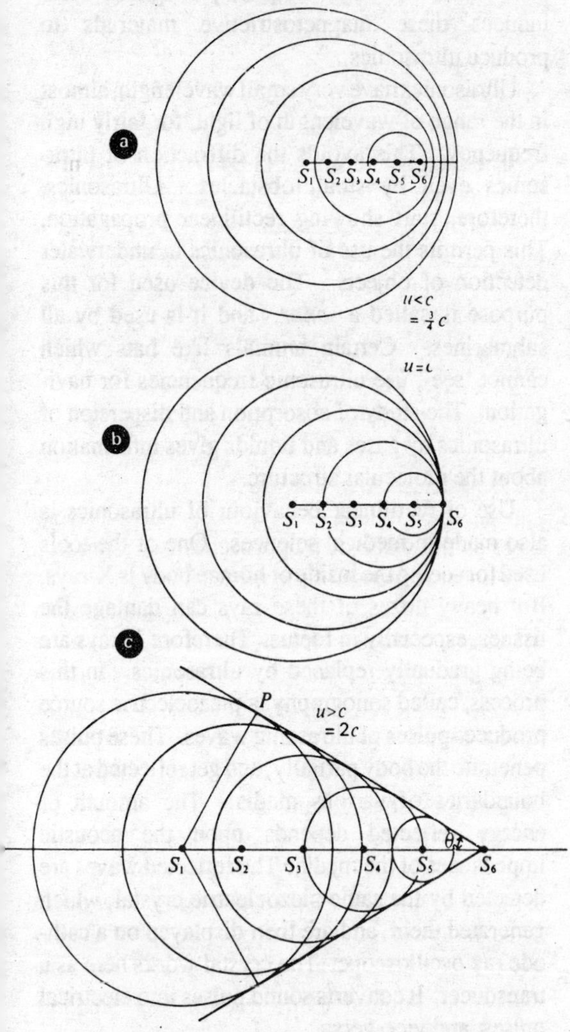

Fig. 11.12 ***Shift in the relative positions of the wavefronts as a function of velocity of the source:*** **(a)** ***Source velocity less than the velocity of sound*** **(b)** ***Source velocity equal to the velocity of sound*** **(c)** ***Source velocity greater than the velocity of sound***

Question: Why is the crossing of the sound barriers so critical for a supersonic plane? This situation arises when *u* crosses *c* [Fig. 11.12 (b)].

The shock wave is also accompanied by sonic boom at supersonic velocity. Consider a supersonic plane approaching from a great distance. Near horizon the plane is almost at the line of sight with observer O (Fig. 11.13). As it is moving faster than sound, a sound produced at a later time can reach O earlier. Consider the position of the plane at P_1 at an instant t_1. When the plane comes overhead at position P_2 at instant t_2, the direction of sound and the motion of plane are at right-angles. Sound, therefore, travels to O as if there is a stationary source. Suppose :

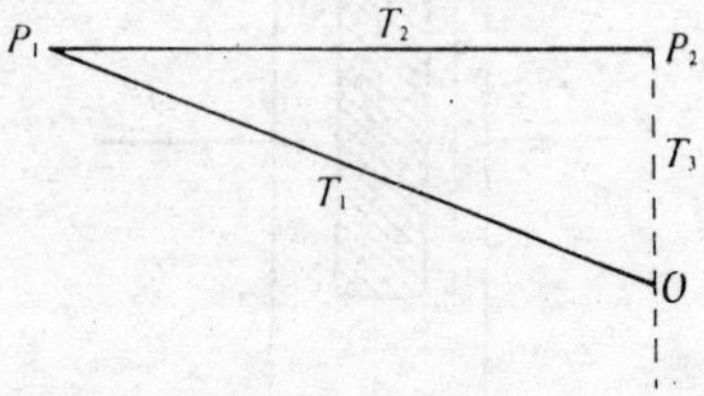

Fig. 11.13 ***Production of a sonic boom***

(*i*) the time taken by sound to cover the distance P_1O is T_1,

(*ii*) the time taken by the plane to reach P_2 from P_1 is T_2, and

(*iii*) time taken by sound to cover the distance P_2O is T_3.

If

$$T_1 = T_2 + T_3 \tag{11.49}$$

the sounds emitted by the plane at positions P_1 and P_2 reach O together. The level of sound, therefore, increases four times, causing an increase in the sound intensity. There may be many positions like P_1 and P_2 for which Eq. (11.49) holds. Thus a continuous boom may be heard.

11.5 Ultrasonics

The range of audible sound lies between 20 Hz and 20,000 Hz. Sound waves with frequency below 20 Hz are called *infrasonic*. (The tidal waves caused in oceans by the movements of the earth are an example of very low frequency waves.) The frequencies higher than 20,000 Hz are called *ultrasonic*. We can calculate roughly the limiting ultrasonic frequency in air at STP. It was pointed out in Section 11.1 that the regions of compression and rarefaction should exceed the mean free path of the particles of the medium. In the limit they can equal the mean free path. Taking the value of mean free path equal to λ, the limiting ultrasonic frequency ν_L can be found. For air at STP, the mean free path is of the order of 10^{-7} m and $c = 331$ m/s. Therefore:

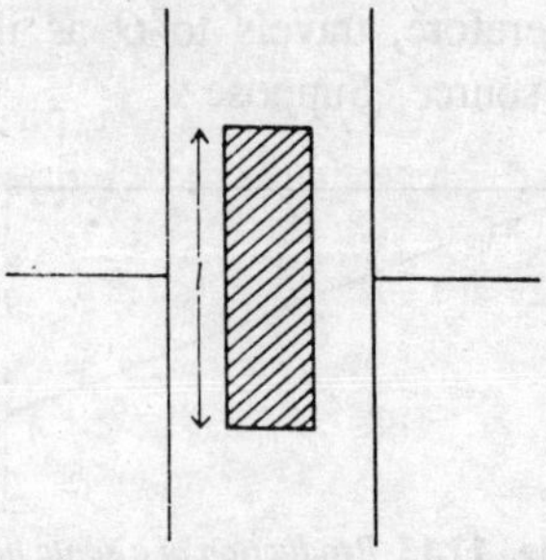

Fig. 11.14 *Generation of ultrasonic waves with the help of piezoelectric crystals*

$$\nu_L \sim \frac{10^2}{10^{-7}} \sim 10^9 \text{ Hz} \qquad (11.50)$$

The frequencies of this order have been produced in the laboratory. Previously, ultrasonic waves were produced by a mechanical arrangement called Galton's whistle. Nowadays the high frequency oscillators using electronic circuitry are used to excite oscillations in piezoelectric crystals and magnetostriction materials. Piezoelectric crystals change in size on applying an electric field. In Fig. 11.14 a rhombic piece of crystal is placed between the plates of a parallel plate capacitor. On applying a high frequency electric field to the plates rapid variations in the length and breadth of the crystal make the crystal oscillate. This gives rise to an ultrasonic wave. The amplitude of the oscillation of the crystal depends upon the strength of the electric field. To get maximum amplitude the frequency of the electric field has to be equal to the natural frequency of the crystal. This frequency is given by

$$\nu = \frac{2725}{l} \text{ kHz} \qquad (11.51)$$

where l is the thickness of the crystal, in cm, in the direction in which the waves are produced.

Another method of producing ultrasonics is to use materials which change their length in magnetic field. A high frequency magnetic field induces these magnetostrictive materials to produce ultrasonics.

Ultrasonics have very small wavelength, almost in the range of wavelength of light, for fairly high frequency. This avoids the diffraction of ultrasonics even by small obstacles. Ultrasonics, therefore, start showing rectilinear propagation. This permits the use of ultrasonics in underwater detection of objects. The device used for this purpose is called a *sonar,* and it is used by all submarines. Certain animals like bats which cannot 'see', use ultrasonic frequencies for navigation. The study of absorption and dispersion of ultrasonics by gases and liquids gives information about the molecular structure.

Use of rectilinear behaviour of ultrasonics is also made in medical sciences. One of the tools used for seeing the inside of human body is X-rays. But heavy dozes of these rays can damage the tissues, especially in foetus. Therefore X-rays are being gradually replaced by ultrasonics. In this process, called sonography, a piezoelectric source produces pulses of ultrasonic waves. These pulses penetrate the body partially, and get reflected at the boundaries of various media. The amount of energy reflected depends upon the acoustic impedances of the media. The reflected waves are detected by the same piezoelectric crystal, which generated them, and are then displayed on a cathode ray oscilloscope. The crystal works here as a transducer. It converts sound pulses into electrical pulses, and vice-versa.

Ultrasonics are also used for probing the interior of those media which absorb electromagnetic waves heavily. These waves can travel through such media upto very large distances because of much smaller attenuation. This property has applications in geology, oceanography, metallurgy, etc.

Acoustic holography is a fast developing field. Ultrasonic sources are used here as coherent sources of sound.

11.6 Architectural Acoustics

Architectural acoustics is the branch of study in which the behaviour of sound in an enclosed medium is studied. In an enclosed volume the sound emitted at a place gets reflected a number of times from walls, etc. The sonic effect at any point is due to the combined effect of the direct and a number of reflected waves. If an auditorium is not properly constructed the combined effect can be undesirable. For example, some of the frequencies of a complex note may diminish in magnitude and the others may increase by the destructive and constructive interference between direct and reflected waves. This will alter the quality of the note, and the listener may not enjoy the music. We may also get zones of silence. Superposition of echoes of the sounds produced earlier with the sounds produced later may also give undesirable effects.

Sabine studied this branch in great detail. He found that maintenance of sound by reflections upto a certain intensity level for an optimum period improves the quality of sound. The optimum period is called the reverberation period. Maintaining the reflected sound waves is not enough because they may interfere destructively to produce zones of silence and may also affect the quality of sound. On the other hand, the absence of reflected sound makes the music concerts in open a dull affair. This happens because in open sound moves in all directions and, from inverse square law, the intensity of sound waves becomes very low even at a small distance. Since our ears are not equally sensitive to all the frequencies, some of the frequencies may go below the range of audibility. This makes the sound flat. This defect can be remedied by using amplifiers in open concerts.

Sabine Formula : The reverberation period is the time in which the level of sound falls to 10^{-6} times its initial value. As sound falls on the walls of the auditorium, a part of it is absorbed. Suppose the average absorption coefficient is $\langle\alpha\rangle$. Then:

$$\langle\alpha\rangle = \frac{A_1\alpha_1 + A_2\alpha_2 + A_3\alpha_3 + A_4\alpha_4 + \dots}{A_1 + A_2 + A_3 + A_4 + \dots} \qquad (11.52)$$

where $\alpha_1, \alpha_2, \alpha_3, \dots$, etc., are the absorption coefficients of areas $A_1, A_2, A_3, \dots$, respectively. Consider an enclosed volume with average sound energy density E. The amount of energy escaping per second, i.e., power, from a window of unit area in a wall is equal to the energy contained in a cylinder of unit area and length c (provided all the energy is moving towards the window). Calling it P_1, it is:

$$P_1 = Ec \qquad (11.53a)$$

But since sound is spreading in all directions, instead of Eq. (11.53 a) we get:

$$P_1 = \frac{1}{4}Ec \qquad (11.53b)$$

Since no energy is reflected from a window, the escaped energy can be treated as energy absorbed by a material of coefficient of absorption equal to unity. If instead of window various absorbing materials are there, then Eq. (11.53b) becomes:

$$P = \frac{1}{4}Ec\Sigma A_n\alpha_n$$

Using Eq. (11.52):

$$P = \frac{1}{4}EcA\langle\alpha\rangle \qquad (11.54)$$

For sound falling normally on the walls, the distance between the walls is the distance travelled by sound between two reflections. (In practice, due to the presence of various reflecting objects, this is not true.) The average distance between the reflections is $4V/A$, where V is the volume enclosed and A is its surface area. In time t the number of reflections is, therefore, $Atc/4V$. The average power absorbed per unit area in each reflection is $\langle\alpha\rangle$. If the initial intensity is I_0, the intensity after time t is given by:

$$I = I_0(1 - \langle\alpha\rangle)^{Atc/4V} \qquad (11.55)$$

or

$$\ln\frac{I}{I_0} = \frac{Atc}{4V}\ln(1 - \langle\alpha\rangle) \qquad (11.56)$$

In the reverberation period T:

$$I = 10^{-6} I_0 \qquad (11.57)$$

Substituting in Eq. (11.56):

$$T = \frac{(\ln 10^{-6})\, 4V}{Ac \ln(1 - \langle\alpha\rangle)} \qquad (11.58)$$

If $\langle\alpha\rangle$ is small then $\ln(1 - \langle\alpha\rangle) \cong -\langle\alpha\rangle$

Therefore, under these conditions T becomes:

$$T = \frac{(-6)(2.3) \times 4V}{-A\langle\alpha\rangle\, c} \qquad (11.59)$$

If V and A are measured in cubic meter and square meter, then using c equal to 333 m/s, Eq. (11.59) becomes:

$$T = \frac{0.165V}{A\langle\alpha\rangle} \qquad (11.60)$$

Relation (11.60) is called Sabine's formula for finding out the reverberation period. For a desirable sonic response, T should lie between 1 and 2 seconds depending on the size of the auditorium and the purpose (speech, music, orchestra, group song, etc.).

We see from the above that T is directly proportional to the volume of the hall, and inversely proportional to the absorption by the wall materials. Thus T can be adjusted to the desired value by properly choosing the various parameters. Echoes can be reduced by properly designing the shape of the hall, its size and by coating its walls with absorbing materials such as felt, coarse cloth, etc. Zones of silence can also be eliminated by properly shaping the building.

Before a building is erected, its scale model is many times prepared. The reflections, interference, zones of silence, etc., are studied by putting the model in water, and by exciting water waves at places where speakers are to be located. Sound and light waves are also used.

Discussion and Problems

11.1 How do you establish that sound is a wave motion?

11.2 The string of a sonometer makes transverse vibrations. How does it produce sound which is a longitudinal wave?

11.3 Does the velocity of sound increase with altitude?

11.4 Does the velocity of sound in a solid increase on heating the solid?

11.5 Are there any dispersive media for sound? Is air a dispersive medium for sound?

11.6 What is the principle of stethoscope?

11.7 Explain why sound can be heard more distinctly at greater distances over water surfaces.

11.8 Discuss an experiment to demonstrate interference of sound in the laboratory.

11.9 There are zones of silence in tuning forks. Explain this in the light of interference of sound waves.

11.10 Show that the amplitude of vibration for the loudest painless sound at 60 Hz is 1.2×10^{-5} m.

11.11 A sound differs by 6 dB from a sound of intensity equal to 10 nanowatt/cm^2. Find the absolute value of intensity for it.

11.12 A person riding on a merry-go-round emits a sound of a certain frequency. Does a person at the centre observe a Doppler shift?

11.13 What happens to the intensity of sound for a Doppler shift in which the pitch falls?

11.14 What conclusion can be drawn from the observed red-shift of the spectra from galaxies?

11.15 Show that relativistically correct formulae for Eqs. (11.41) and (11.43) are the same. Does this result disprove the existence of ether?

11.16 The crack of a riding whip produces a sonic bang? Explain this.

11.17 While discussing interference it was stated that condition (4) is relaxable for sound waves. Explain this.

11.18 Discuss the use of ultrasonic frequencies.

11.19 Find out the upper limit of ultrasonic frequencies in solids.

11.20 The audibility of the speech of a man standing on the ground is more than that of a man standing at the top of a staircase. Explain this.

11.21 Find out the value of molecular weight for a gas in which the velocity of sound is 1260 m/sec at 0°C. $\gamma = 1.4$.

11.22 A locomotive travelling with a velocity of 108 km hr passes a stationary observer. The locomotive emits a frequency of 2000 c/sec. Find the observed pitch for the locomotive while approaching towards and receding from the observer.

11.23 What is the velocity of the observer for whom a note is 10 per cent lower than the emitted frequency?

11.24 Microwaves from a radar source are reflected from a moving plane. The reflected wave produces 5000 beats with the original radar wave. Find the speed of the plane in Mach units. The wavelength of the radar waves is 0.10 m.

11.25 For two sound waves of equal intensity compare the pressure amplitude of the waves in air and water.

11.26 A bullet is fired with a speed of 700 m/sec. Find the Mach angle.

11.27 Phase velocity of light in glass is 2/3 of the velocity of light in vacuum. Find the velocity of an electron for which the Cherenkov radiation makes a Mach angle of 60°.

11.28 A steel ball drops from a height of 1 m on hard ground. If all the energy of the ball at the point of impact is converted into a sound pulse lasting for 0.1 s, find the intensity of the sound a metre away. The mass of the steel ball is 50 g.

11.29 An observer is moving away from a source of sound with speed v, the source is moving towards the observer with speed u, and the wind is blowing with speed V in the direction from the source towards the observer. All these speeds are measured with respect to the earth. Show that the pitch ν' recorded by the observer is,

$$\nu' = \frac{c+V-v}{c+V-u}\nu$$

where ν is the frequency of the source, and c is the speed of sound in still air.

11.30 In normal conversation the sound energy output of a person is around 10^{-5} J/s, and the area of the mouth opening is about 10 cm^2. What will be the sound intensity at his mouth? How much will this intensity be at 1 m distance?

Chapter 12

Vibrations of Strings and Air Columns

12.1 Boundary Restrictions on SHM

In Section 7.2 there has been a mention of complex harmonic motion. It was also shown in Section 7.7 how the combination of two orthogonal SHMs gives rise to complex motion as demonstrated by Lissajous figures. The presence of two orthogonal SHMs indicates that the vibrating body has two degrees of freedom. A simple pendulum used as a conical pendulum (Problem 2.33) is an example. It has freedom to vibrate in two orthogonal directions. On the other hand, the vibrating pendulum of a clock is confined in one direction only and therefore has one degree of freedom.

Complex motions also arise when a number of oscillators are coupled together. Suppose there are two objects executing linear SHM with certain frequencies. If they are coupled together, it becomes a compound system with two degrees of freedom. The composite system has two modes of vibration having different frequencies. As an example take two identical simple pendulums with their bobs, which are at the same height and have a mass m each, coupled by a spring of spring constant k. When the two bobs always move in the same direction, the spring remains unstretched and the time-period is equal to $2\pi\sqrt{l/g}$. In the second mode the two bobs at any time move in opposite directions and the time-period becomes $2\pi/\sqrt{(g/l + 2k/m)}$.

Exercise : Derive the above relations.

If the restriction of linear vibration is given up and vibration along all the three axes is permissible giving another four degrees of freedom, then the composite body has $3 \times 2 = 6$ degrees of freedom. For any physical object the modes of vibration can, therefore, be known if the number of constituents and the degrees of freedom permitted to each are known*. In 22.4 litre of a monatomic gas at STP there are 6.0×10^{23} atoms, and each atom has three degrees of freedom. Thus the total number of modes of vibration possible for 22.4 litre of the gas can be as large as 18.0×10^{23}. In practice this does not happen. The atoms in the gas are moving about chaotically and the vibration of individual particles is meaningless. Only the vibration of the layers of the gas at least as thick as the mean free path of the atoms has some meaning. This has already been pointed out in Chapter 11.

To make it more clear let us consider the vibration of beads embedded in a cord, along x-axis, fixed between two rigid supports. Also, suppose that only the vibration along the y-axis is permitted (Fig. 12.1). Then as the number of beads increases, the modes of vibration also increase as is apparent from the figure. If we have a very large number of beads, say N, then they will appear to be distributed continuously giving N modes of vibration. A continuous string can be compared

*For details refer to Chapters 1 and 2 of F.S. Crawford, Jr., *Waves : Berkeley Physics Course*-Vol. 3 (McGraw-Hill Book Co.), 1968.

with such a system. We can, therefore, imagine that if a continuous string (or wire) is considered, a large number of modes with a spectrum of frequencies is possible. The modes corresponding to the higher frequencies exist for a short period. They die off leaving only a few dominant modes of the string. Any other freely vibrating body will have a similar behaviour. For example, an infinitely large elastic sheet will also have as many modes as the degrees of freedom of the component oscillators in it. But on excitation only a few dominant modes persist longer.

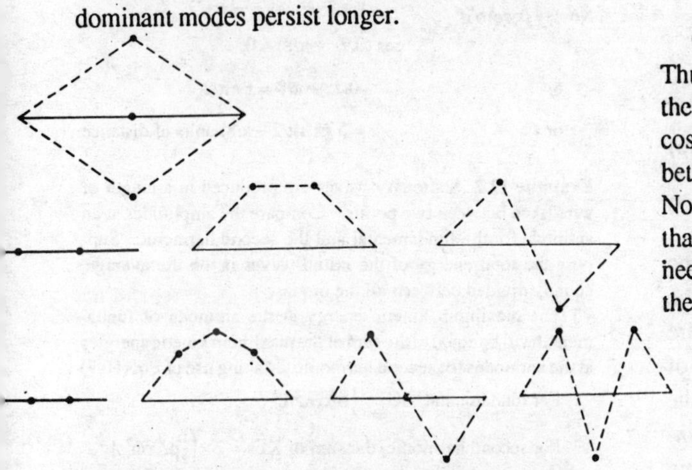

Fig. 12.1 *Transverse vibrations of beads embedded in a cord. When vibrating only along one direction:* (a) *one bead has only one mode of vibration,* (b) *two beads have two modes of vibration, and* (c) *three beads have three modes of vibration*

If there exist certain boundary restrictions, the situation becomes quite different. After all, we cannot have infinitely long strings and infinitely extending elastic sheets. Suppose we have a string AB fixed at both the ends having x-coordinates, $x = 0$ and $x = L$ (Fig. 12.2). The string is picked at certain point C, stretched and then released. Obviously, the point C will start vibrating and waves of transverse motion will start travelling both to the left and the right. These waves meet discontinuities at A and B, and if the supports are rigid enough, these will get reflected. We have, therefore, two sets of waves moving in opposite directions. There may be many modes of vibration and waves of many frequencies. Let us take the waves of one particular frequency. Since the two waves are moving in opposite directions, their wave equations will be:

$$y_1 = A \sin(\omega t - kx), \tag{12.1}$$

and $$y_2 = A \sin(\omega t + kx + \phi) \tag{12.2}$$

Here, ϕ is the relative phase angle between the two waves at point x. The resultant at x is, therefore,

$$y = y_1 + y_2$$

$$= A \sin(\omega t - kx) + A \sin(\omega t + kx + \phi),$$

or $$y = 2A \cos(kx + \phi/2) \sin(\omega t + \phi/2) \tag{12.3}$$

Thus the resultant wave has the same frequency as the component waves, but the amplitude has a cosinusoidal term. Therefore, the amplitude varies between $-2A$ and $2A$ as we move along the string. Now, the boundary conditions for the string AB are that y should be zero both at $x = 0$ and $x = L$. We need only the amplitude part of Eq. (12.3) to apply these conditions. Putting $x = 0$, we get,

$$\text{Amplitude} = 2A \cos \phi/2 \tag{12.4}$$

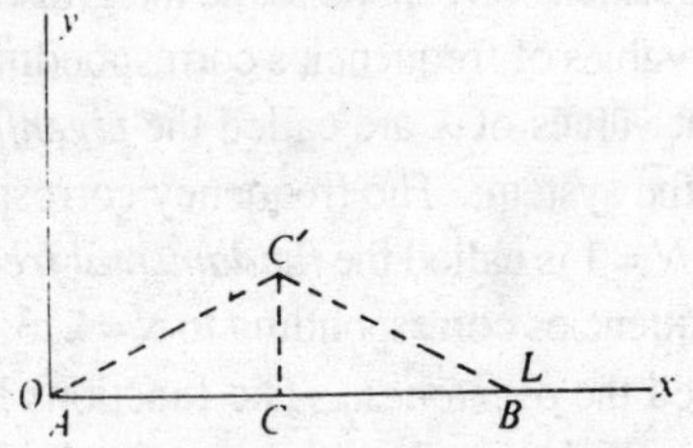

Fig. 12.2 *An infinitely long string may be considered to be made of an infinite number of beads, and thus having an infinite number of modes with continuously varying frequency. But when it is fixed at two ends (or it becomes a bound system) the frequency spectrum becomes discrete.*

If $\phi/2 = \pi/2$, the amplitude becomes zero and y also becomes zero as desired. Putting $\phi/2 = \pi/2$ in Eq. (12.3)

$$y = -2A \sin kx \cos \omega t \tag{12.5a}$$

The negative sign in Eq. (12.5a) is irrelevant because the amplitude part will still vary from $+2A$ to $-2A$. We can, therefore, absorb the negative sign in the amplitude and write Eq. (12.5a) as

$$y = 2A \sin kx \cos \omega t \tag{12.5b}$$

Let us now apply the second boundary condition, i.e. $y = 0$ at $x = L$. Thus:

$$2A \sin kL = 0$$

This condition will be satisfied only if

$$kL = \pm N\pi,$$

or
$$\lambda = 2L/N \qquad (12.6)$$

Here, N is an integer. As λ cannot be negative, N will have only positive values. The various modes of vibration can be obtained by giving N the values 1, 2, 3, ..., and so on.

Question: Is the value zero for N permissible?

Substituting for λ in Eq.(12.5b) from Eq.(12.6)

$$y = 2A \sin\frac{N\pi x}{L}\cos \omega t \qquad (12.7)$$

Thus we see that by bounding the string system the number of modes of vibration of the string can be limited. Further, only discrete wavelengths are possible which correspond to the integral values of N. The values of frequencies corresponding to the different values of λ are called the *eigenfrequencies* of the system. The frequency corresponding to λ for $N = 1$ is called the *fundamental frequency*. The frequencies corresponding to $N = 2, 3, ...$, etc., are called the *overtones*. The functions like sin $(N\pi x/L)$ which describe the system for the different modes of the vibrating system are called the *eigenfunctions* of the system. The corresponding energies are called the *eigenvalues* of the system.

The overtones for the uniform string are integral multiples of the fundamental. In practice the systems are not simple and the eigenfrequencies are fractional multiples of the fundamental frequency. The eigenfrequencies of a stretched, bounded membrane are fractional multiples of the fundamental frequency. An example of this is the musical instrument —drum. The frequency spectrum for kettle drum contains fraction overtones. It is found that the quality of sound is not musical if the harmonics present in the sound are not integral multiples of fundamental. The sound of the kettle drum is, therefore, not very musical. It is interesting to note that if a stretched membrane is loaded about the centre, the harmonics become nearly integral multiples. *Tabla* is such an arrangement, and it emits a musical sound.

It may be noted in passing that at the fixed end of the string, no work can be done on the support because this end cannot move. Thus, energy cannot leave the string. To explain the form of a reflected wave, we can imagine here the interference of the incident wave with its own reflection.

Example 12.1 Two progressive waves $y_1 = 5\sin(20t + 0.2x + \pi/4)$, and $y_2 = 5\sin(20t - 0.2x)$ travel on a string. At what values of x the displacement will be zero?

The signs of the coefficients of x for y_1 and y_2 are opposite. The waves are, therefore, travelling in opposite directions. Let

$$y = y_1 + y_2 = 5\sin(20t + 0.2x + \pi/4) + 5\sin(20t - 0.2x)$$

$$= 10\cos(0.2x + \pi/8)\sin(20t + \pi/8)$$

Now, y is zero if

$$\cos(0.2x + \pi/8) = 0,$$

or
$$0.2x + \pi/8 = \pm n\pi/2,$$

or
$$x = 5(\pm n\pi/2 - \pi/8) \text{ units of distance}$$

Example 12.2 Stationary waves are produced in a length of wire fixed between two points. Compare the amplitudes at an antinode for the fundamental and the second harmonic. Suppose the total energy of the initial waves is, on the average, equally divided between all the modes.

The maximum kinetic energy at the antinode of fundamental will be equal to the sum of the maximum kinetic energies at the antinodes for second harmonic. Making use of Eq. (10.7)

For fundamental: $\text{KE} = \frac{1}{2}\rho\Delta x\omega^2 A_0^2$

For second harmonic, the sum of KEs = $2\times\left(\frac{1}{2}\rho\Delta x\omega_1^2 A_1^2\right)$,

or
$$A_1 = A_0/2\sqrt{2}$$

Here $\omega_1(= 2\omega)$ is the angular frequency of the second harmonic, ρ is the mass of the wire per unit length, and Δx is the length of a small element of the wire at each antinode.

The amplitudes of the harmonics go on decreasing justifying our statement that vibrations of higher modes are transitory. This is because the small energy associated with antinodes of the higher modes is quickly dissipated by frictional mechanism.

12.2 Stationary Waves

The value of $N = 1$ gives the fundamental mode of vibration. For the higher values of N we get the harmonic frequencies, also called overtones. Let us now see how AB will look for the fundamental mode. If we take high speed snapshots of the string vibrating in fundamental mode we will see the string go through the various dotted phases to the maximum giving a sort of bowed shape [Fig. 12.3 (a)]. The amplitudes of various points on the string vary, but any point of the string goes on vibrating simple harmonically with an angular frequency ω_0 corresponding to $\lambda_0 = 2L$. The points of zero amplitude are called *nodes*, and the points of maximum amplitude are called *antinodes*. For a fundamental mode the two nodes are at the two fixed ends and an antinode is in between. It becomes very easy to visualise the various harmonics in terms of nodes and antinodes. For example, for the second harmonic the possibility is to have one more node at the middle, apart from the two compulsory nodes at the clamped ends A and B. Therefore, the second harmonic will look like the pattern shown in Fig. 12.3(b).

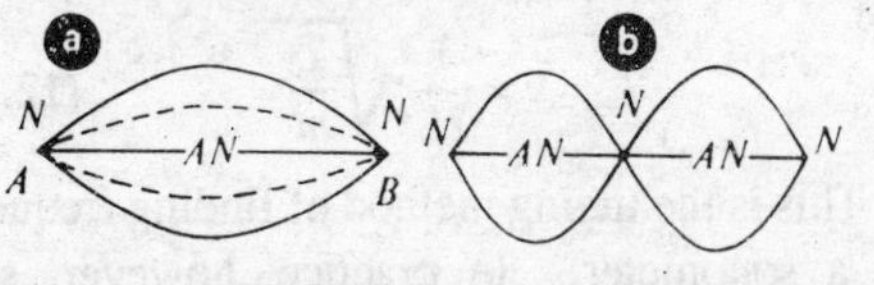

Fig. 12.3 (a) *A bound string vibrating in its fundamental mode* (b) *The string vibrating in the second mode (second harmonic) or first overtone*

Question: (1) What is the distance between two adjacent nodes and between adjacent node and antinode in terms of wavelength? (2) What is the wavelength of the second harmonic on a string of length L which is clamped at both the ends? What is it for the third harmonic?

It may be pointed out that as the points at the nodes are always at rest, energy cannot pass through them. Thus, there is no propagation of energy from one place to another. The pattern just stands and waves. But in a progressive wave, nothing of this sort happens, and there is a transfer of energy from one place to another.

If we divide ω by $2\pi/\lambda$, i.e. k, we get,

$\omega/k = \nu\lambda =$ velocity of sound in the medium,

$$\text{or} \quad \omega/k = \sqrt{T/m} \tag{12.8}$$

Here, T is the tension in the string, and m its mass for unit length. This relation is called a dispersion relation. We have discussed dispersive and non-dispersive media while discussing phase velocity and group velocity. If ω/k is a constant we call the medium to be nondispersive. In such a medium the phase and group velocities are the same. If ω/k is not a constant but changes with k, then we call the medium a dispersive medium. It is the dispersive medium which throws the various frequencies of a complex wave apart. We will discuss this while studying dispersion of light waves by a prism.

Question: Does a string form a dispersive medium?

It does not need any telling that the waves moving on the string AB in Fig. 12.2 are transverse. The waves are moving along the x-axis. The normal vibration of the string could have been in any direction in the yz-plane. But in the above case we have considered the vibration only along y-axis, i.e. in the yx-plane which is the plane of the paper (Fig. 12.4). If, out of all the possible waves only those in one particular plane are present, e.g. the waves in the yx-plane discussed above, such waves are called *polarized waves*. We will discuss them in detail while studying optics.

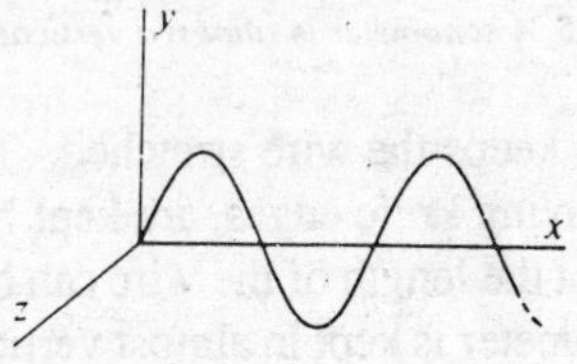

Fig. 12.4 *Polarized vibrations*

Example 12.3 Show that the period of the fundamental mode in a stationary wave is equal to the double of the time the component waves take in traversing the distance between the fixed ends.

Let the component waves be

$$y_1 = A \sin(\omega t - kx),$$

$$\text{and} \quad y_2 = A \sin(\omega t + kx)$$

The phase velocity of the waves is $\omega/k = c$. Let the length of the string between the fixed ends be L. From Eq. (12.6):

$$\lambda \text{ (fundamental)} = 2L$$

$$\therefore \quad T\text{(fundamental)} = 2L/c$$

= twice the time taken for the component waves to travel between the fixed ends

12.3 Sonometer

Sonometer is a simple arrangement to find the frequency of an object. The apparatus consists of a hollow wooden box over which a wire is placed with one end anchored to one end of the box. The string is then passed over a pulley fixed at the other end of the box, and a weight is attached to the free end of the wire (Fig. 12.5). The force exerted by

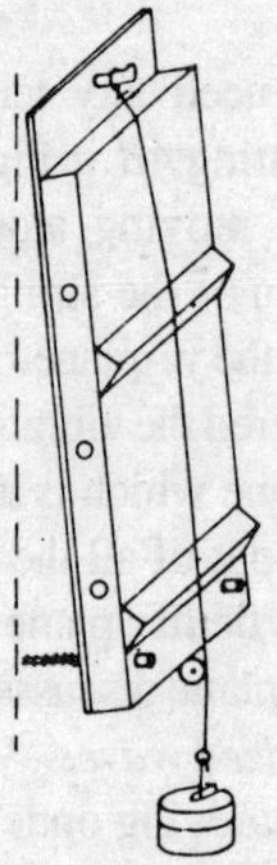

Fig. 12.5 *A sonometer in almost a vertical position*

the weight keeps the wire stretched. Two sliding bridges, having knife edges, are kept beneath this wire so that the length of the wire can be adjusted. If the sonometer is kept in almost vertical position and the string is plucked at the centre, it will vibrate in its most dominant fundamental mode. Since it rests on the knife edges placed over the sounding board, vibrations are clamped down here. The air in the hollow box receives these vibrations through the sounding board. The transmission of the vibrations of the string through the sounding board to the air makes the sound loud enough to 'beat' with the sound emitted by the tuning fork. Sometimes the hollow box has holes in its sides. The presence of these holes makes the energy of the vibrating air inside the box dissipate earlier. The resonance due to the natural frequency of the hollow box is, therefore, suppressed. If the object whose frequency we wish to determine is struck lightly, it will also emit a note. The distance between the knife edges is so adjusted that the string length between them vibrates with the same frequency as that of the vibrating object. This can be judged by the absence of beats between the two. If the string length is L, then for the Nth mode

$$L = N\lambda/2 \tag{12.9}$$

From relations (12.8) and (12.9)

$$\nu = \frac{1}{\lambda}\sqrt{\frac{T}{m}},$$

or

$$\nu = \frac{N}{2L}\sqrt{\frac{T}{m}} \tag{12.10a}$$

For fundamental mode, $N = 1$. Therefore,

$$\nu = \frac{1}{2L}\sqrt{\frac{T}{m}} \tag{12.10b}$$

This is the tuning method of finding frequency by a sonometer. In practice, however, some persons are not trained to hear the beats. For them the easier method is to place the sonometer horizontally. Now, a light-weight wire is placed on the confined string. When the tuning occurs, then due to resonance, the string vibrates violently and the piece of wire is thrown out. But in this case the tension T of the string is not Mg, due to the frictional drag in the sonometer wire by the pulley if it does not operate smoothly.

Example 12.4 A wire stretched between a fixed end and a pulley of a sonometer is kept in tension by a mass of 1 kg. The length of the wire is adjustable and it is placed between the poles of a horseshoe magnet. Alternating current of 100 c/sec is sent through the wire. Find the length of the wire for which the resonance will occur. Mass of the wire is 0.040 g/cm.

The interaction between the magnetic field and ac will set the wire into vibrations. The resonance will occur when the natural frequency of the wire becomes 100 c/sec. Using Eq. (12.10b):

$$100 = \frac{1}{2L}\sqrt{\frac{1000 \times 981}{0.040}},$$

or

$$L = 25 \text{ cm}$$

An instrument for measuring ac frequency can be designed on this principle. It is called the frequencymeter.

12.4 Melde's Experiment

In a sonometer the dominant mode of vibration is the fundamental one. Melde proposed an experiment in which higher modes of vibration can be demonstrated. The experimental arrangement consists of a long string attached to a prong of an electrically driven tuning fork at one end, and to a weight at the other end (Fig. 12.6). The string passes over a pulley, and its length can be adjusted by a support S.

As the tuning fork is excited, the string also vibrates. When the length of the string (or the tension in it) is adjusted, then at one stage the fundamental frequency of the string becomes the same as that of the tuning fork and a loop is formed between AB. If the length AB (or tension) is now so adjusted that the string vibrates in the second mode, two loops will be observed. We can increase the number of loops by suitable adjustments. Thus it gives us a glimpse of the various modes of vibration. We can also use this experiment to find the frequency of a tuning fork of unknown frequency. In the sonometer the tuning length used to calculate frequency is found between the two knife edges. At edges the discontinuity does not allow a very accurate location of nodes and, hence, there is a likelihood of an error in determining the tuning length. In Melde's experiment a large number of loops permit us to obtain a better estimate of the position of nodes and the length needed to calculate the frequency.

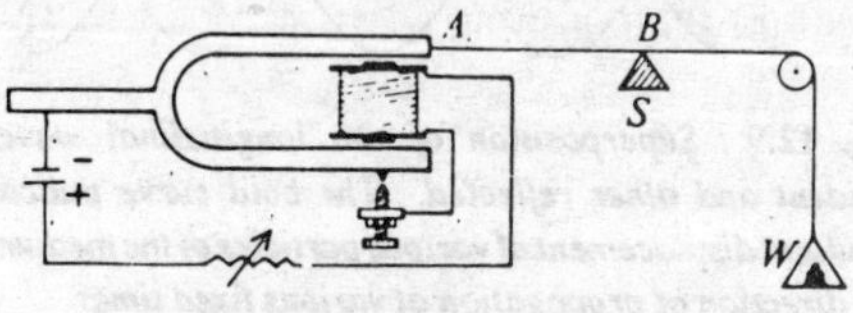

Fig. 12.6 *Melde's apparatus*

The string in this experiment can be excited in two ways (Figs. 12.7 and 12.8). In Fig. 12.7, the tuning fork vibrates at right-angles to the string and the frequency of the string is equal to the frequency of the fork. This is because when the prong, to which the string is attached, goes up to its maximum position, it takes the string with it to the maximum upward position. When the prong comes down to the maximum position on other side, string also comes with it. Thus the string makes the same number of vibrations as the tuning

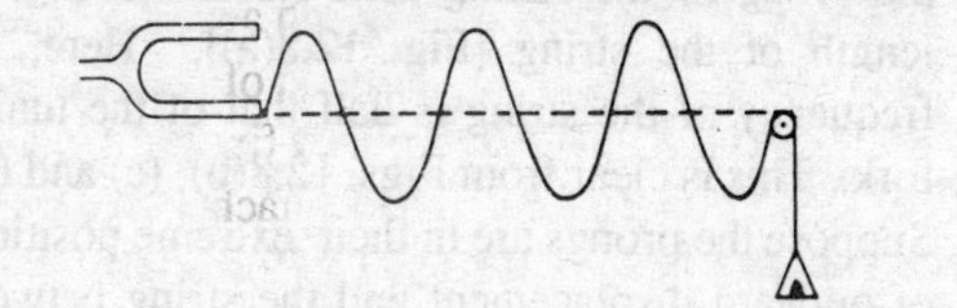

Fig. 12.7 *When tuning fork prong vibrates perpendicular to the length of the string both the fork and the string have the same frequency of vibration.*

fork makes. Suppose n loops are formed for a particular string length and tension. Let the length of the string be L metres, and the tension in the string be Mg newton. Since the length of n loops is equal to L metres, one loop is equal to L/n metres.

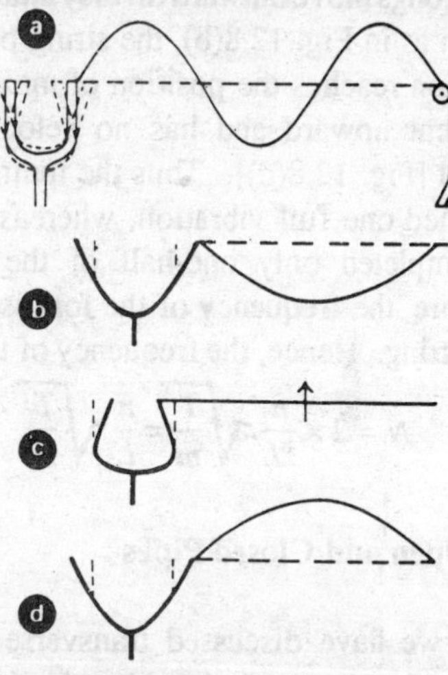

Fig. 12.8 (a) ***Tuning fork prong vibrating along the length of the string.*** (b) ***When the prong goes out, the string gets loosened and goes down.*** (c) ***When the prong goes in the string gets tightened, comes back to the original position and has a velocity upward.*** (d) ***When the prong now goes out the string goes up***

$$\therefore \quad \frac{\lambda}{2} = \frac{L}{n} \tag{12.11}$$

or

$$\lambda = 2L/n \tag{12.12}$$

Therefore, the frequency of the tuning fork, N, which is equal to the frequency of vibration of the string, is:

$$N = \frac{n}{2L}\sqrt{\frac{T}{m}} \qquad (12.13)$$

In the second case, called the longitudinal mode, the prong of the tuning fork vibrates along the length of the string [Fig. 12.8(a)]. Here, the frequency of the string is half that of the tuning fork. This is clear from Figs. 12.8(b), (c) and (d). Suppose the prongs are in their extreme positions of outward displacement and the string between the prong and the pulley is slackened [Fig. 12.8(b)]. At this moment the prongs as well as the string have no velocity. Then the prongs start coming near to each other till they reach the extreme position of inward displacement. The string straightens up and is horizontal [Fig. 12.8(c)]. The string has no displacement from its mean position, but has maximum velocity in the upward direction. Next, when prongs move outward till they attain the same position as in Fig. 12.8(b), the string by virtue of its motion reaches the position of maximum displacement upward and has no velocity at that moment [Fig. 12.8(d)]. Thus the tuning fork has performed one full vibration, whereas the string has completed only one-half of the vibration. Therefore, the frequency of the fork is twice that of the string. Hence, the frequency of the fork is:

$$N = 2\times\frac{n}{2L}\sqrt{\frac{T}{m}} = \frac{n}{L}\sqrt{\frac{T}{m}} \qquad (12.14)$$

12.5 Open and Closed Pipes

So far we have discussed transverse stationary waves. Let us now consider the longitudinal stationary waves. It has been shown in Section 10.1 that there can also be a transverse representation of a longitudinal wave, in which the parts of the curve having positive slopes represent a rarefaction, and the parts of the curve with negative slopes represent a compression.

Now, suppose such a longitudinal wave is produced at S (thin solid curve in Fig. 12.9) and after moving through a certain distance it falls on a rigid wall. As the reflection coefficient is -1 (Eq. 11.37), the displacement wave suffers a phase reversal of π. In other words, the reflected wave has the same amplitude as the incident wave but is always dead out of phase with the latter. Let us assume that at time $t = 0$ a compression falls on the wall. We have now to trace the wave motion through the periodic time T. It is easier to compute the reflected wave at various times if it is assumed that a wave is coming from the wall towards left (dotted line in Fig. 12.9). At this time the two waves will have shapes as shown by the solid and

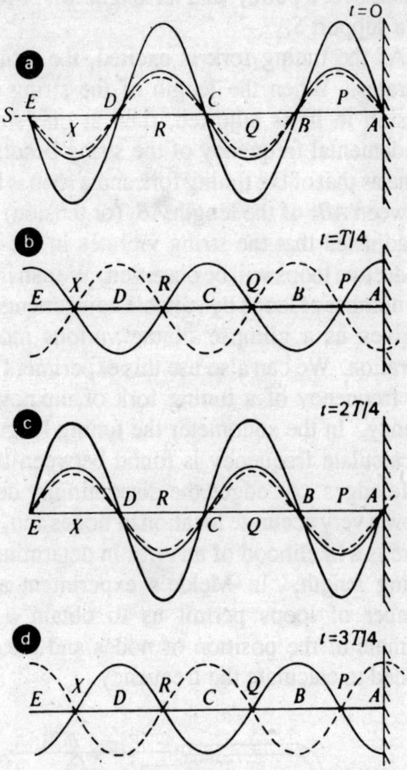

Fig. 12.9 *Superposition of two longitudinal waves—one incident and other reflected. The bold curve indicates the resultant displacement of various particles of the medium along the direction of propagation at various fixed times*

broken curves in the figure, and will be moving in opposite directions. These waves reinforce each other at the points P, Q, R and X and produce maxima, while the displacements at the points A, B, C, D, and E are zero. At time $t = T/4$, the two waves produce zero displacement everywhere. At time $t = T/2$, the wave patterns will be such as to produce maxima again at points P, Q, R and X, but this time in directions opposite to the ones produced at $t = 0$. The displacements at points A, B, C, D and E will be zero even now. At time $t = 3T/4$, the two waves again produce zero displacement everywhere. The pattern at time $t = T$ will be the same as in the beginning and will repeat thereafter.

From the above we can conclude the following:

(*i*) The particles at points A, B, C, D and E are always at rest. These points are called *nodes*.

(*ii*) The particles at the points P, Q, R and X suffer displacements, reaching maximum positive or maximum negative after an interval of time $T/2$. These points are called *antinodes*.

(*iii*) The displacement amplitudes of the particles between the nodes and the antinodes lie between zero and that at the antinodes.

(*iv*) From the pattern at $t = 0$, we notice that at some nodes the resultant curve, shown by a thicker line, has positive slope, e.g. at B and D, representing a rarefaction. But at $t = T/2$ these points represent a compression. Thus the nodes are the points with maximum pressure variation, though the displacement variation is zero there.

(*v*) At antinodes though the displacement variation is maximum, the pressure variation is zero. This is because the slope of the curve is always zero at these points.

(*vi*) The nodes and antinodes alternate.

(*vii*) Half a wave lies between two consecutive nodes or antinodes.

looking at the above characteristics of standing waves, we find that they are certainly different from the progressive waves. It will be instructive for you to write the differences in the tabular form.

Let us now study the propagation of waves through pipes. Suppose we have a pipe open at both the ends and longitudinal vibrations are excited in the medium (air) of the tube (Fig. 12.10). At the open ends there is a complete freedom of vibration. Therefore, there will be antinodes here. To find out the modes of the possible vibration in the hollow pipe, we have to take care of only this boundary condition. Let us now calculate the modes.

First mode: In Fig. 12.10(a), the length L of the tube is equal to $\lambda_0/2$, or $\lambda_0 = 2L$. Therefore, the fundamental frequency is

$$\nu_0 = c/2L \qquad (12.15a)$$

where c is the velocity of sound in the medium inside the pipe.

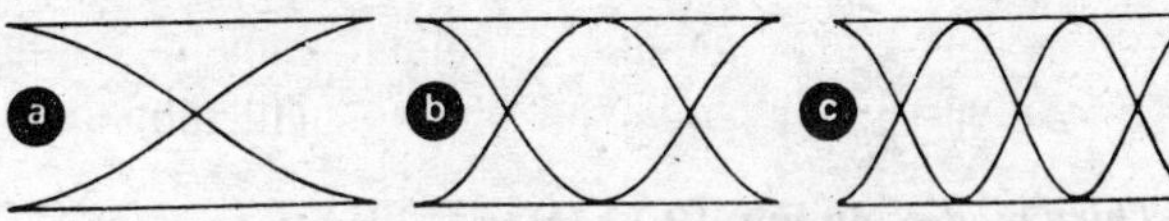

Fig. 12.10 *Standing waves in a pipe open at both the ends:* (a) *First mode* (b) *Second mode* (c) *Third mode*

Second mode: In this mode apart from two antinodes at the ends, there is one more antinode at the centre of the tube [Fig. 12.10(b)].

$$\therefore \qquad L = \lambda_1,$$

or

$$\nu_1 = \frac{c}{L} = \frac{2c}{2L} \qquad (12.15b)$$

Third mode: In Fig. 12.10(c), it is clear that

$$L = \frac{3}{2}\lambda_2, \text{ or } \lambda_2 = \frac{2L}{3}$$

$$\therefore \quad \nu_2 = \frac{3c}{2L} \qquad (12.15c)$$

and so on.

Comparing Eqs. 12.15(a), 12.15(b), ..., etc., we see that:

$$\nu_0 : \nu_1 : \nu_2 : \nu_3 : \nu_4 \ldots :: 1 : 2 : 3 : 4 : 5\ldots$$

All the integral harmonics are, therefore, possible in a hollow tube with both ends open.

If we consider a tube closed at one end (Fig. 12.11), the boundary conditions now demand an antinode always at the open end and a node always at the closed end. Let us now discuss the various modes, which have been shown in Figs. 12.11 (a), 12.11 (b), ...

Fig. 12.11 *Standing waves in a pipe closed at one end and open at the other :* (a) *First mode* (b) *Second mode* (c) *Third mode*

First mode: In Fig. 12.11 (a)

$$L = \frac{\lambda_0}{4}, \text{ or } \lambda_0 = 4L$$

$$\therefore \quad \nu_0 = c/4L \tag{12.16a}$$

Second mode: In Fig. 12.11 (b)

$$L = \frac{3\lambda_1}{4}, \text{ or } \lambda_1 = \frac{4L}{3}$$

$$\therefore \quad \nu_1 = \frac{3c}{4L} \tag{12.16b}$$

Third mode: In Fig. 12.11 (c)

$$L = \frac{5\lambda_2}{4}, \text{ or } \lambda_2 = \frac{4L}{5}$$

$$\therefore \quad \nu_2 = \frac{5c}{4L} \tag{12.16c}$$

and so on. Therefore,

$$\nu_0 : \nu_1 : \nu_2 : \nu_3 \ldots :: 1 : 3 : 5 : 7 \ldots$$

Thus only odd integral harmonics are possible in this case.

12.6 Kundt's Tube

You might have already learnt to measure the velocity of sound by the resonance column. Let us now study another method, the Kundt's method, to measure this velocity. The apparatus consists of a hollow glass tube with one end closed by a rigid cork (Fig. 12.12). At the other end there is a piston carried by a metallic rod, which is fixed at its centre by a clamp. The tube is first washed, dried, evacuated, and then filled with the desired gas. Lycopodium powder is sprinkled along the length of the tube. The free end of the metallic rod is now rubbed by a piece of chamois leather to excite vibrations in it. These vibrations are released into the medium of the tube through the piston attached to the rod. The waves in the medium of the tube are reflected from the other end and a standing wave is set up on properly adjusting the distance between the piston and the other closed end. This adjustment can be done by sliding the glass tube to left or right. We know that antinodes have maximum displacement. Therefore, the antinodes of the standing waves in the tube throw off the lycopodium powder and heaps of the powder collect at nodes, where the displacement is zero. The distance between the heaps, i.e. the nodes, is measured. Twice this distance is λ. If the frequency of the wave is known, then we can find the velocity of sound. The velocity of sound in the material of the rod must be known. Since the rod is clamped at its centre the centre is a node and the free ends of the rod are antinodes. Let the length of the rod be L.

$$\therefore \quad L = \frac{\lambda_{\text{rod}}}{2} \tag{12.17}$$

Let the velocity of sound in the rod be u, and the distance between the two consecutive lycopodium powder heaps be $l = \lambda/2$. Then:

$$\nu = \frac{u}{\lambda_{\text{rod}}} = \frac{u}{2L}$$

and the velocity of sound in the medium of the tube is

$$v = \frac{u}{2L} \times 2l = \frac{ul}{L} \tag{12.18}$$

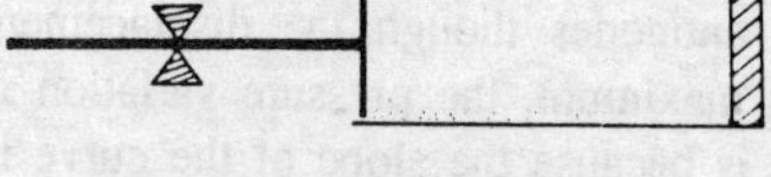

Fig. 12.12 *Kundt's tube*

Suppose u is not known. Then first we fill the tube with air, the velocity of sound in which is known, say c. Using Eq. (12.18)

$$c = \frac{ul_{\text{air}}}{L} \tag{12.19}$$

The value of u calculated from Eq. (12.19) is then used to find v from Eq. (12.18). If we use liquid medium, iron filings are used instead of lycopodium powder.

12.7 Musical Scales

Some sounds reaching our ears give a pleasant feeling, while some others are irritating. On analysis it has been found that various frequencies and their intensities in the former type, called *musical sounds,* follow a certain order. The following are the three main characteristics of any sound.

(i) *Pitch:* It denotes that sensation produced in our ears which depends on frequency of sound. The shrillness of sound increases with its frequency. The first key of harmonium emits a grave tone, while as the upper keys are played the shrillness of the tones increases. Alternatively, we say that the pitch of the sound emitted by harmonium increases as the higher keys are played. A student can experience this feeling with the help of tuning forks.

(ii) *Loudness:* This is a measure of magnitude of auditory sensation produced by sound. If for the same musical note, say of a violin, volume of the amplifier is increased the loudness of the sound increases while the degree of shrillness remains the same. Physically, the loudness of a note is measured by its intensity (see Section 11.2).

(iii) *Quality:* This is a characteristic which differentiates two notes of the same pitch and loudness produced by two sources, say, a sitar and a violin. Here we must differentiate between a tone and a note. The *tone* is a single frequency, whereas the *note* has fundamental mixed with its *harmonics* or *overtones.* Difference between the two notes is caused by the presence of higher harmonics in different degrees. The sensation of sweetness or harshness of a note depends on the presence or absence of various overtones and their relative strengths. Quality is also sometimes called *timbre.* A tuning fork is supposed to produce only the tone of frequency engraved on it.

When two or more frequencies are sounded it is termed a *cord.* If the combined effect of these frequencies is pleasant then the sound produced is called *concordant,* and if the combined effect is unpleasant then it is called *discordant.* The frequencies are said to be in *consonance* and *dissonance,* respectively. The unpleasantness in the ear is due to beats that are produced by the component tones. These beats produce unpleasantness in the same manner as the flickering of light in the eye. When the flicker is either very low or very rapid, it is not unpleasant to the eye, but when it is within a given range it becomes very unpleasant. Musical compatibility of two notes is found to depend upon the ratio of their frequencies. The ratio between two frequencies is called *interval**. Musical intervals involving small integers are consonant, while those involving large integers are dissonant. The simplest ratio is 1 : 2. In this interval a note having as fundamental the lower frequency is called *tonic,* and the one having its fundamental twice as large is called the octave of the first. The two notes together form the octave. This interval being pretty large, other notes are inserted in between to fill the gap.

Musical Scale: A number of notes emitted together or one after another to produce a pleasant effect is called a musical scale. If the notes are sounded together, as is western music, then they are said to be in *harmony.* In Indian music *melody* is more important which is obtained by producing notes in succession. When we want to construct a musical scale, first we have to select a base note, i.e., the starting note having the lowest pitch. Once the base note is selected, the other notes are then selected to form consonant intervals with the base note.

One of the early musical scales is *True* or *Diatonic Scale.* In this scale an octave is divided into intervals as given below:

Notation in western music	**C**	**D**	**E**	**F**	**G**	**A**	**B**	**C**
Notation in Indian music	**Sa**	**Re**	**Ga**	**Ma**	**Pa**	**Dha**	**Ni**	**Sa**
Relative frequency	**1**	**9/8**	**5/4**	**4/3**	**3/2**	**5/3**	**15/8**	**2**
Interval between successive notes		**9/8**	**10/9**	**16/15**	**9/8**	**10/9**	**9/8**	**16/15**

It may be noted that if these intervals are multiplied together the result is 2, the interval of an octave.

***The reason for taking ratio, and not difference, is the same as for intensities (Section 11.2).**

As an example if the base note has the fundamental frequency 240 Hz, fundamental frequencies of other notes would successively be 270 Hz, 300 Hz, 320 Hz, 360 Hz, 400 Hz, 450 Hz and 480 Hz. The intervals inserted between the tonic and the octave are: major tone (9/8), minor tone (10/9) and semi-tone (16/15). Thus this ordinary diatonic scale is an octave which is the sum of three major tones, two minor tones and two semi-tones. These intervals can be obtained by the sub-division of a string.

Exercise : Take 256 Hz as the base note and find the others.

(Ans. 288, 320, 341.33, 384, 426.67, 480, 512 Hz)

To make matters more regular and easy a tempered scale has been designed where the same interval $2^{1/12}$ is taken between successive notes. Thus it is constructed in the following manner.

C	Cs	D	Ds	E	F	Fs	G
$2^{0/12}$	$2^{1/12}$	$2^{2/12}$	$2^{3/12}$	$2^{4/12}$	$2^{5/12}$	$2^{6/12}$	$2^{7/12}$

Gs	A	As	B	C′
$2^{8/12}$	$2^{9/12}$	$2^{10/12}$	$2^{11/12}$	$2^{12/12}$

Main advantage of the tempered scale is that the interval of successive notes being the same, a musician can choose any note as his base note.

Exercise: Frequency of the key-note *C* is 256 Hz. Find the frequency of the note *F*. (Ans. 341.72 Hz)

Discussion and Problems

12.1 Find the frequency of the third harmonic (second overtone) for a wire fixed at two ends 0.5 m apart. Tension in the wire is 10 N. Linear density of the wire is 4×10^{-3} kg/m.

12.2 A cable of length 30 m is stretched between two poles with a force of 20 kg. If a pole is struck with stick, how long will it take to hear the sound created at the other end? The mass of the cable is 4.5 kg.

12.3 Explain what is meant by tuning a stringed instrument.

12.4 Will a wire vibrate for a longer time or a shorter time, if it is mounted on a sounding board?

12.5 Show that no distinguishable beats are possible for two frequencies in the ratio 1/3 if the harmonics of the frequencies also beat against each other.

12.6 How are the different notes produced in a flute by closing the holes at the lower end?

12.7 What is the function of silencers in cars, scooters, etc.?

12.8 A forward and a reflected wave superpose to form a stationary wave. What happens if their amplitudes are not equal?

12.9 Draw a stationary wave pattern like the one in Fig. 12.9 for the reflection coefficient equal to one, e.g. when the wave enters from a denser to a rarer medium.

12.10 Sometimes in a stringed musical instrument a thick wire is wrapped by a thin wire.. Why?

12.11 The dust piles in a Kundt's tube are 0.10 m apart in air, and 0.13 m apart in methane gas. Find the velocity of sound in methane. The velocity of sound in air is 330 m s^{-1}.

12.12 What type of waves are produced in the rod of the Kundt's tube?

12.13 What points of the stretched string between two fixed points must be plucked and touched to make: (*i*) the fundamental predominant, and (*ii*) the first overtone predominant.

12.14 What is the function of holes in a sonometer box? How are the vibrations transmitted from the tuning fork to the box?

12.15 The tuning length of the wire held in tension on a sonometer for a note emitted by a tuning fork is *l*. A second resonance is obtained at *l*/2. Explain why the second resonance is obtained?

12.16 How will you compare the γ for two gases by Kundt's tube ($\gamma = C_P/C_V$)?

12.17 Two objects of mass *m* each have been placed on a frictionless horizontal surface and are coupled with the help of three identical springs of spring constant *k* (Fig. 12.13). Describe the modes of vibration along the length of the springs and calculate their frequencies.

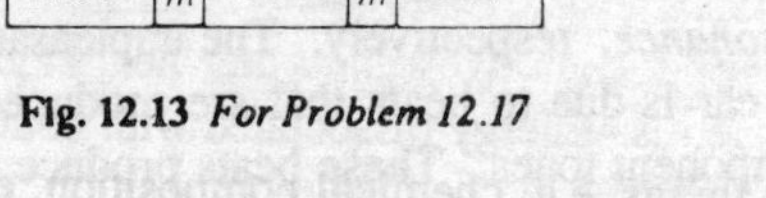

Fig. 12.13 *For Problem 12.17*

Chapter 13

Temperature

13.1 Introduction

While studying the motion of a particle and a system of particles the principle of 'conservation of energy' is evolved. We observe in our every day experiences bodies moving even on horizontal surfaces slowing down and ultimately coming to a stop. Apparently the energy, as a sum of potential and kinetic energies, is not conserved. Take another example of a ball falling vertically. After a few rebounds of decreasing heights the ball stops. Does this mean that the law of conservation of energy is not a universal law? You have every right to raise this question and you must have an open mind. Once Bohr, known as father of modern physics, had also doubted the sacred (at present) law of conservation of momentum at the atomic scale. But before we do this, we have to be doubly sure that we may not be missing something.

On referring to Chapter 5, you will find that the law of conservation of mechanical energy holds good only if the friction is negligible. To appreciate this restriction, rub your palms and see. They get warmed up. Let a large number of iron balls fall from a certain height on an iron plate—the iron balls and iron plate get warmer. Why do they warm up? We may guess that the energy, which apparently seems to be lost, has heated the palms, the iron plate and the balls. Our guess is substantiated when we see locomotives and other heat engines working for us. Energy (associated with hotness and other things, e.g. chemical composition, etc.) which is present inside a body is known as *internal energy*. Energy associated with hotness alone is called thermal energy.

It is a common experience that if a hot and a cold body are brought into contact, the hotter body keeps on cooling and the colder body becomes hotter and hotter, till both the bodies are equally hot. If hotness is a measure of energy, we can conclude that energy has flown from the hotter body to the colder body. This energy in transfer is called *heat*. In Chapters 13-17 we will talk about it in detail.

We know that if position and velocity, or position and momentum, of a particle (or a system of particles) are known, we can find out its mechanical energy. Position, velocity, etc., are called mechanical coordinates. Let us now seek the type of information we need to know about internal energy. We have already noted the effect of one parameter, i.e. hotness. Other things remaining the same, the hotter the body the more the internal energy it has. Similarly, pressure, volume, etc., also give information about the internal energy. All such macroscopic parameters (parameters which can be measured), which determine the internal energy of a system, are known as *thermodynamic coordinates*. Alongwith all these, the phase of a system is also important. We know that when we heat ice, or just rub a slab of ice, some of it melts. Here, the effect of transfer of energy is not to heat up the ice but to change its phase from solid to liquid.

All the experiments to verify the law of conservation of momentum are done on a horizontal plane because if the momentum of a certain system is to be conserved, it should be isolated. The same holds for energy. Eliminating friction amounts to isolating the system. Many times we need this type of isolation in the study of heat also. Take a gas and a burner. Cover the gas from all sides with a thick asbestos wall [Fig. 13.1(a)]. We cannot now change the state (thermodynamic coordinates) of the gas. If the bottom layer is replaced by an iron plate [Fig. 13.1(b)], the volume of the gas starts increasing. If the volume of the gas is to be kept constant, the pressure on it will have to be increased. In any case, the state of the system (gas) will change. The walls of the former type are called *adiabatic walls,* and the walls of the latter type are known as *diathermic walls.* Adiabatic walls* isolate the system from the surroundings, while diathermic walls establish thermal interaction between the two.

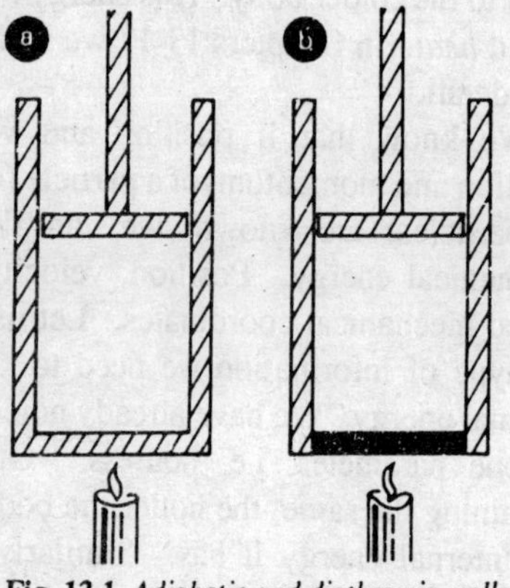

Fig. 13.1 *Adiabatic and diathermic walls*

13.2 Concept of Temperature

In common language we understand what we mean by low and high temperatures. Ice is cold and steam is hot. We say steam is at a higher temperature than ice. By using the sensation of touch we can compare two temperatures qualitatively. But this method can prove deceptive sometimes. The response of our senses depends upon their previous state. Another drawback is that qualitative analysis does not help us much. We have to answer 'how much'? Therefore, we need some operational definition of temperature.

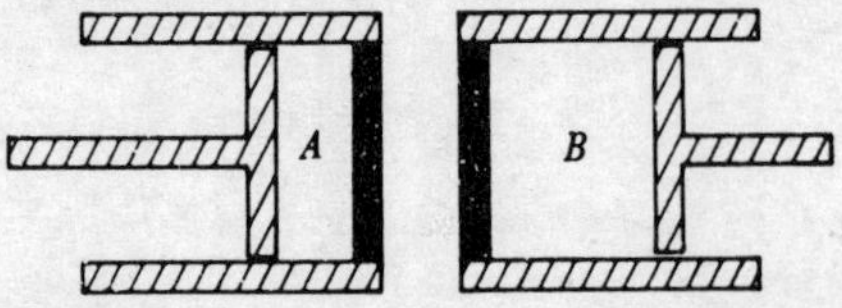

Fig. 13.2 *Thermal contact between two systems*

Let us take two bodies *A* and *B*, as in Fig. 13.2. To simplify our discussion we can take two gases. Let us bring them into thermal contact with the help of a diathermic wall. If *A* is hotter than *B* then, pressures remaining constant, the volume of *A* will decrease and the volume of *B* will increase; thus changing the thermodynamic coordinates of both the systems *A* and *B*. We may say that the two systems are not in 'thermal equilibrium'. After some time, the volumes cease to change with time, and the systems achieve thermal equilibrium. We say that now both the systems are at the same temperature. Thus temperature is a parameter which tells us whether a system will be in thermal equilibrium with another system or not when brought into contact through a diathermic wall. If the thermodynamic coordinates (say pressure and volume in the case of gases) of two bodies after thermal contact remain the same as they were before this, we say the two bodies were at the same temperature.

It is also our experience that if two systems *A* and *B* in thermal equilibrium with another system *C* separately (Fig. 13.3) are brought into contact by a diathermic wall, their coordinates will not change. This means they will also be in thermal equilibrium with each other. This experimental law is often known as 'zeroeth law of thermodynamics'. This law is the basis of the existence of the concept of temperature.

*'Adiabatic wall' is an idealization and, strictly speaking, cannot be had in practice.

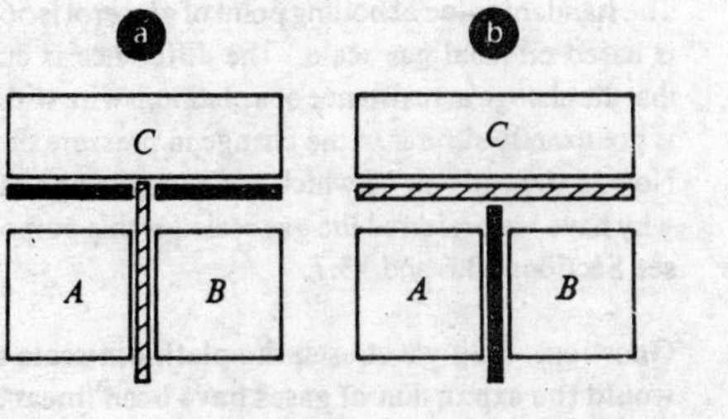

Fig. 13.3 *If two systems A and B are in thermal equilibrium with system C separately, they will be in thermal equilibrium with each other also.*

Questions: (1) There are three bodies *A*, *B* and *C*. If there is no force (neglecting gravitation) between *A* and *B* (i.e. they are in mechanical equilibrium) and between *A* and *C*, is it necessary that there should be no force between *B* and *C*? (2) If a body *A* attracts bodies *B* and *C* separately, is it necessary that bodies *B* and *C* should also attract each other?

After discussion of the above questions one will realize that the zeroeth law of thermodynamics is not self evident.

13.3 Measurement of Temperature

Let us take a system A (say a gas in a cylinder) which is in thermal equilibrium with another system B (say ice). If x and y are the thermodynamic coordinates of A, an infinite number of paired values of x and y are possible for which A is in thermal equilibrium with B. Plot all such values in xy-plane, as in Fig. 13.4, and join the points with a line. This line is called an isotherm. Let us number it '1'. According to zeroeth law of thermodynamics, all states represented by this isotherm are in thermal equilibrium with each other. Now take another system C (say boiling water). Let all the states of A which are in thermal equilibrium with C be represented by isotherm '2'. Similarly, we can draw other isotherms. It is clear that any state on any one isotherm is not in equilibrium with any other state on any other isotherm. Thus we say that various isotherms are at different temperatures. What happens when we go from one isotherm to another? Naturally x and y will change. If we move along the dashed line in Fig. 13.4, y will remain constant and x will be different for different isotherms*. Thus change in x from isotherm to isotherm can be taken in some way to represent the change in temperature. Then x is said to be a *thermometric property*. Familiar examples of thermometric property are length, l, in mercury thermometer; pressure, P, in constant volume gas thermometer; resistance, R, in platinum resistance thermometer; and thermo emf, ε, in a thermocouple.

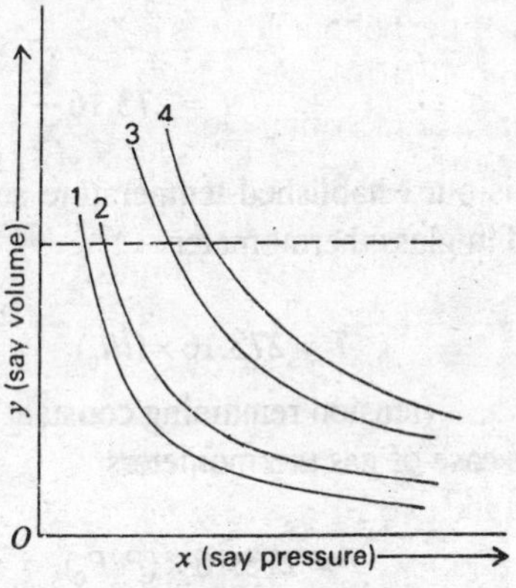

Fig. 13.4 *Isotherms*

13.4 Choosing Temperature Scale

It has been seen above that temperature is a function of thermodynamic coordinates. If one of these is kept constant, temperature becomes a function of the second alone. Mathematically, if y remains constant,

$$T(x) = f(x) \tag{13.1}$$

Here we can take any convenient form of $f(x)$ to represent T. The simplest form of $f(x)$ is x itself or some multiple of x. Therefore, we take T directly proportional to x. In other words,

$$T(x) = ax \tag{13.2}$$

where a is a constant. Thus, if x_1 and x_2 are the values of x at two isotherms, then the ratio of their temperatures is,

$$\frac{T_1}{T_2} = \frac{x_1}{x_2} \tag{13.3}$$

---

*Alternatively, if we move vertically, then x will stay constant and y will change.

One can see from Eq. (13.3) that if one temperature is known, the other can be found out by measuring x_1 and x_2. The present practice is to treat the triple point of water as a standard fixed point of thermometry and assign its temperature a numerical value of 273.16.* Let subscript '0' denote various values at this temperature. Thus,

$$\frac{T}{T_0} = \frac{x}{x_0},$$

or

$$T = 273.16 \frac{x}{x_0} \quad (13.4)$$

This is our established temperature scale. For a liquid in glass thermometer

$$T = 273.16 \times (l/l_0) \quad (13.5)$$

(tension remaining constant)

In the case of gas thermometers

$$T = 273.16 \times (P/P_0) \quad (13.6)$$

(volume remaining constant)

If we use resistance elements

$$T = 273.16 \times (R/R_0) \quad (13.7)$$

(tension remaining constant)

For a thermocouple

$$T = 273.16 \times (\varepsilon/\varepsilon_0) \quad (13.8)$$

(tension remaining constant)

Example 13.1 Resistance of a platinum resistance thermometer at triple point of water is 26.22 Ω. When kept in boiling glycerol, this resistance becomes 54.41 Ω. Find out the boiling point of glycerol on this scale. Compare it with the standard value of boiling point of glycerol and explain the difference, if any.

Let T_p be the boiling point of glycerol on platinum scale. Using Eq. (13.7)

$$T_p = 273.16 \times \frac{54.41}{26.22}$$

$$= 567.0 \text{ degrees}$$

The standard value of boiling point of glycerol is 563.2 K, which is based on ideal gas scale. The difference is due to the fact that the change in resistance of a platinum wire with temperature is not exactly similar to the change in pressure of an ideal gas. Now, it depends on us which scale to take as a standard. But why have we preferred the gas scale for this purpose? For this, see Sections 13.6 and 13.7.

Question: Had we chosen the platinum scale as standard, would the expansion of gases have been linear?

13.5 Triple Point of Water

To produce the standard temperature of the triple point of water in the laboratory, the apparatus shown in Fig. 13.5 is used. Distilled water is taken into this vessel and it is connected to an exhaust pump. All the air is pumped out and the vessel is sealed. Now this vessel contains water and water vapour only (no air). The whole apparatus is cooled with the help of a freezing mixture till inside this vessel we have water, water vapour and ice in equilibrium. This temperature is known as triple point of water, which is a standard fixed point of thermometry.

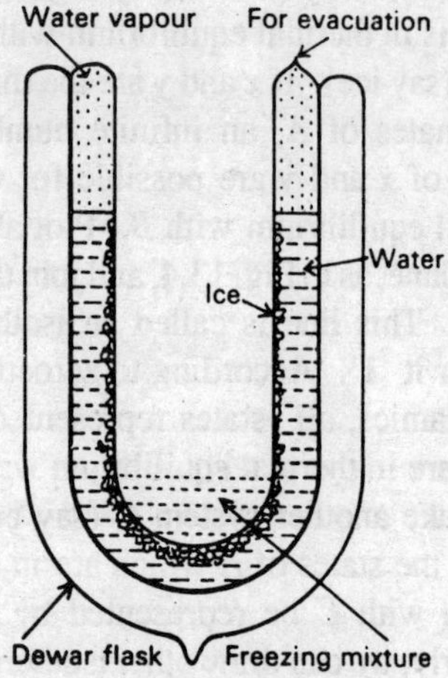

Fig. 13.5 *Triple point bath*

13.6 Constant Volume Gas Thermometer

Now there are so many thermometers at our disposal according to Eqs. (13.5) to (13.8). If we find out various normal boiling points (nbp) and normal

*Agreed upon at the Tenth General Conference on Weights and Measures held in 1954 at Paris. This numerical value is taken to make the unit of temperature on Kelvin scale equal to one degree Celsius.

melting points (nmp) with each of these thermometers, the values vary from thermometer to thermometer. So we have to make a choice. Take any one thermometer as a standard and calibrate the rest in terms of this standard. We all know about the requirements and considerations to be kept in mind while choosing a standard. The main considerations are that it should be reproducible with time and should be easily replaceable if destroyed due to some accident. The disagreement between any two gas thermometers is negligible at low pressures, and can be removed by a procedure described in Section 13.7. Moreover, the change in thermometric property (here pressure) with temperature is quite large and, thus, can be measured accurately. Therefore, gas thermometer has been chosen as the standard thermometer. It is this thermometer which defines the empirical temperature scale.

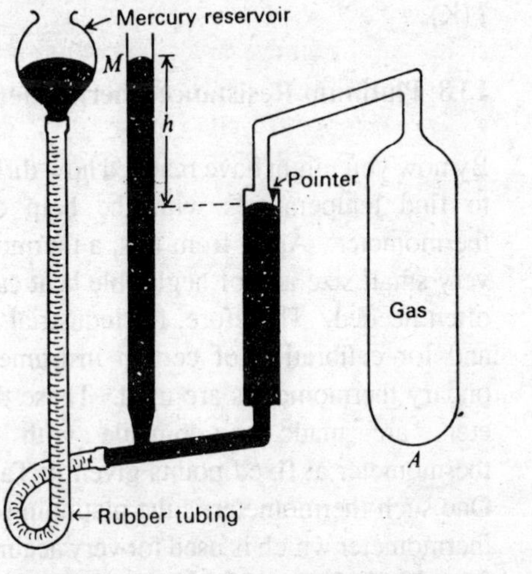

Fig. 13.6 *Constant volume gas thermometer*

The construction of this thermometer is shown in Fig. 13.6 which is self explanatory. The material of the bulb A and the gas inside it differ from laboratory to laboratory. Their nature also depends upon the temperature range we are interested in. Generally used gases are hydrogen and helium. The bulb may be of Pyrex glass for ordinary temperatures and that of platinum, or some alloy of platinum, for very high temperatures.

First of all the bulb is kept in the triple point bath discussed earlier. The mercury is made to touch the pointer and the height h is noted which gives the pressure of the gas inside the bulb. Next the bulb is immersed in the bath, of which the temperature is to be found out. Volume of the gas in the bulb is kept constant by again making the mercury to touch the pointer. The new pressure of the gas is noted. By using Eq. (13.6), the temperature of the bath is determined.

13.7 Ideal Gas Temperature Scale

We have mentioned above that various gas thermometers do not quite agree amongst themselves. But it has been seen experimentally that this small disagreement reduces further as we reduce the pressure of the gas in the bulb. Let us use the following procedure. Take a gas thermometer. Measure the pressures at triple point of water and at the unknown temperature. Find out the unknown temperature using Eq. (13.6). Now take out some gas from the bulb of the thermometer thus reducing the pressure. Find out the new value of the unknown temperature. Repeat this procedure many times, and then plot a graph between the gas pressure and unknown temperature. Such results are shown in Fig. 13.7 for nbp of water. From

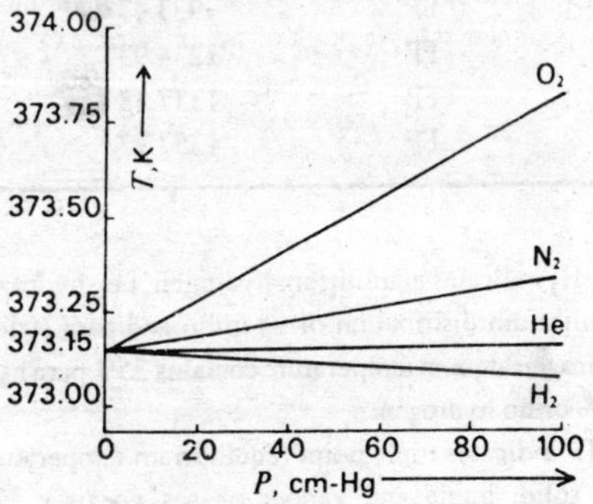

Fig. 13.7 *Under the limit when pressure approaches zero, all gas thermometers read the same temperature for a particular bath*

this figure one can see that the temperature extrapolated to zero pressure is the same for all gases. Temperatures found in this manner are known as ideal gas temperatures. Some such temperatures are given in Table 13.1. These are called the *fixed points*. A fixed point is an easily reproducible state of an arbitrarily chosen system.

These temperatures are used to calibrate the secondary thermometers as discussed later. It may be noted that though these temperatures are independent of a particular gas, these do depend on one property of gases, i.e. the change in pressure. Thus the perfect gas temperature scale is not an absolute one. An absolute temperature scale would be that

Table 13.1 *Fixed Points of International Temperature Scale of 1990 (ITS - 90)*

Material[a]	Equilibrium State[b]	Temperature K	Temperature °C
e-H_2	TP	13.8033	−259.3467
Ne	TP	24.5561	−248.5939
O_2	TP	54.3584	−218.7916
Ar	TP	83.8058	−189.3442
Hg	TP	234.3156	−38.8344
H_2O	TP	273.16	0.01
Ga	MP	302.9146	29.7646
In	FP	429.7485	156.5985
Sn	FP	505.078	231.928
Zn	FP	692.677	419.527
Al	FP	933.473	660.323
Ag	FP	1234.93	961.78
Au	FP	1337.33	1064.18
Cu	FP	1357.77	1084.62

a e-H_2 indicates equilibrium hydrogen, i.e., hydrogen with the equilibrium distribution of its ortho and para states. Normal hydrogen at room temperature contains 25% para hydrogen and 75% ortho hydrogen.

b TP indicates triple point (equilibrium temperature at which the solid, liquid and vapour phases coexist); FP indicates freezing point, and MP indicates melting point (the equilibrium temperature at which the solid and liquid phases coexist under a pressure of 101325 Pa, one standard atmosphere). The isotopic composition is that naturally occurring.

which is independent of any property of any substance. There exists one. It was given by Lord Kelvin and is based on the second law of thermodynamics. He also showed that so long as gases remain in the gaseous form, the two scales coincide. Therefore, the name 'kelvin', symbol 'K', has been adopted for the unit of temperature on this scale.

The two temperature scales we are already familiar with are the Celsius and the Fahrenheit scales. On the Celsius temperature scale (previously known as 'Centigrade' scale), a degree (the unit of temperature) is of the same magnitude as that of the ideal gas or Kelvin scale. But the zero of this scale is the ice point (equilibrium of ice and air-saturated water) which is 273.15 K. Thus,

$$t°C = T\,K - 273.15 \qquad (13.9)$$

establishes a relationship between the Celsius temperature t (° C) and the absolute temperature T(K).

13.8 Platinum Resistance Thermometer

By now you might have realized how difficult it is to find temperatures with the help of a gas thermometer. Apart from this, a thermometer of very small size and of negligible heat capacity is often needed. Therefore, for industrial purposes and for calibration of certain instruments, secondary thermometers are used. These thermometers are made to coincide with the gas thermometer at fixed points given in Table 13.1. One such thermometer is the platinum resistance thermometer which is used for very accurate work from 13.8033 K to 1234.93 K(ITS-90). It is shown schematically in Fig. 13.8. R is the platinum resistance element, LL are the leads and CC are the wires the resistance of which compensates for the resistance of the leads.

This thermometer is first immersed in the triple point bath and its resistance is measured by using the Callender and Griffiths bridge* which is a

*See Problem 21.19. Also refer to some book on experimental physics, e.g. B.L. Worsnop and H.T. Flint, *Advanced Practical Physics for Students* (Methuen & Co., London), 1951.

modified version of a post-office box. Then the thermometer is put into the bath the temperature of which is unknown, and the resistance again measured. Any unknown temperature found by using Eq. (13.7) is called the platinum temperature, which, as mentioned earlier, will be different from the ideal gas temperature. Therefore, this thermometer is calibrated by measuring the resistance at various ideal gas temperatures, and then representing the results by an empirical formula. In small ranges, the following quadratic relation is generally employed.

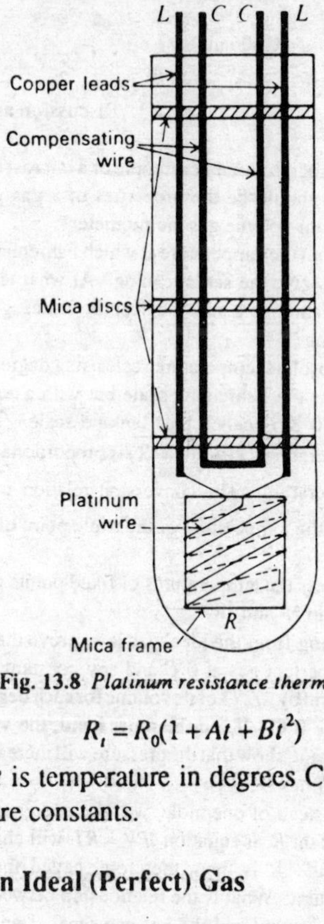

Fig. 13.8 *Platinum resistance thermometer*

$$R_t = R_0(1 + At + Bt^2) \qquad (13.10)$$

where t is temperature in degrees Celsius, and A and B are constants.

13.9 An Ideal (Perfect) Gas

We have seen that when pressure approaches zero, then for all gases

$$T = 273.16\frac{P}{P_0} \text{ (at constant volume),}$$

$$\text{or} \quad \frac{P}{P_0} = \frac{T}{T_0} \qquad (13.11)$$

There is one more similar quantity which is independent of a gas under the limit pressure approaches zero. This is the product of the pressure and the volume for a given amount (in moles) of a gas when temperature is kept constant. This, like temperature, is independent of a particular gas. Some such experimental results have been plotted in Fig. 13.9 at triple point. It is further seen that this product (PV) changes with temperature. Thus,

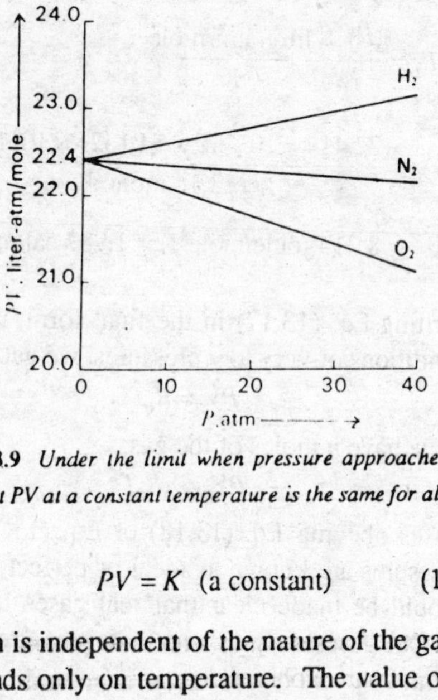

Fig. 13.9 *Under the limit when pressure approaches zero, product PV at a constant temperature is the same for all gases*

$$PV = K \text{ (a constant)} \qquad (13.12)$$

which is independent of the nature of the gas and depends only on temperature. The value of this constant at triple point of water is $K_0 = 22.414$ litre atm/mole. From Eq. (13.12),

$$\frac{PV}{(PV)_0} = \frac{K}{K_0} \qquad (13.13)$$

Eq. (13.13) is a general relation for any two points on the isotherms at temperatures T and T_0. If we take two points, one on each isotherm, such that $V = V_0$, we get

$$\frac{P}{P_0} = \frac{K}{K_0} \qquad (13.14)$$

Further, using Eq. (13.11) we get

$$\frac{K}{K_0}=\frac{T}{T_0} \qquad (13.15)$$

Hence, $$\frac{PV}{(PV)_0}=\frac{T}{T_0} \qquad (13.16)$$

Thus, $$\frac{PV}{T}=\frac{(PV)_0}{T_0} \qquad (13.17)$$

The term $(PV)_0/T_0$ is called the universal gas constant (or simply gas constant) because it is the same for all gases. We give it a symbol R_0. Thus,

$$R_0=\frac{(PV)_0}{T_0}\ \frac{\text{litre-atm/mole}}{\text{K}}$$

$$=\frac{22.414\times10^{-3}\ \text{m}^3\times1.01325\times10^5\ \text{Nm}^{-2}}{273.16\ \text{mole K}}$$

$$=8.314\ \text{joule/mole-K}=1.987\ \text{cal/mole-K}$$

Writing Eq. (13.17) in the final form, under the conditions of very low pressures, we get

$$PV=R_0T \qquad (13.18)$$

If we have n moles of the gas,

$$PV=nR_0T \qquad (13.19)$$

Gases obeying Eq. (13.18) or Eq. (13.19) at all pressures are known as ideal or perfect gases. It should be made clear that real gases behave as perfect gases only at very low pressures (below about 2 atmospheres).

Example 13.2 A cylindrical vessel contains an ideal gas at a temperature of 27°C and a pressure of 1 atm. The volume of the gas is ten litre. The gas is compressed with the help of a piston to half of its original volume. The temperature rises to 47°C. What is the new pressure? How many moles of the gas are there?

In this process the mass of the gas remains constant. Therefore, using Eq. (13.19), we get,

$$\frac{P_iV_i}{T_i}=\frac{P_fV_f}{T_f}$$

where the subscripts i and f refer to initial and final values of various parameters respectively. From the given data,

$$P_i=1\ \text{atm},\ V_i=10\ \text{litre},\ T_i=27+273=300\ \text{K}$$

$$P_f=?,\ V_f=5\ \text{litre},\ T_f=47+273=320\ \text{K}$$

Therefore,

$$\frac{P_f\times5}{320}=\frac{1\times10}{300},$$

or $$P_f=\frac{320\times10}{300\times5}=2.13\ \text{atm}$$

Also, $$PV=nR_0T,$$

or $$n=\frac{P_iV_i}{R_0T_i}\ \text{(say)},$$

or $$n=\frac{1.013\times10^5\ \text{Nm}^{-2}\times10\times10^{-3}\ \text{m}^3}{8.314\ \text{J K}^{-1}\ \text{mole}^{-1}\times300\ \text{K}}$$

$$=\frac{1.013\times10}{8.314\times3}\ \text{mole}$$

$$=0.406\ \text{mole}$$

Discussion and Problems

13.1 Is temperature a macroscopic or a microscopic concept?

13.2 What should be the properties of a gas you use for a constant volume gas thermometer?

13.3 Find out the temperature at which Fahrenheit and Celsius scales give the same reading. At what temperature do the Fahrenheit and the Kelvin scales give the same reading?

13.4 An absolute temperature scale using degrees of the same size as the Fahrenheit scale but with a zero coinciding with 0 K is called the Rankine scale. Thus Rankine temperature T_R (written °R) is proportional to the Kelvin temperature. The conversion relation is $T_R=(9/5)T$. Find the temperature of the triple point of water on this scale.

13.5 Express the temperatures of fixed points given in Table 13.1 in °F and °R.

13.6 Starting from the ideal gas law, prove that a volume of any perfect gas at 0°C and any constant pressure, will expand by 1/273 of its volume for each degree it is heated above 0°C. If, on the other hand, the volume is held constant, show that the pressure will increase by the same amount.

13.7 If, instead of one mole, we take one gram of a gas, the constant R in equation $PV=RT$ will change from gas to gas. It is then, therefore, called the specific gas constant. What is the relationship between the specific gas constant and the gas constant? Density of helium under standard conditions ($t=0°C$, $P=760$ mm-Hg) is 0.178 g/litre. What are the specific and the universal gas constants?

13.8 A litre of nitrogen has a temperature of 27°C and a pressure of 850 mm-Hg. Find the mass of the gas if the density of nitrogen at STP is 1.25 g/litre.

13.9 The atmospheric pressure at a place is 74 cm-Hg. But a mercury barometer, after introducing some oxygen into it, gives a pressure of 70 cm-Hg. Calculate the amount of oxygen introduced when the temperature of the place is 30°C, area of cross-section of the barometer tube is 1 cm^2, and the height of the barometer tube above the mercury level in the outer vessel is 100 cm.

13.10 In the J-shaped tube of Fig. 13.10 mercury is poured into the open end in such a way that air gets trapped into it. How much mercury can be poured before it starts overflowing. Atmospheric pressure is 74 cm-Hg. Volume of the bend may be neglected.

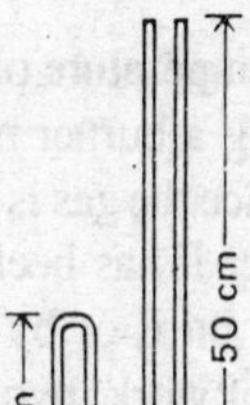

Fig. 13.10 *For Problem 13.10*

13.11 At −20° C and a pressure of 740 mm-Hg, a gas has a volume of 1 litre. At what Celsius temperature will it occupy a volume of 1.5 litre, if the pressure rises to 800 mm-Hg?

13.12 A litre of hydrogen at a pressure of 1 atm and at a temperature of 20°C, is heated till both the volume and the pressure are doubled. What is the amount of hydrogen and what is its final temperature?

13.13 An air bubble of diameter 1 mm is at the bottom of a sea 5 km deep, where the temperature is 4°C. The bubble rises to the surface which is at 27°C. Find the diameter of the bubble just before it reaches the surface, assuming that the air has attained the temperature of the surrounding water. Also assume that the density of water is 1g/cm^3 throughout.

13.14 A vessel of capacity 10 litre contains carbon dioxide at a pressure of 4 atm and temperature 37°C. Because of leakage some gas escapes out. If the pressure of the remaining gas at 27°C is 2 atm, find out the amount of CO_2 which has leaked.

13.15 We have two vessels of equal volume, one containing hydrogen and the other equal mass of helium. The temperature of both the containers is 27°C. (*i*) What is the relative number of molecules in two vessels? (*ii*) If pressure of hydrogen is 2 atm, what is the pressure of helium? (*iii*) If the temperature of helium is kept 27°C and that of hydrogen is changed, at what temperature will its pressure become equal to that of helium?

Note: For problems after Problem 13.6 assume all gases to be perfect. Any other data you need can be had from the tables of constants.

Chapter 14

Laws of Thermodynamics

14.1 Introduction

If the law of conservation of energy is to hold good, then 'something' (called heat) which is shared between two bodies at unequal temperatures is also energy (Sec. 13.1). In the early eighteenth century this 'something' was called 'caloric fluid'. It was further believed that the caloric fluid is conserved, i.e. the loss in the caloric fluid of the hotter body is equal to the gain in the caloric fluid of the colder body. This concept can be easily disproved. If a girder is cut with a saw and water is continuously poured at the place of cutting to keep it cool, the water continuously gains in the caloric fluid while the girder remains in the same state. One can conjecture that the continuous dissipation of work done on the saw may be directly responsible for heating. It was shown by Rumford and Joule that when a definite amount of work was expended a definite amount of heat was produced. That is in mathematical language

$$W = JQ \tag{14.1}$$

In MKS system work W is in joules and heat Q is in kilocalories (heat needed to raise one kg of water from 14.5° to 15.5°C). Thus unit of J, the mechanical equivalent of heat, will be joule/kcal. The value of J is 4184 joules/kcal, or 4.184 J/cal. In SI the unit used for both Q and W is the same, i.e. joule. In such a case, J reduces to unity.

Heat engines function on conversion of heat into work. From this we can conclude that work and heat are interchangeable. Let us take a gas in a cylinder having diathermic walls. On compression the gas gets heated. The daily experience is a cycle pump. This method of heating the gas involves mechanical work. The temperature of the gas can also be raised by bringing a burner near it. This process involves heat. Once the gas is hot, we have no method to tell whether it has been brought to this state by doing work on it or by heating. In mechanics, the concept of work involves transfer of energy other than due to temperature difference. Similarly, the concept of heat also involves transfer of energy due to difference in temperature. Processes which involve work and heat constitute the subject matter of thermodynamics. Here we are concerned about the internal state of a system.

Example 14.1 A waterfall is 100 m high. Assuming that all the heat generated stays in water, how much warmer the water at the bottom will become as compared with the water at the top?

Let us take 1 kg of water. When it falls through a height of 100 m, the decrease in its potential energy is

$$\Delta E = 1 \text{ kg} \times 9.81 \text{ ms}^{-2} \times 100 \text{ m} = 981 \text{ J}$$

This decrease in potential energy will first appear as kinetic energy of water as it falls. When it strikes the bottom, this kinetic energy reduces to zero. This now appears as thermal energy. Therefore, the thermal energy produced is

$$Q = 981 \text{ J} = 981/4.184 \text{ cal} = 234.5 \text{ cal}$$

Heat input required to raise the temperature of 1 kg of water through 1°C is 1 kcal. Therefore

Rise in temperature $= 234.5/1000\ °\text{C} = 0.23°\text{C}$

14.2 External Work Done by a System

Let the gas in a cylinder (Fig. 14.1) expand quasi-statically. By the term quasi-static we mean that the gas should change its state so slowly that the temperature throughout the gas should remain uniform. This is because we are assigning one temperature to the whole system. It is possible only when the unbalanced force responsible for expansion is negligibly small. In other words, the difference in force on both the sides of the piston is very small.

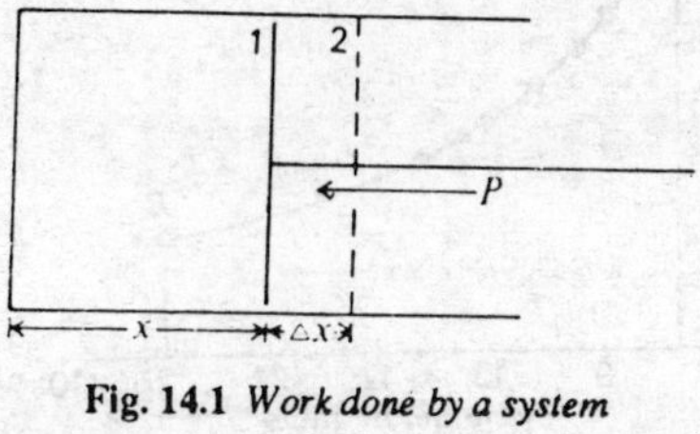

Fig. 14.1 *Work done by a system*

If A is the area of cross-section of the piston, the force F exerted by the gas on it is,

$$F = PA \qquad (14.2)$$

where P is the pressure of the gas. Thus work done by the gas in going from position (1) to position (2) is

$$\Delta W = F \times \Delta x$$

$$= P(A\Delta x),$$

or

$$\Delta W = P\Delta V \qquad (14.3)$$

where ΔV is the increase in volume of the gas.

On the PV-diagram given in Fig. 14.2, this work has been shown by the heavily shaded area. If the system goes from initial state i to the final state f, the work done will be equal to the sum of infinitesimal works like ΔW. This work has been shown in Fig. 14.2 by the area *iabfi*.

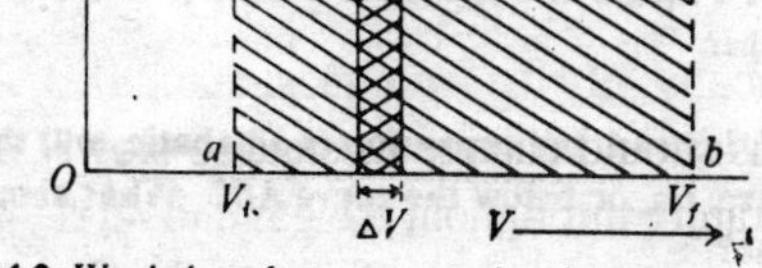

Fig. 14.2 *Work done by a system when it expands from state i to state f is given by the area iabfi*

In mechanics work done on a system is taken as positive, while here it is just the opposite. Here, we concentrate on the forces exerted by the system on the surroundings. Thus in the case we have discussed just now, force and displacement are in the same direction and, hence, work is done by the system, and it is taken as positive. On the other hand, heat transferred to the system is taken as positive. This sign convention stems from the theory of heat engines, where heat is always taken by the engine and work is done by it.

Question: When a wire is pulled with a force F its length increases by an amount x. How much is the work done by the wire?

One thing must be noted here. This work done depends upon the path traversed by the system in going from i to f. Let us take two cases. Let the system first expand from i to m in Fig. 14.3 isobarically (iso = same, bar = barometer = pressure); and then go from m to f isochorically (chor = volume). The work done is the area *imbai*. In the second process let it first go from i to n isochorically; and then from n to f isobarically. This time the work done by the system is only *nfban*. If the system completes one full cycle going to f via m and coming back to i via n, the net work done by the gas is the area of the loop *imfni*.

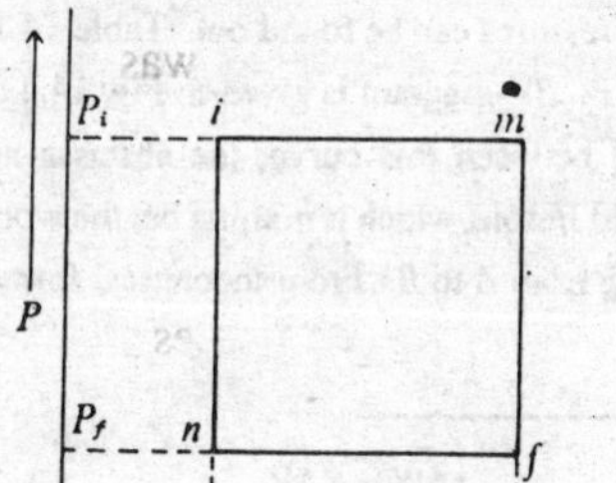

Fig. 14.3 Work done by a system in going from state i to state f depends upon the path taken.

A similar conclusion will be drawn if instead of doing work, the system is given heat to take it from state i to state f. Let T_i and T_f be the temperatures of the system in states i and f, respectively. Let us first reduce the pressure to P_f and then heat the gas at this constant pressure till we get the temperature T_f. We can also first heat the gas at pressure P_i to reach the temperature T_f and then reduce the pressure at constant temperature till we get the pressure P_f. The heat given to the system in each case will be different. The same will hold for other paths followed.

Example 14.2 One mole of a perfect gas is allowed to expand from a pressure of 4 atm to a pressure of 1 atm keeping the temperature constant at 27°C (an isothermal process). Draw the *PV*-diagram (or indicator diagram) and from this find out the work done by the gas. Compare it with the theoretical value given by the formula* $W = -R_0T \log_e(P_f/P_i)$, where P_i and P_f refer to the initial and the final pressures, respectively.

For one mole of an ideal gas, the equation of state is,

$$PV = R_0T$$

$$\therefore \quad V = \frac{R_0T}{P}$$

$$= \frac{8.314\ \text{J K}^{-1}\ \text{mole}^{-1} \times 300\ \text{K}}{P}$$

$$P, 1\ \text{atm} = 1.013 \times 10^5\ \text{N m}^{-2}$$

Substituting various values of P in the above equation, molar volumes at these pressures can be found out. Table 14.1 gives V as a function of P. *PV*-diagram is given in Fig. 14.4.

Area enclosed between this curve, the abscissa and the broken lines is 3420 J/mole, which is nothing but the work done by the gas in going from A to B. From theoretical formula,

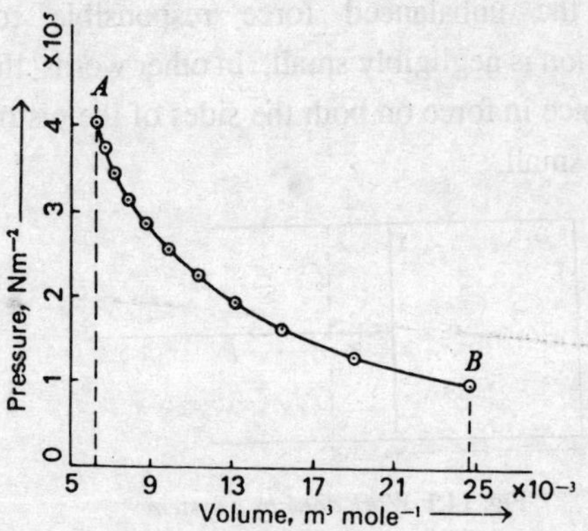

Fig. 14.4 *For Example 14.2*

$$W = -R_0T \log_e \frac{P_f}{P_i} = R_0T \log_e \frac{P_i}{P_f}$$

$$= R_0T \times 2.303 \times \log_{10} 4$$

$$= 8.314\ \text{JK}^{-1}\ \text{mole}^{-1} \times 300\ \text{K} \times 2.303 \times 0.6021$$

$$= 3458\ \text{J/mole}$$

Table 14.1 *Volume as a Function of Pressure*

Pressure		Volume
Atm	*Nm*$^{-2}$	*m*^{3}*mole*$^{-1}$
4.0	4.052×10^5	6.155×10^{-3}
3.7	3.748×10^5	6.655×10^{-3}
3.4	3.444×10^5	7.241×10^{-3}
3.1	3.141×10^5	7.941×10^{-3}
2.8	2.836×10^5	8.792×10^{-3}
2.5	2.532×10^5	9.849×10^{-3}
2.2	2.228×10^5	11.19×10^{-3}
1.9	1.925×10^5	12.96×10^{-3}
1.6	1.621×10^5	15.39×10^{-3}
1.3	1.317×10^5	18.94×10^{-3}
1.0	1.013×10^5	24.62×10^{-3}

The difference in the two results is due to some error involved in plotting the graph and counting squares. We take more than half a square as one and leave out less than half a square altogether.

Question: If, instead, the expansion is adiabatic, will the curve be above, on, or below the curve *AB*? What about the work done?

$$^*\Delta W = P\Delta V$$

$$\therefore \quad W = \int_i^f PdV$$

But $\quad PV = R_0T$

Therefore for isothermal expansion:

$$PdV + VdP = 0$$

$$\therefore \quad W = -\int_i^f VdP = -R_0T \int_i^f \frac{dP}{P} = -R_0T \log_e \frac{P_f}{P_i}$$

14.3 First Law of Thermodynamics

We have seen above that if a system goes from one state to another, work and heat taken separately depend upon the path followed by the system. But we have learnt in Chapter 5 that work done in going from one point to the other (without friction) in gravitational field is independent of the path. When this is the case, the work done can be expressed as a difference of the value of a function which depends only on position. This function (for unit mass) was called gravitational potential energy function. You will come across a similar situation in the electric field. There too the work done in carrying an electric charge in a static electric field will be independent of the path and we will, through this, introduce electric potential energy function. Here also it has been seen experimentally that if in going from state i to state f a system does a work W and Q is the heat transferred to the system, then $(Q-W)$ is independent of the path. We thus conclude that there exists a function of the thermodynamic coordinates (analogous to functions of space coordinates mentioned above), whose value at the final state minus its value at the initial state is equal to $(Q-W)$. This function we decide to call *internal energy function* and give it a symbol U. Expressed in mathematical language

$$Q-W=U_f-U_i \qquad (14.4)$$

Now Q is the energy given to the system by way of heat, and W is the energy taken out of the system by way of work. Thus, RHS represents increase in internal energy of the system. As in the case of gravitational and electrical potential energies, it is the change in internal energy which matters. Thus, Eq. (14.4) is the law of conservation of energy, and is called the *first law of thermodynamics*. If the changes are infinitesimal, then

$$\Delta Q-\Delta W=\Delta U \qquad (14.5)$$

It may be interesting to point out that through the zeroeth law of thermodynamics we have introduced a thermodynamic parameter called *temperature*, and through this (first) law we have brought out the existence of a thermodynamic function *internal energy*. Another thermodynamic function, entropy, will be introduced through second law of thermodynamics.

Example 14.3 When one gram of boiling water is converted into steam at a pressure of one atmosphere, the steam occupies a volume of 1671 cm^3. The heat (or the latent heat) of vaporization of water at 1 atm is 9.70 kcal/mole. Compute the external work done by the system, and the increase in its internal energy.

$$\text{Heat input} = 9.70 \text{ kcal/mole} \times 1 \text{ g}$$

$$= \frac{9.70\times10^3}{18} \text{ cal/g} \times 1 \text{ g}$$

$$= 539 \text{ cal}$$

$$= 539\times4.184 \text{ J} = 2255 \text{ J}$$

Now, in the process of formation of steam the external pressure remains constant at 1 atm. Therefore, work done by the system

$$W=\Sigma P\Delta V = P\Sigma\Delta V = P(V_f-V_i)$$

$$= 1.013\times10^5 \text{ N m}^{-2}\times\frac{(1671-1)}{10^6}\text{m}^3$$

$$= \frac{1.013\times1670}{10}\text{ N m}$$

$$= 169 \text{ J}$$

Using Eq. (14.5), increase in internal energy of the system:

$$\Delta U = \Delta Q - \Delta W,$$

or

$$U_f-U_i = (2255-169) \text{ J}$$

$$= 2086 \text{ J} = 499 \text{ cal}$$

14.4 Principle of Calorimetry

Let us take two systems A and B [Fig. 14.5(a)]. System A may be a solid or any other thing of which the specific heat is to be measured; and system B may be a water bath. Let these two systems be brought into thermal contact by a diathermic wall (or by mixing) and be surrounded by an adiabatic wall [Fig. 14.5(b)]. For system A alone:

$$Q_A-W_A=U_{fA}-U_{iA} \qquad (14.6)$$

and for system B alone :

$$Q_B-W_B=U_{fB}-U_{iB} \qquad (14.7)$$

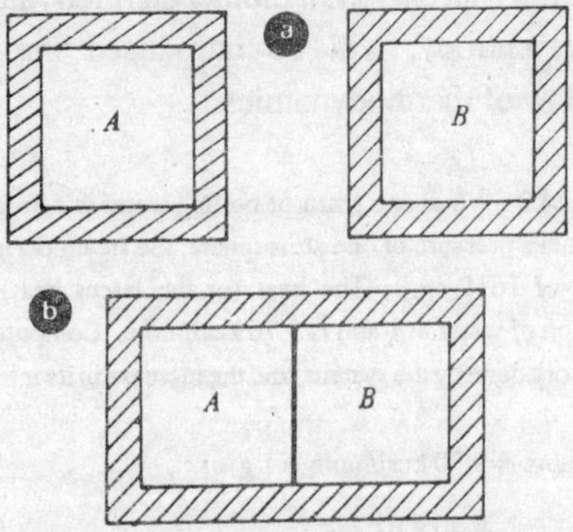

Fig. 14.5 *When two systems in* (a) *are brought into thermal contact, but are otherwise isolated from the surroundings, as in* (b), *the heat gained by one equals the heat lost by the other*

From Eqs. (14.6) and (14.7) we get,

$$(Q_A + Q_B) - (W_A + W_B)$$

$$= (U_{fA} + U_{fB}) - (U_{iA} + U_{iB}) \qquad (14.8)$$

Now, the right-hand side of Eq. (14.8) is the increase in internal energy of the composite system $(A + B)$. Also, on the left-hand side, $(W_A + W_B)$ is the work done by this composite system. Therefore, from the first law of thermodynamics, $(Q_A + Q_B)$ is the heat supplied to the composite system. But since this composite system is surrounded by an adiabatic wall, heat can neither go in nor go out of this system. Therefore

$$Q_A + Q_B = 0,$$

or

$$Q_A = -Q_B \qquad (14.9)$$

Thus, under the conditions given above, heat gained by system A equals heat lost by system B, or vice versa. It is this basic principle which works behind all calorimetric measurements.

Example 14.4 By definition, the specific heat of a substance is the amount of energy required to increase the temperature of its unit mass by one degree. In an experiment to find the specific heat of tin, a 200.0 g piece of tin at 98.0°C is put into a copper calorimeter containing 100.0 g of water at 20.0°C. The temperature of the mixture becomes 27.1°C. Find out the specific heat of tin. The mass of the calorimeter is 102.0 g and the specific heat of copper is 0.094 kcal/kg-K. The specific heat of water is 1 kcal/kg-K.

Let the specific heat of tin be S.

$$\text{Heat lost by tin piece} = \frac{200.0}{1000}\ \text{kg} \times S \times (98 - 27.1)\ \text{K}$$

$$= 0.2 \times S \times 70.9\ \text{kg K}$$

$$\text{Heat gained by water} = \frac{100.0}{1000}\ \text{kg} \times 1\ \text{kcal/kg-K} \times (27.1 - 20.0)\ \text{K}$$

$$= 0.1 \times 7.1\ \text{kcal}$$

$$\text{Heat gained by calorimeter} = \frac{102.0}{1000}\ \text{kg} \times 0.094\ \text{kcal/kg-K} \times (27.1 - 20.0)\ \text{K}$$

$$= 0.102 \times 0.094 \times 7.1\ \text{kcal}$$

Now, for the composite system of tin and water in calorimeter, no heat has either gone out or come in. Therefore, applying the above calorimetry principle

$$0.2 \times S \times 70.9\ \text{kg K} = 0.1 \times 7.1\ \text{kcal} + 0.102 \times 0.094 \times 7.1\ \text{kcal}$$

or,

$$14.18\,S = (0.71 + 0.068)\ \text{kcal/kg-K}$$

$$= 0.778\ \text{kcal/kg-K},$$

or

$$S = 0.778/14.18 = 0.055\ \text{kcal/kg-K}$$

$\therefore$ Specific heat of tin = 0.055 kcal/kg-K

14.5 Specific Heats of a Gas

Let us give an amount of heat ΔQ to a system. If its temperature rises by an amount ΔT (it is not necessary that temperature should always rise on heating), then the heat capacity* of that system is defined as

$$C = \Delta Q/\Delta T \qquad (14.10)$$

Heat capacity per unit mass is known as specific heat, which is a characteristic of the material of the body. C varies with temperature around which ΔT is taken. Apart from this, we have to specify some other factors. As we have already discussed, a system having an initial temperature T_i can be taken to a final temperature T_f in an infinite number of ways. Further, final temperature alone does not specify the final state. There can be an infinite number of states having the temperature T_f. Thus

*The term *heat capacity* is erroneous. It gives the impression as if we are talking about the capacity of the system to hold heat. Heat in a body has no meaning. Actually what it means is simply the amount of heat added for unit rise of temperature.

the simple relation (14.10) can give any value for the heat capacity, depending upon the path taken to achieve the final temperature. For solids and liquids the specification of the final state does not matter much because their volumes change by a very small amount. But in the case of gases one has to be specific about it.

In the determination of heat capacities, it is decided that we will not do work on the system. We will simply supply it heat. What remains to be specified is whether the system does work or not. As we have already pointed out, the work done ($\Sigma P\,\Delta V$) by solids and liquids can be neglected because ΔV is very small. But for gases one cannot neglect the change in volume because it is quite appreciable. For gases we are generally concerned with two heat capacities : one in which pressure is kept constant, and the second in which volume is kept constant. Thus,

$$C_P = \left(\frac{\Delta Q}{\Delta T}\right)_P, \qquad (14.11)$$

and

$$C_V = \left(\frac{\Delta Q}{\Delta T}\right)_V \qquad (14.12)$$

The amount of gas generally taken is one mole. Thus, C_P and C_V will be in joules per mole per kelvin.

Let us take one mole of a perfect gas* in a cylinder having a diathermic bottom (Fig. 14.6). When the cylinder is heated at the bottom the gas pressure increases. Increase pressure on the piston accordingly, so that outside pressure is always equal to the gas pressure. Thus gas will not expand and, hence, will do no work. If ΔQ is the heat thus supplied to the gas and ΔT is the increase in its temperature, then according to the first law of thermodynamics,

$$\Delta Q = \Delta U + \Delta W \qquad (14.13)$$

Using the defining relation (14.12) and putting $\Delta W = 0$,

$$\Delta U = C_V\,\Delta T \qquad (14.14)$$

*For a perfect gas U depends only on temperature (see Ch. 15).

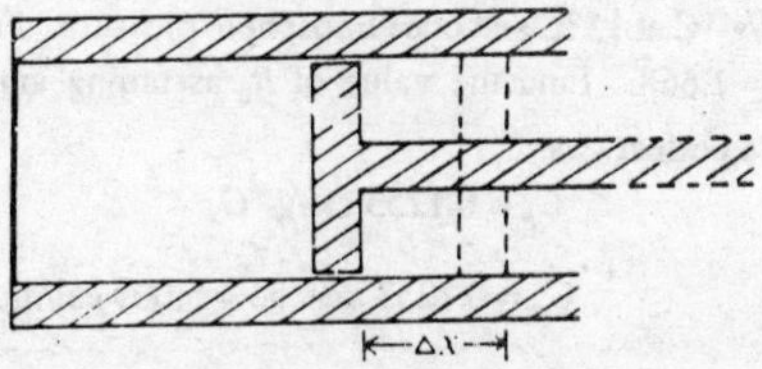

Fig. 14.6 ***If a gas is heated at a constant pressure then, apart from heating, heat will have to be supplied even for the work done by the gas***

In the second process, let us heat the gas slowly. As its temperature rises, its pressure will increase and the piston will be pushed out. Let the heating be so slow that the gas pressure is always almost equal to the external pressure P. In other words, the expansion of the gas should be quasi-static. In Fig. 14.6, the final position of the piston is shown by dashed lines. As already shown in Section 14.2, Eq. (14.3) will represent the work done by the gas. Applying first law of thermodynamics to this process,

$$\Delta Q = \Delta U + P\,\Delta V \qquad (14.15)$$

By the defining relation (14.11), LHS of Eq. (14.15) equals $C_P \Delta T$. Therefore,

$$C_P\,\Delta T = \Delta U + P\,\Delta V,$$

or

$$C_P\,\Delta T = C_V\,\Delta T + P\,\Delta V \qquad (14.16)$$

Now, for a perfect gas,

$$PV = R_0 T \qquad (14.17)$$

If, pressure remaining constant, volume changes from V to $(V+\Delta V)$ when temperature changes from T to $(T+\Delta T)$, then,

$$P(V+\Delta V) = R_0(T+\Delta T) \qquad (14.18)$$

From Eqs. (14.17) and (14.18), by subtraction, we get,

$$P\,\Delta V = R_0 \Delta T \qquad (14.19)$$

Using Eq. (14.19), Eq. (14.16) becomes,

$$C_P \Delta T = C_V \Delta T + R_0 \Delta T,$$

or

$$C_P - C_V = R_0 \qquad (14.20)$$

In Eq. (14.20) C_P, C_V and R_0 are in the same units.

Eq. (14.20) shows that molar specific heat of a gas at constant pressure is always greater than that at constant volume.

Example 14.5 Specific heat of argon at constant pressure is 0.1253 cal/g-°C at 15°C and one atmosphere pressure. Further $C_P/C_V = \gamma = 1.668$. Find the value of R_0 assuming argon to behave as a perfect gas.

$$C_P = 0.1253 \text{ cal/g-°C},$$

or
$$C_P = 0.1253 \times 39.95 = 5.005 \text{ cal/mole-°C}$$

$$C_V = 5.005/1.668 = 3.000 \text{ cal/mole-°C}$$

Thus
$$R_0 = C_P - C_V$$
$$= 2.005 \text{ cal/mole-°C}$$

Compare it with the standard value of R_0, i.e. 1.987 cal/mole-°C. The difference is quite small. Therefore, we can conclude that, at moderate pressures, argon behaves more or less as an ideal gas.

14.6 Carnot Cycle – Second Law of Thermodynamics

From calorimetric measurements it can be easily found that the work done can be completely converted into heat. Let us see whether the reverse can also be done. If a contrivance, called heat engine, takes heat energy Q from some source and out of it converts W into work, then its efficiency η is defined as

$$\eta \equiv W/Q \qquad (14.21)$$

The efficiency of actual heat engines varies from about 5 to about 40 per cent. Carnot undertook theoretical investigations regarding improvements in the efficiency of a heat engine and found that even an ideal engine cannot convert all the input of heat energy into work. In any heat engine we require four things: (1) A source of heat (e.g. a furnace), (2) a working substance (e.g. steam or air), (3) a machine, and (4) a heat sink at a temperature lower than that of the source. For an engine to have maximum efficiency, Carnot suggested the following procedure. For the machine, let us take a cylinder of adiabatic walls and an adiabatic end cap (Fig. 14.7). This will not allow any heat loss by way of conduction, and maximum heat will be converted into work. Also assume that there is no friction between the piston and the cylinder. Take a perfect gas* as a working substance and operate the engine in the following four steps.

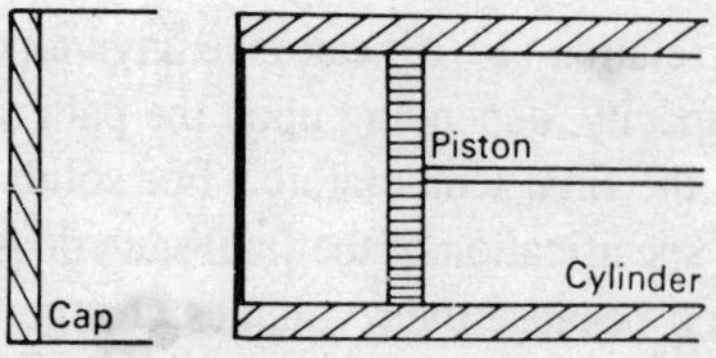

Fig. 14.7 *A heat engine*

(1) First remove the cap from the cylinder and put its conducting end in thermal contact with the heat reservoir of Fig. 14.8, which is at temperature T_1 K. Let the piston of the engine move out slowly in such a way that the temperature of the gas always remains only infinitesimally less than T_1. During this isothermal expansion the point representing the state of the gas moves from A to B in indicator diagram (Fig. 14.9). Work W_1 done by the gas in

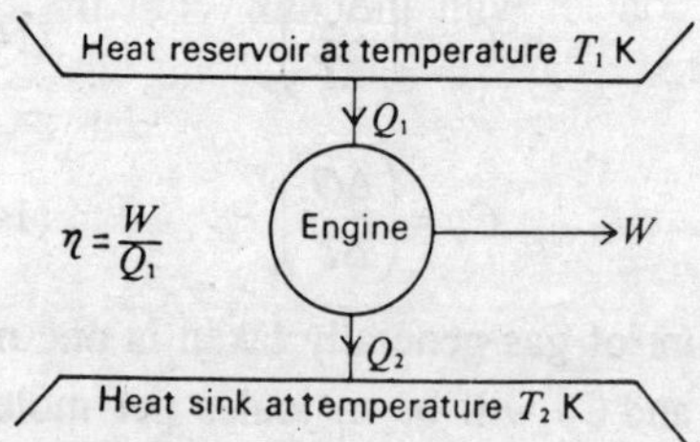

Fig. 14.8 *Working of a heat engine*

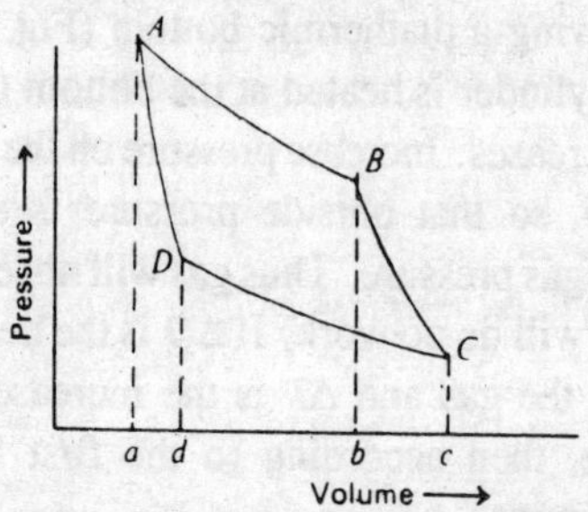

Fig. 14.9 *Carnot cycle using perfect gas*

*We will show later that the efficiency of a Carnot engine is independent of a working substance. We take here a perfect gas because its equation of state is known and, thus, theoretical calculations are possible.

this process equals the heat Q_1 taken by it from the source.* Thus,

$$W_1 = Q_1 = R_0 T_1 \ln\left(\frac{V_B}{V_A}\right) \qquad (14.22)$$

where V_A and V_B are the volumes of the gas in the cylinder at points A and B, respectively. This work is also represented by the area $AabB$.

(2) Now replace the adiabatic cap of the cylinder and allow the gas to expand (due to inertia) adiabatically. This process is to be quasi-static such that the pressure inside the cylinder is infinitesimally greater than the outside pressure throughout this expansion. The point on indicator diagram moves from B to C. Stop at C where the temperature of the gas has fallen to T_2 K, which is the temperature of the sink. The work done by the gas in this process is the area $BbcC$.

(3) Remove the end cap and put the cylinder in thermal contact with the sink. Let the gas be compressed quasi-statically keeping its temperature infinitesimally greater than T_2. The work W_2 thus expended appears as heat Q_2, which is given to the sink. Let the point on the indicator diagram move from C to D. Therefore,

$$W_2 = Q_2 = -R_0 T_2 \ln\left(\frac{V_D}{V_C}\right) = R_0 T_2 \ln\left(\frac{V_C}{V_D}\right) \qquad (14.23)$$

which is equal to the area $CcdD$. Here V_C and V_D are the volumes of the gas at points C and D, respectively.

(4) Stop at a point D such that if the adiabatic cap is now replaced the gas, in adiabatic compression due to inertia, should be able to go back to the point A. Work done on the gas during this process is equal to the area $AadD$.

In the adiabatic expansion from B to C, as $\Delta Q = 0$, work done by the gas $= -\Delta U = U_B - U_C$. Similarly, work done by the gas in going from D to $A = -\Delta U = U_D - U_A$. Since $U_A = U_B$, and $U_C = U_D$, the work done by the gas from B to C (area $BbcC$) equals the work done on it from D to A (area $AadD$). Hence in one complete cycle the engine takes heat Q_1 from a source at temperature T_1, rejects heat Q_2 to a sink at temperature T_2, and converts the difference $(Q_1 - Q_2)$ into work W. This work is represented on the indicator diagram by the area $ABCD$. Now,

$$W = Q_1 - Q_2$$

$$\therefore \qquad \eta = \frac{W}{Q_1} = \frac{Q_1 - Q_2}{Q_1}$$

$$= 1 - \frac{Q_2}{Q_1} \qquad (14.24)$$

Using Eqs. (14.22) and (14.23), and isotherms and adiabatics of Fig. 14.9, it can be shown that

$$\frac{Q_1}{T_1} = \frac{Q_2}{T_2} \qquad (14.25)$$

Hence $$\eta = 1 - \frac{T_2}{T_1} \qquad (14.26)$$

which is less than unity and depends on the two temperatures only.

The above engine has been operated in such a way that it can retrace its path at every step. Therefore, the above operation is called a *reversible cycle*. Though the Carnot cycle cannot be realized in practice, it is very useful for theoretical studies. When operated in reverse direction, it works as a refrigerator. It takes heat Q_2 from a lower temperature T_2, work W is done on it, and it rejects heat $Q_1 (= Q_2 + W)$ at a higher temperature T_1.

---

*As the gas is perfect and there is no change in its temperature, $\Delta U = 0$. Therefore, from first law of thermodynamics :

$$\Delta Q = \Delta W = \int_A^B P dV = \int_A^B \frac{R_0 T_1}{V} dV$$

$$= R_0 T_1 [\ln V]_{V_A}^{V_B} = R_0 T_1 \ln\left(\frac{V_B}{V_A}\right)$$

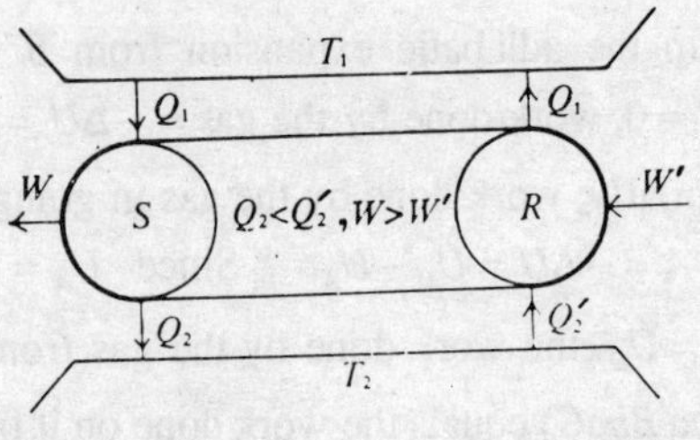

Fig. 14.10 *A composite device consisting of a revrsible engine and an irreversible engine*

The Carnot cycle described above gives maximum efficiency to a heat engine. To prove this we take two engines R and S which are reversible and irreversible, respectively. Let them work between two fixed temperatures T_1 and T_2 (Fig. 14.10). Let S be more efficient than R. If both of them take the same amount of heat, say Q_1, from the source, work output of S will be more than that of R. Therefore, heat Q_2 rejected by S to the sink will be less than Q_2', the heat rejected by R. Let us now reverse R and make it work as refrigerator. It will now take heat Q_2' from the sink, work W' will be done on it, and it will reject heat Q_1 to the source. Let R now be coupled with S. As $W > W'$, S will be able to run R, and apart from it, work $(W - W')$ will be available for use elsewhere. The combination of S and R jointly will take heat $(Q_2' - Q_2)$ from the sink and the whole of it will be converted into work $(W - W')$, and no net heat will be taken from the source. Thus if we enclose the above combination in a box, we will have an engine which is hundred per cent efficient, and takes heat energy from sources at lower temperatures. But our experience tells us that this is impossible. If it were possible, mankind could use huge amounts of internal energy stored in atmosphere, sea, and earth. This way the perpetual motion machines could be made.

If we assume that a reversible engine using a particular working substance is more efficient than another using different working substances, we reach the same conclusion as above. Thus we conclude that *among several engines working between two given temperatures all reversible engines have the same efficiency and this is the highest limit for the efficiency of any heat engine.* This is Carnot's theorem. Thus the low efficiency of a heat engine is not an engineering deficiency of designing and fabrication, but is something in the very nature of the things.

We have given the above discussion to make you appreciate that physical laws evolve from realities of life. Our experience with the problem of converting heat into work has given rise to the second law of thermodynamics. It states that *it is impossible to construct an engine which, operating in a cycle, will produce no effect other than the extraction of heat from a source and the conversion of this heat completely into work.*

In Carnot cycle, Q_1 is the heat taken by the working substance at temperature T_1 and Q_2 is the heat rejected by it at temperature T_2. If, (Q_1/T_1) represents the gain in 'something' of the working substance, then (Q_2/T_2) will represent loss in it. Therefore, from Eq. (14.25) we see that this 'something' of the working substance is conserved in a reversible cycle. Such quantities are very important to us. Therefore, we give it a name *entropy* and a symbol S. We can now say that the entropy of a system after undergoing a reversible cycle remains unchanged. If, during a reversible process, all interacting systems are taken into consideration, the entropy of all systems taken together does also not change. Let us, for example, take the process from A to B in Fig. 14.9. During this process, the reservoir loses heat Q_1 at temperature T_1. Therefore, the decrease in its entropy is Q_1/T_1, which is equal to the increase in entropy of the working substance.

After one complete cycle of a Carnot engine, the entropy of the reservoir decreases by an amount (Q_1/T_1), and the entropy of the sink increases by an amount Q_2/T_2. One can see from Eq. (14.25) that these two changes are equal. For all other engines, as the heat rejected is comparatively more, the increase in entropy of the sink is greater than the decrease in entropy of the source. Thus, the overall effect is an increase in entropy. Hence, an alternate statement of the second law is that *the entropy cannot decrease.* You know from daily experience that heat always flows from hotter bodies to colder bodies. It also amounts to increase in entropy of hot and cold bodies taken together. This process of heat transfer is irreversible. The above conclusion that entropy cannot decrease is very general and holds for any process occurring by itself. This general experience can be enunciated as : *Every physical or chemical process in nature takes place in such a way that the sum of the entropies of all systems taking part in the process increases. In the limiting case of a reversible process, the sum of the entropies remains constant.* This is also called the principle of increase of entropy and is the most general statement of the second law of thermodynamics. This does not hold for engines only. It holds for all processes taking place in this universe.

We can now differentiate clearly between the first and the second laws of thermodynamics. The first law states that if a process takes place, the energy can only be transformed from one form to the other, but it has to be conserved. The second law tells us whether that process will take place or not. According to Clausius the first law says that the energy of the world remains constant, and the second law says that the entropy of the world tends to be maximum. Thus the first law denies the possibility of creating or destroying energy, and the second law denies the possibility of using this energy in a particular way.

If we have two engines having a common sink, the efficiency of that engine will be higher which takes heat from a source at higher temperature. [See Eq. (14.26)]. It has also been mentioned above that the flow of heat from a higher temperature to a lower temperature amounts to increase in entropy. Therefore, the principle of increase of entropy implies the increase in unavailability of internal energy for useful purposes.

Entropy is also a measure of disorder or randomness in the molecular arrangement of the system. We know that when we give heat to a system its entropy increases. We also know from the kinetic theory that the addition of heat to a system produces a more violent agitation of its molecules. Opposite to it, the continuous subtraction of heat from a system amounts to decrease in its entropy. Due to this, the molecular motions also tend to vanish and the molecular arrangement attains a state of greater order. Hence, the increase in entropy implies a transition from an ordered state to a less ordered state. In other words, we can say that the principle of increase of entropy implies an increase in disorder. Therefore, we can say that according to the second law of thermodynamics *the disorder in nature is increasing.* It might be your experience also that the natural tendency of the things is to go into a disordered state. One has to put in a lot of effort to either maintain, or create order. According to kinetic theory, therefore, the entropy is a statistical concept. It can be shown that it is a logarithmic measure of the number of states accessible to the system. Or, it is a logarithmic measure of the degree of randomness of the system.

Question: A perfect gas expands freely from a volume V_1 to a volume V_2. (i) Is there any change in the entropy of the gas? (ii) If so, how much?

Question: Starting from the principle of increase of entropy, show that : (i) No engine can convert the heat taken from one source into work without affecting the surroundings. (ii) It requires a sink to reject some of the heat, and the temperature of this sink is to be lower than that of the reservoir. (iii) Eq. (14.26) is an upper limit to the efficiency of any engine. (iv) No living being, if isolated, can remain alive.

Hint: Principle of increase of entropy is that ***the entropy of an isolated system cannot decrease.*** This system can contain two or more sub-systems. The entropy of the engine (including the working substance) remains unchanged after one cycle. Output of the engine is used to lift a stone.

14.7 Steam Engine

Carnot cycle, though most efficient, cannot be realized in practice. Heat intake being at constant temperature, the process is very slow and the engine becomes very bulky and heavy in relation to the power output. The practical engines, on the other hand, must be quick-acting, i.e., a cycle ought to be completed as quickly as possible.

Steam engines, though not light, are quite fast. In these engines a mixture of water and water vapour is used as the working substance. Four steps of the *Rankine cycle*, the most commonly used cycle in steam engines, are as follows :

1. Superheated steam (at about 300°C) from a separate boiler is injected into the cylinder (at point *A* in Fig. 14.11). The piston is moved outwards and the substance expands isothermally and isobarically at temperature T_1 and pressure P_1 taking the heat from the source. P_1 is the pressure which saturated steam would exert at temperature T_1. This part corresponds to the conversion of water into steam in the boiler and its admission into the cylinder, and is represented by *AB*.

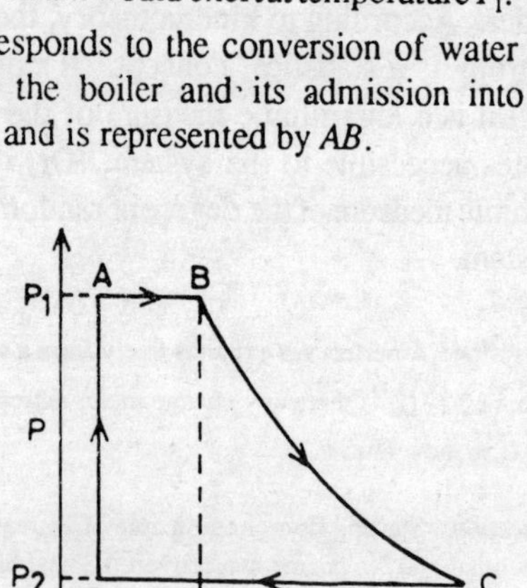

Fig. 14.11 *Rankine cycle*

2. Next the steam expands adiabatically (quasistatically in the ideal cycle) doing work on the piston (and hence on the machine coupled to it) till its temperature falls to T_2, the temperature of the condenser. This is represented by *BC* in Fig. 14.11.

3. In the third stage the substance is compressed isothermally till all the steam is condensed at pressure P_2 and temperature T_2. P_2 is the pressure of saturated steam at temperature T_2. This conversion of 'wet' steam into saturated water, and its isobaric isothermal transfer from cylinder to condenser is represented by *CD*.

4. The condensed water is transferred to the boiler by means of a separate feed-pump and the pressure is thereby increased from P_2 to P_1. This compression, assumed to be adiabatic, is shown by *DA*.

In the above adiabatic compression there is practically no change in temperature. At *A* the water is first heated from temperature T_2 to T_1 in the boiler and subsequently converted into steam. The process *AB* in the ideal case includes: (i) reversible isobaric heating of water to the boiling point, (ii) reversible isobaric isothermal vaporization of water into saturated steam.

The work done by the steam engine following the Rankine cycle will be equal to the area of the indicator diagram. Efficiency in ideal Rankine's cycle is less than in Carnot's cycle and is about 30%. The efficiency of actual steam engines is about 60-70% that of the ideal Rankine engine.

14.8 Internal Combustion Engines

In steam engine heating of the working substance is done outside the cylinder. As a result heat is lost during the transfer of the hot working substance to the cylinder. No separate furnace is used in internal combustion engines, but heating is done inside the cylinder itself*. Further, as the working substance can be heated to much higher temperatures in this way, internal combustion engines are much more efficient than the steam engines.

The Otto Cycle: In this engine, which is quite fast and comparatively light, heat is absorbed at constant volume. (It may be recalled that in Carnot cycle heat is absorbed at constant temperature.) The four strokes of this cycle are:

1. In the *charging stroke* the inlet valves are opened and a suitable mixture of air and gasoline vapour is sucked into the cylinder. This process is shown as *EC* in Fig. 14.12.

*By contrast, the steam engine may be called an external combustion engine, as heating is done in the boiler.

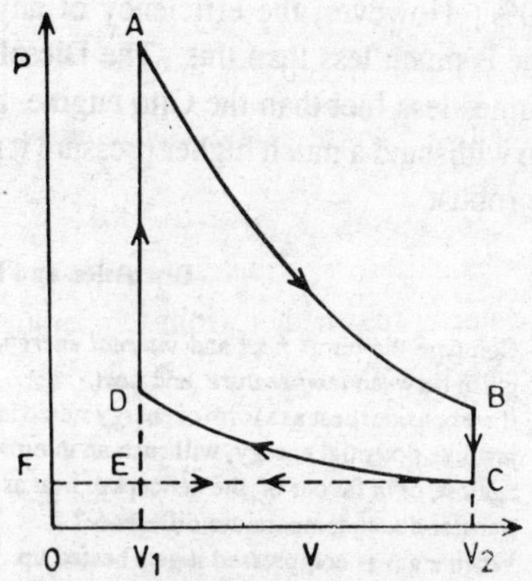

Fig. 14.12 *Otto cycle*

2. In the *compression stroke* all the valves are closed and the mixture is compressed adiabatically to about 1/9th of its original volume. In the process the temperature is raised to about 700°C. This is represented by *CD* in Fig. 14.12.

At *D* the mixture is fired by a spark, and the resulting combustion raises the temperature of the gas to more than 2000°C, and takes it from *D* to *A*. The chemical energy is thus converted into thermal energy *at a constant volume*. It may be noted that the transfer of heat here is almost instantaneous.

3. The piston is now thrown out with great force due to the high pressure developed, thus moving the machine coupled to it. This *working stroke* (an adiabatic expansion) is shown by curve *AB* in the figure.

4. At the end of this working stroke, i.e., at *B*, the exhaust valves open, pressure in the cylinder drops rapidly to atmospheric pressure and the inward moving piston pushes out the 'burnt' gases. After scavenging is complete, a fresh charge of air and gas is sucked in and the fresh cycle begins.

For the purposes of computation the above cycle is approximated by ***air standard*** or *Otto cycle* given in Fig. 14.12. Section *BC* represents cooling at constant volume. The ratio $V_2/V_1 = \rho$ is called the *compression ratio*. The work output is again given by the area of the closed curve *ABCD*. Heat input Q_1 is the heat supplied at constant volume along the line *DA*, and is given by:

$$Q_1 = C_V (T_A - T_D) \tag{14.27a}$$

Here C_V is the heat capacity of air at constant volume and is assumed to remain constant. It may be remembered that air is the working substance in this cycle, the function of gasoline being merely to heat the air by combustion. The heat exhaust Q_2 is along BC. Therefore,

$$Q_2 = C_V (T_B - T_C) \tag{14.27b}$$

Therefore, the efficiency of the Otto cycle becomes,

$$\eta = 1 - \frac{Q_2}{Q_1} = 1 - \frac{T_B - T_C}{T_A - T_D} \tag{14.28}$$

From the relation $TV^{\gamma-1}$ = constant for an adiabatic, we get

$$T_A V_1^{\gamma-1} = T_B V_2^{\gamma-1}, \tag{14.29}$$

and

$$T_D V_1^{\gamma-1} = T_C V_2^{\gamma-1}, \tag{14.30}$$

Therefore,

$$\frac{T_A}{T_B} = \frac{T_D}{T_C} = \left(\frac{V_2}{V_1}\right)^{\gamma-1} = \rho^{\gamma-1} \tag{14.31}$$

$$\frac{T_B}{T_C} = \frac{T_A}{T_D}, \quad \frac{T_B - T_C}{T_C} = \frac{T_A - T_D}{T_D} \tag{14.32}$$

and

$$\frac{T_B - T_C}{T_A - T_D} = \frac{T_C}{T_D} = \left(\frac{1}{\rho}\right)^{\gamma-1} \tag{14.33}$$

Hence, the efficiency becomes,

$$\eta = 1 - \left(\frac{1}{\rho}\right)^{\gamma-1} \tag{14.34}$$

Taking $\rho = 9$, $\gamma = 1.4$, we get

$$\eta = 1 - \left(\frac{1}{9}\right)^{0.4} = 1 - 0.42 = 0.58 = 58\%$$

This is the efficiency of an ideal Otto cycle. The efficiency of actual engines is much less due to friction, thermal conduction, turbulence, acceleration, etc., which are usually present. It may be noted that higher the ρ, the compression ratio, the higher the efficiency of the engine.

The Diesel Cycle: Diesel, not satisfied by the low efficiency of the Otto cycle, tried to find other cycles with which one could reach the efficiency of the Carnot cycle. In the process he invented another engine which, for some purposes, offers certain advantages over the Otto engine.

It has been seen that in Otto cycle if the compression ratio, ρ, is pushed up the efficiency can be further increased. But ρ cannot be increased beyond a certain value, otherwise the mixture would explode prematurely during compression before the spark passes.

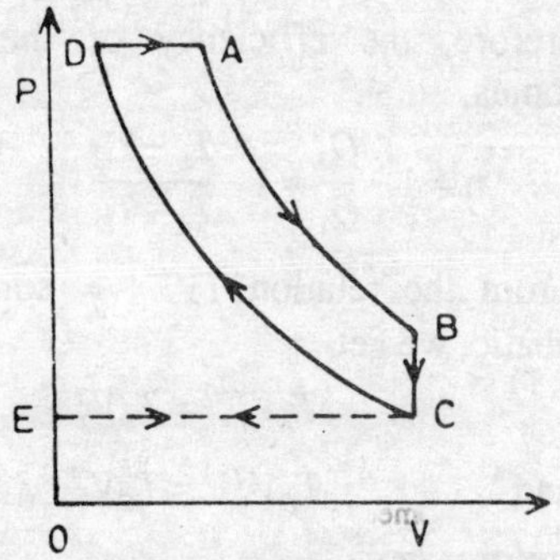

Fig. 14.13 *Diesel cycle*

In the Diesel cycle no fuel is introduced during compression. An ideal Diesel cycle (all processes are quasistatic) is shown in Fig. 14.13. First, pure air is sucked isobarically (path *EC*) and compressed adiabatically to about 1/20th of its volume (path *CD*). A valve is now opened and fuel injected into the cylinder under pressure. The fuel catches fire instantaneously as the temperature in the cylinder is already about 1000°C, above the ignition point of the fuel. Pressure during the combustion is kept constant as piston moves out (path *DA*) by regulating the supply of fuel. Thus here the *heat is absorbed at constant pressure*. The supply of fuel is cut-off after the temperature reaches a certain maximum value (point *A*). The piston is now allowed to move forward adiabatically upto *B*. At *B* a valve is opened and the pressure drops to *C*. *CE* is the exhaust stroke.

The efficiency of this cycle will clearly be,

$$\eta = 1 - \frac{C_V(T_B - T_C)}{C_P(T_A - T_D)} \qquad (14.35)$$

The efficiency of the idealized Diesel cycle is about 65-70%. However, the efficiency of any actual engine is much less than this. The Diesel engine consumes less fuel than the Otto engine, but as it has to withstand a much higher pressure it must be more robust.

Discussion and Problems

14.1 Compare the terms *heat* and *internal energy*. Distinguish between *temperature* and *heat*.

14.2 If we consider heat as a form of energy stored in a system, just like potential energy, will such an interpretation be against, or in favour of, the concept of heat as energy in transfer due to temperature difference?

14.3 When a gas is compressed it gets heated up. Has there been a transfer of heat?

14.4 When we rub on a piece of ice, it melts without change of temperature. Does this process involve transfer of heat? If so, does it violate the concept of heat as energy in the process of transfer because of temperature difference?

14.5 The internal energy of a body can be increased in two ways—by performing work on it or by heating it. Can you distinguish whether the internal energy has been acquired by doing work, or by transfer of heat?

14.6 On how many thermodynamic coordinates does the internal energy of a system, say a gas, depend? Two objects have the same temperature. Do they necessarily have the same internal energy?

14.7 In Fig. 14.14, vessel *A* of volume V_A contains air at a high pressure. Another vessel *B* is connected to it through a capillary and a stop-cock. *P* is a piston, and the other end of *B* is open to atmosphere. To start with, *P* is at the extreme left. Now, the stop-cock is opened slowly so that the pressure of the gas in *B* which pushes the piston out is only slightly greater than the atmospheric pressure. After the stop-cock is opened completely and the equilibrium is reached, volume V_B of the vessel *B* gets filled. (*i*) What is the work done? Assume the piston to be frictionless and leakproof. (*ii*) Is this process quasistatic? (*iii*) How much work will be done if air is allowed to leak directly into the atmosphere?

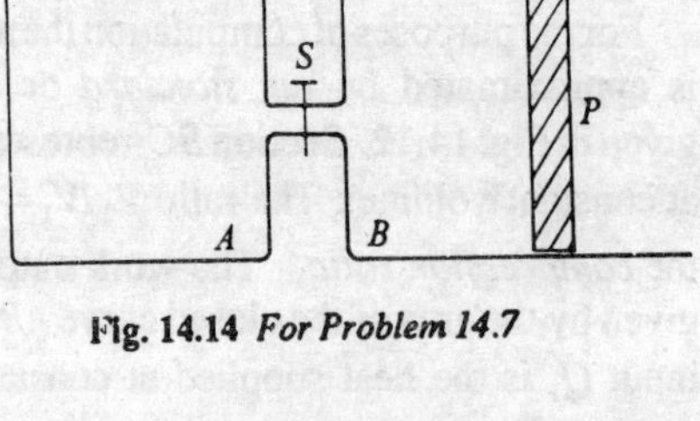

Fig. 14.14 *For Problem 14.7*

14.8 A vessel, having adiabatic walls, consists of two compartments—one full of a gas and the other having vacuum. The partition is suddenly broken. What is the work done during any infinitesimal portion of this process ? This process is called free expansion. Does internal energy of the gas change?

14.9 A length of a wire, in which there is a tension τ, is changed quasi-statically from length L to $L + \Delta L$. What is the amount of work done? Take care of the sign. If the process is adiabatic, what is the change in the internal energy of the wire?

14.10 Let us have a soap film as shown in Fig. 14.15. The film is pulled out with the help of wire W. If the wire moves by a distance Δx, what is the work done? Let the coefficient of surface tension of the film be σ N m^{-1}. Assume that the wire has been pulled quasi-statically and isothermally.

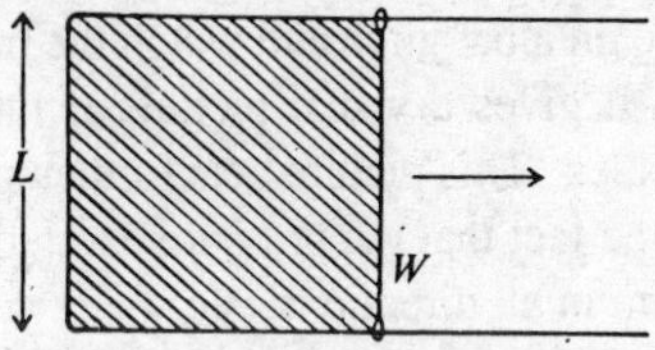

Fig. 14.15 *For Problem 14.10*

14.11 A trunk of mass 100 kg is pulled along a horizontal surface for 30 m. If the coefficient of sliding friction is 0.30, how much thermal energy is produced?

14.12 A 3000-lb automobile, travelling at 60 mph on a level road, is stopped by applying brakes. How much is the thermal energy produced?

14.13 A copper block slides down an incline 10 m long. Due to friction its velocity remains constant. The angle of inclination with the horizontal plane is 45°. If the block absorbs 80 per cent of the thermal energy produced, determine its temperature rise. The specific heat of copper is 0.094 kcal/kg-°C.

14.14 A 20 g lead bullet, having a horizontal velocity of 500 m/s, strikes an object of mass 480 g and gets studded into it. The target is lying on a frictionless horizontal table. (*i*) What is the amount of thermal energy produced? (*ii*) If 30 per cent of this thermal energy goes into the bullet, calculate the rise in its temperature. The specific heat of lead is 0.031 cal/g-deg. (*iii*) Is this temperature rise due to the heat given to the bullet, or due the work done on it?

14.15 When a thermodynamic system is taken from state 1 to state 3 via path 1-2-3 in Fig. 14.16, 100 cal are given to the system, and 40 cal worth of work is done. (*i*) How much heat flows into the system if it is taken along the path 1-4-3, and if the work done in 14 cal. (*ii*) When the system is brought back to state 1 from state 3 along a straight line path 3-1, how much work is done? How much heat the system either absorbs or rejects? (*iii*) Take internal energy at state 1, U_1, to be zero. What is U_3? (*iv*) If $U_4 = 10$ cal, how much is the heat involved in the process 1-4? And in the process 4-3?

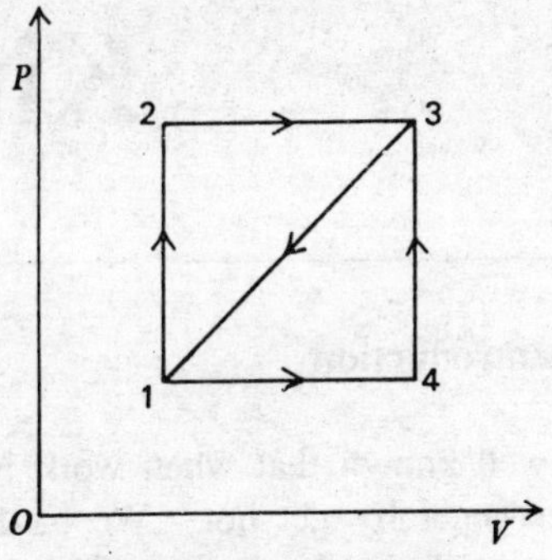

Fig. 14.16 *For Problem 14.15*

14.16 A thermometer of mass 10 g is used to measure the temperature of 100 g of ethyl alcohol. At steady state it reads 50°C. What was the temperature of the alcohol before the insertion of the thermometer? The initial reading of the thermometer is 20°C, and its average specific heat is 0.15 cal/g-°C. The specific heat of alcohol is 0.58 cal/g-°C.

14.17 In Joule's experiment to find a relationship between work done and heat produced, water was churned with a paddle wheel driven by two loads, each of mass 14 kg, and each falling vertically 2 m. Joule had 7 kg of water to be heated. After each churning he lifted his loads up and let them fall again. After twenty such falls, he found the water temperature to rise by 0.38°C. From this data establish a relationship between a calorie and a joule. Neglect heat capacity of the container.

14.18 Calculate the difference between two molar specific heats of hydrogen, being given that one litre of this gas weighs 0.0899 g at STP.

14.19 From Eqs. (14.3), (14.5) and (14.14),

$$\Delta Q = C_V \Delta T + P \Delta V$$

From Eq. (14.17),

$$P \Delta V + V \Delta P = R_0 \Delta T$$

(i) Show that,

$$\Delta Q = C_P \Delta T - V \Delta P$$

(ii) Assuming γ to stay constant during quasi-static adiabatic expansion of an ideal gas, show that,

$$PV^{\gamma} = \text{constant}$$

Chapter 15

Kinetic Theory of Gases

15.1 Introduction

It is well known that when work is expended, bodies generally get hot. We have studied in mechanics that in the absence of friction the effect of doing work on a body is to increase its velocity. Now, when we do work and it gets lost, it may be that the velocity of something gets increased. In other words this hotness may be due to the increased motion of some material particles which we cannot see. Laws of chemical combination support this atomistic or particle nature of matter. This atomistic (granular, particle-like, bundle-like, or discrete) nature is also exhibited by charge (Sec. 19.7), as well as by radiation (Sec. 17.9). Such a model for gases will be developed in this chapter. We have chosen gases because their behaviour is simpler than that of liquids and solids, and their properties are better known.

15.2 Model for a Gas

Whatever model we develop, it should be based on the experimental observations made on the gases. We describe the model in the form of the following postulates.

(*i*) A gas consists of small particles which we call molecules. This is based on the arguments described in Section 15.1.

(*ii*) If we release some gas in a chamber, the gas, after a short time, fills the whole space. From this one can say that particles move very fast and move in all directions in a random way. We neglect the force of gravity on the molecules, which means that the density of molecules is uniform and there is no preferred direction for any velocity. Throw a stone on a bee-hive and watch the motion of the bees. It gives a visual picture of motion of the molecules. This picture gets further substantiated from the fact that the pressure exerted by gases is uniform in all directions.

(*iii*) Now, when particles strike a curtain, the curtain gets stretched. If the gas particles strike a curtain or a wall at finite time intervals, one must observe fluctuations in the pressure, which we never do. Therefore, the bombarding particles should be very light and their number should be very large. Individual impulses will, then, be too small to be detected. Further, the number being too large, the bombardment will be almost uniform.

(*iv*) Gas molecules obey Newtonian mechanics. One may be surprised over this, and may feel that it is well established experimentally. But we have done it with the systems which we can see. The size of the molecules comes out to be extremely smaller than the objects which we have studied (about 10^{-8} times in linear dimensions). To assume the validity of Newtonian mechanics at atomic scale is nothing but extrapolation, which may or may not be correct. The extrapolation to very high velocities has failed.

When we make a model, it should be as simple as possible to start with. We study it and reach certain conclusions. If these results agree with the experiment well and good, otherwise we will improve upon the model. A simple model will atleast provide guidelines. It will tell us whether we are going in the right direction. So, we make the following simplifying assumptions along with the above postulates.

(*v*) We assume that there are no intermolecular forces. In other words, the molecules move in straight lines in between two collisions.

(*vi*) The collisions amongst the molecules and between the molecules and the walls are elastic.

(*vii*) The size of the molecules is extremely small compared to the size of the container in which they move. This assumption is plausible because we know that gases are very compressible. Further, if a gas is liquefied, the volume of the liquid is thousands of times smaller than that of the gas. It has two consequences : First, almost the whole of the volume of the container is available to the molecules for movement; second, the duration of the collision is negligible as compared with the duration of free movement. Thus, averaged over time, the kinetic energy converted into potential energy during an elastic collision is extremely small. In other words, the energy of the gas is wholly kinetic.

The gases obeying the last three conditions, along with the first four, are called *ideal gases* or *perfect gases*. Now our model is ready. To see whether it is good or bad we have to put it to test. The method is to see whether it fits into the experiment or not.

15.3 Pressure Exerted by a Perfect Gas

Let us take one mole of a perfect gas enclosed in a cubical vessel (Fig. 15.1) of dimensions *a*, *b*, and *c*, in the directions of *x*, *y*, and *z*, respectively. Consider a molecule having a velocity $\mathbf{v}_i$. Let the components of this velocity in the directions *x*, *y* and *z* be v_{ix}, v_{iy} and v_{iz}, respectively. We consider the motion of the molecule in the direction of *x*. When it strikes the wall *ABEF*, its momentum is mv_{ix} (*m* is the mass of each molecule of the gas). Now the collision is normal and elastic. Therefore, the molecule will, after collision, rebound with a velocity v_{ix}. The momentum of the molecule after collision is again mv_{ix} but its direction is reversed. Therefore, the change in momentum is $2mv_{ix}$ in the negative direction of *x*. According to the principle of conservation of momentum, this will be the momentum transferred to the wall in the positive direction of *x*. After this collision the molecule will strike the wall *OCDG* and will come back to strike* the wall *ABEF* again. This will apparently be after a time $2a/v_{ix}$. Hence it will strike this wall $v_{ix}/2a$ times per second and will thus impart it a momentum mv_{ix}^2/a, in one second. Thus total momentum transferred to the wall *ABEF* per second due to all molecules is $\Sigma mv_{ix}^2/a$, where the summation is over all the molecules. Since pressure is force per unit area, and force further equals the rate of change of momentum, the pressure exerted on this surface is

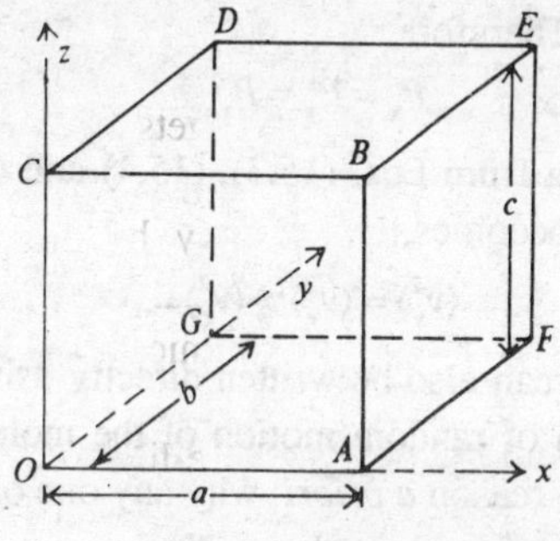

Fig. 15.1 *A container having a gas in it*

$$P_x = \frac{1}{bc}\Sigma\frac{mv_{ix}^2}{a}$$

$$= \frac{m}{abc}\Sigma v_{ix}^2 = \frac{m}{V}\Sigma v_{ix}^2 \qquad (15.1)$$

where *V* is the volume of the container. The same will hold good for the face *OCDG*.

Now, *N* (Avogadro's number) is the total number of molecules in the vessel. Thus by definition,

*Here we have not considered the collision of this molecule with other molecules during its flight from one wall to the other. But it will not affect our results. If it collides with any other molecule, it will interchange with it its *x*-component of velocity. This is because both the molecules are point particles without any intermolecular force, have equal masses and the collision is elastic. As no time is lost in this collision, the second molecule will cover the rest of the distance in the same time.

$$\langle v_x^2 \rangle \equiv \frac{\Sigma v_{ix}^2}{N} \tag{15.2}$$

where $\langle v_x^2 \rangle$ represents the average over all the molecules. Thus,

$$P_x = \frac{mN}{V} \langle v_x^2 \rangle \tag{15.3}$$

Similarly for y- and z-directions

$$P_y = \frac{mN}{V} \langle v_y^2 \rangle, \tag{15.4}$$

and
$$P_z = \frac{mN}{V} \langle v_z^2 \rangle \tag{15.5}$$

Now, it is an experimental fact that the pressure exerted by a gas on all the sides of a container is the same. Therefore,

$$P_x = P_y = P_z \tag{15.6}$$

Substituting from Eqs. (15.3), (15.4) and (15.5), Eq. (15.6) becomes,

$$\langle v_x^2 \rangle = \langle v_y^2 \rangle = \langle v_z^2 \rangle \tag{15.7}$$

This result can also be written directly using our assumption of random motion of the molecules. There is no reason *a priori* why any one of these should be preferred to others. Now,

$$\Sigma v_i^2 = \Sigma v_{ix}^2 + \Sigma v_{iy}^2 + \Sigma v_{iz}^2 ,$$

or
$$\frac{\Sigma v_i^2}{N} = \frac{\Sigma v_{ix}^2}{N} + \frac{\Sigma v_{iy}^2}{N} + \frac{\Sigma v_{iz}^2}{N},$$

or
$$\langle v^2 \rangle = \langle v_x^2 \rangle + \langle v_y^2 \rangle + \langle v_z^2 \rangle \tag{15.8}$$

Therefore, from Eqs. (15.7) and (15.8),

$$\langle v_x^2 \rangle = \langle v_y^2 \rangle = \langle v_z^2 \rangle = \frac{1}{3} \langle v^2 \rangle \tag{15.9}$$

Hence, the pressure exerted by a gas is,

$$P = P_x = P_y = P_z = \frac{mN}{3V} \langle v^2 \rangle \tag{15.10}$$

Putting $(mN/V) = \rho$, the density of the gas,

$$P = \frac{1}{3} \frac{mN}{V} \langle v^2 \rangle = \frac{1}{3} \rho \langle v^2 \rangle \tag{15.11}$$

Example 15.1 The density of nitrogen at 0°C and 76 cm-Hg pressure is 0.00125 g/cm^3. Find the root mean square velocity of the molecules.

By definition, root mean square (rms) velocity is:

$$v_{rms} \equiv \sqrt{\langle v^2 \rangle}$$

Thus, in this problem

$$v_{rms} = \sqrt{\frac{3 \times 76 \times 981 \times 13.6}{0.00125}}$$

$$= 4.93 \times 10^4 \text{ cm/s}$$

Now, the kinetic energy of one molecule is $mv^2/2$. Thus the average kinetic energy per molecule would be,

$$\frac{\Sigma \frac{1}{2} m v_i^2}{N} = \frac{1}{2} m \langle v^2 \rangle$$

and total kinetic energy,

$$U = \frac{1}{2} mN \langle v^2 \rangle \tag{15.12}$$

$$\therefore \quad P = \frac{2}{3}(U/V) = \frac{2}{3} u \tag{15.13}$$

where u is the energy density.

Thus, Eq. (15.13) tells us that pressure is related to the kinetic energy of the molecules. But one thing should be remembered very carefully. This kinetic energy is the kinetic energy associated with the centres of mass of the molecules. If a molecule is rotating or vibrating, the kinetic energy associated with these motions is not to be considered. It should be clear from the derivation of the expression for pressure. There, v was the velocity of the molecule as a whole.

15.4 Deduction of Gas Laws

Let us first see microscopically the mechanism of heat transfer when two systems at unequal temperatures are brought into thermal contact, as in Fig. 15.2. Let the system A be at a higher temperature than the system B. We know that heat goes from A to B. When molecules of A strike the wall W, they make its molecules vibrate faster and these molecules in turn make their neighbours do so. This way, these vibrations reach the other side of the wall. When molecules of B strike the wall, the net result is the transfer of energy to the molecules of B. From this we can conclude that kinetic energy of the molecules of A (at higher temperature) is greater than that of B(at lower temperature). When A and B attain the same common temperature, there is no further transfer of heat. We can say that, on the average, molecules of A have the same kinetic energy as the molecules of B.

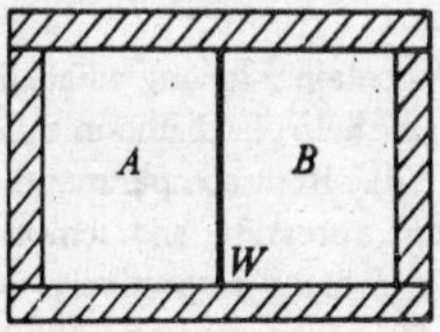

Fig. 15.2 *Two systems at different temperatures brought into thermal contact*

Avogadro's law: If we have two gases A and B at the same temperature, from the above discussion we can say that,

$$\frac{1}{2}m_A\langle v_A^2\rangle = \frac{1}{2}m_B\langle v_B^2\rangle \qquad (15.14)$$

Further, if both gases are at the same pressure,

$$\frac{1}{3}\frac{m_A N_A}{V_A}\langle v_A^2\rangle = \frac{1}{3}\frac{m_B N_B}{V_B}\langle v_B^2\rangle \qquad (15.15)$$

From Eqs. (15.14) and (15.15), if $V_A = V_B$,

$$N_A = N_B \qquad (15.16)$$

In other words, if two gases have the same pressure and temperature, then equal volumes contain equal number of molecules. This is the Avogadro's law.

Boyle's law: Let us rewrite Eq. (15.11) as

$$PV = \frac{1}{3}mN\langle v^2\rangle \qquad (15.17)$$

In this equation we see that for a given amount of gas the right-hand side will remain constant if $\langle v^2\rangle$ remains constant. This holds true provided temperature remains constant as shown above. This is the Boyle's law.

15.5 Kinetic Theory Interpretation of Temperature

From Eq. (15.13) we get $PV = \frac{2}{3}U$. Experiment tells us that

$$PV = R_0 T \qquad (15.18)$$

Thus,
$$T = \frac{2}{3}\frac{1}{R_0}U \qquad (15.19)$$

Example 15.2 Calculate the rms speed of a nitrogen molecule at 0°C.

From the above we see that the translational kinetic energy associated with one mole of a gas is, $U = (3/2)R_0T$. Therefore, the average translational kinetic energy associated with one molecule is

$$\frac{3R_0}{2N}T = \frac{3}{2}kT,$$

or

$$\frac{1}{2}m\langle v^2\rangle = \frac{3}{2}kT$$

$$\therefore \qquad v_{rms} = \sqrt{\frac{3kT}{m}} = \sqrt{\frac{3R_0T}{M}}$$

where m is the mass of one molecule, and M the molecular weight of the gas. Therefore,

$$v_{rms} = \sqrt{\frac{3\times 8.314\ \text{J K}^{-1}\ \text{mole}^{-1}\times 273\ \text{K}}{28\ \text{g mole}^{-1}}}$$

$$= \sqrt{\frac{3\times 8.314\times 273\ \text{J}}{28/1000\ \text{kg}}}$$

$$= \sqrt{\frac{3\times 8.314\times 10^3\times 273\ \text{kg m}^2\ \text{s}^{-2}}{28\ \text{kg}}}$$

$$= 493\ \text{m s}^{-1}$$

We thus see that molecular speeds are extremely high, ~10^3 km/hr. With such high speeds a molecule should cross an average sized room in about one-hundredth of a second. But if a scent is sprayed at one end of a room, it takes a few seconds to reach the other end. This is because of molecular collisions. The scent molecules collide with air molecules and get deflected from their straight line paths. Such a migration of gas particles from one place to another is called *diffusion*. On the average the transfer of the substance is from a region of higher concentration to a region of lower concentration. This motion is assisted by thermal motions of the molecules, and is impeded by molecular collisions. Equilibrium rate of transfer from higher to lower concentration depends upon the concentration gradient.

Eq. (15.19) tells us that temperature is proportional to the average kinetic energy (associated with the centres of mass) of the molecules. This kinetic energy is not the kinetic energy associated with the motion of the gas as a whole. If a container having a gas inside it starts moving, it will not increase the temperature of the gas. Thus, temperature is a measure of the thermal kinetic energy, i.e. the energy associated with the random motion of the centres of mass of the molecules. If we have only one molecule, the concept of temperature has no meaning. It comes into picture when we have a collection of a large number of molecules, that too in random motion. In other words, the concept of temperature is a statistical concept.

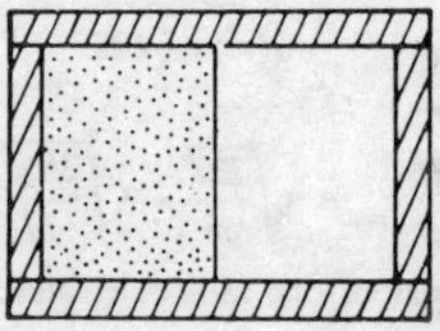

Fig. 15.3 *Free expansion of a perfect gas*

From Eq. (15.19) we see that, for perfect gases, U depends only on temperature. As long as temperature remains constant, U does not change with volume or pressure. Let there be a gas in half of the vessel having adiabatic walls (Fig. 15.3). The other half is vacuum. If we break the partition and let the gas occupy the whole volume, then U (and hence T) should remain unchanged because the gas is not doing any external work. In other words, in free expansion, temperature of a perfect gas does not change. This is Joule's law. Though free expansion is also an adiabatic process, the term 'adiabatic expansion' is reserved for a particular process in which the system expands against an external pressure. Thus the expanding gas does work on the surroundings.

Question: What happens to the temperature of a perfect gas when it undergoes an adiabatic expansion?

We can explain Joule's law physically also. Free expansion amounts to increasing the average distance between the molecules. As there are no forces between the molecules, (according to our assumption) no work has been done in pulling the molecules apart. Hence kinetic energy, and thus temperature, remains unchanged.

Note another point. We cannot conclude from Eq. (15.19) that at absolute zero $U = 0$, because all gases get liquefied much earlier than this, and our model stops working.

Example 15.3 A container having adiabatic walls is divided into two parts with the help of a diathermic partition, and is kept horizontal (Fig. 15.4). Both compartments are filled with an ideal gas. Volume, pressure and temperature in the left compartment are V, P and T respectively; while these parameters for the gas in the right compartment are $2V$, $2P$ and T. The partition is now let go, and it moves without friction.

(*i*) What are the relative amounts of the gas in the two compartments?

(*ii*) What are the temperatures on both sides?

(*iii*) What are the pressures on both sides?

(*iv*) What are the final volumes?

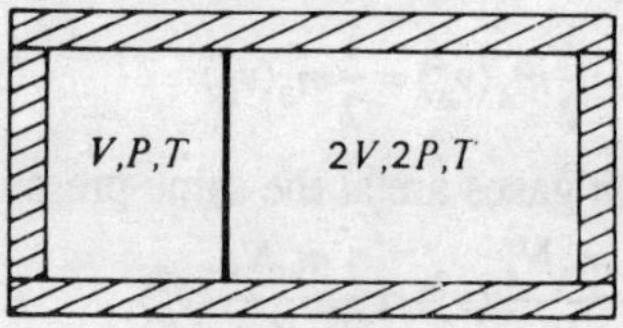

Fig. 15.4 *For Example 15.3*

(*i*) For the left compartment,

$$\frac{\text{Pressure} \times \text{Volume}}{\text{Temperature}} = \frac{PV}{T}$$

For the right compartment,

$$\frac{\text{Pressure} \times \text{Volume}}{\text{Temperature}} = \frac{4PV}{T}$$

Therefore, the amount of gas in the right-hand compartment is four times greater than that in the left compartment.

(*ii*) When the partition is let go, it will move to the left. Therefore, the gas on the left of it will temporarily get heated up, and that on the right will get cooled. After some time both temperatures will become equal.

If we take whole of the gas as one system, we have neither given any heat to it, nor any work has been done. Therefore, according to the first law of thermodynamics, its internal energy will remain unchanged. But we have seen above that, for an ideal gas, internal energy is a function of temperature only. Hence the final temperature will be the same as before, i.e. T K.

(*iii*) As the piston is frictionless, pressure in both the compartments will be the same.

(*iv*) Now the gases in the two compartments have the same pressure and temperature. Therefore the ratio of the volumes occupied by them is equal to the ratio of their masses. Thus,

$$V_{left} : V_{right} :: 1 : 4$$

But the total volume,

$$V_{left} + V_{right} = 3V$$

Therefore volume of the left compartment

$$V_{left} = 3V/5$$

and the volume of the right compartment

$$V_{right} = 12V/5$$

Question: What is the final pressure? What do you expect if the partition is a bad conductor?

15.6 Degrees of Freedom and Principle of Equipartition of Energy

We are considering here smooth and perfectly elastic molecules. Thus, these molecules are capable of undergoing translational motions only. The average energy of one such molecule is

$$\langle E\rangle = \frac{1}{2}m\langle v^2\rangle$$

$$= \frac{1}{2}m\langle v_x^2\rangle + \frac{1}{2}m\langle v_y^2\rangle + \frac{1}{2}m\langle v_z^2\rangle \qquad (15.20)$$

Thus in the energy expression we have three squared terms and, hence, the molecule has three degrees of freedom. The number of independent squared terms in the energy expression is the number of degrees of freedom.

Now, according to Eq. (15.19), this energy is equal to $(3/2)kT$. We have already shown that mean square velocities in the three directions are equal. Thus, on the average, this energy is equally shared by the three degrees of freedom, and energy associated with each degree of freedom is $kT/2$. In a general case it can be proved by classical statistical mechanics that the average energy associated with any squared term in the energy expression of the molecule is $kT/2$. This is the principle of *equipartition of energy*.

If we take a diatomic molecule (Fig. 15.5), which is rigid, i.e. which cannot vibrate, the energy of the molecule is,*

$$E = \left(\frac{1}{2}mv_x^2 + \frac{1}{2}mv_y^2 + \frac{1}{2}mv_z^2\right)$$

$$+\left(\frac{1}{2}I_y\omega_y^2 + \frac{1}{2}I_z\omega_z^2\right) \qquad (15.21)$$

where m is the total mass, and I_y and I_z are the moments of inertia about y- and z-axes, respectively.

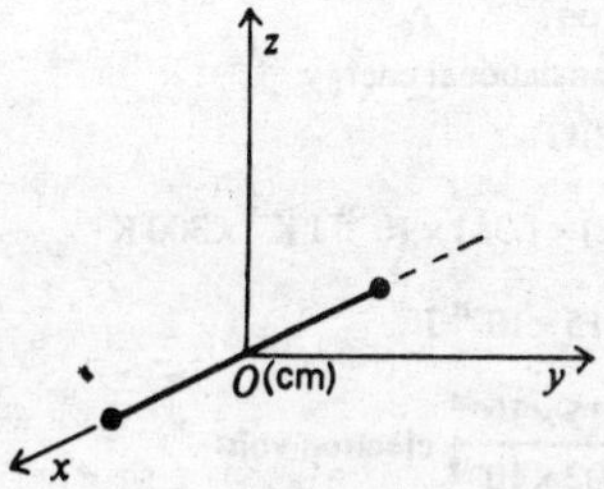

Fig. 15.5 *A rigid rotator*

We have not considered the energy term associated with the rotation of the molecule about x-axis because the moment of inertia about this axis (and hence the kinetic energy associated with it) is negligible. Thus, this molecule has five squared terms in the energy expression and hence five degrees of freedom. Therefore, the energy associated with this molecule is $(5/2)kT$.

If this molecule is non-rigid, it can also oscillate along the x-axis. The energy of such a molecule will be

$$E = \left(\frac{1}{2}mv_x^2 + \frac{1}{2}mv_y^2 + \frac{1}{2}mv_z^2\right)$$

$$+\left(\frac{1}{2}I_y\omega_y^2 + \frac{1}{2}I_z\omega_z^2\right) + \left(\frac{1}{2}\mu\dot{\xi}^2 + \frac{1}{2}K\xi^2\right) \quad (15.22)$$

Here all symbols are as explained above and in addition μ is the reduced mass, K is the restoring force per unit displacement, ξ is the sum of the displacements of the atoms from their equilibrium

***Eq. (15.21) is based on the fact that total kinetic energy of a system is equal to the sum of the kinetic energy of the centre of mass (called orbital kinetic energy) and the kinetic energy of the constituent particles with respect to the centre of mass (called internal kinetic energy). The internal kinetic energy is the kinetic energy observed in the C-frame.**

positions, and ξ is its time derivative. The term $(1/2)K\xi^2$ represents the internal potential energy (the potential energy due to the separation between the atoms). Therefore, such molecules have seven degrees of freedom and hence the energy associated with each molecule is $(7/2)kT$.

Example 15.4 Calculate average translational energy of a mercury atom at 27°C. Is it enough to ionize a mercury atom during a collision?

Average translational energy

$$= (3/2)kT$$

$$= (3/2) \times 1.381 \times 10^{-23} \text{ J K}^{-1} \times 300 \text{ K}$$

$$= 6.215 \times 10^{-21} \text{ J}$$

$$= \frac{6.215 \times 10^{-21}}{1.602 \times 10^{-19}} \text{ electron volt}$$

$$= 0.0388 \text{ eV}$$

A mercury atom cannot be ionized because the energy required for this purpose is 10.4 eV. These atoms cannot even be excited because the energy required is 4.9 eV.

Note: See Section 19.6 for the energy unit eV; and see Chapter 31 for what we mean by excitation and ionization of an atom.

15.7 Specific Heats of Gases

Monatomic gas: Each molecule of this gas has three degrees of freedom. Thus, if we take one mole of this gas,

$$U = \langle E \rangle N = (3/2)kTN = (3/2)R_0T \quad (15.23)$$

We know that,

$$C_V = \Delta U/\Delta T$$

= increase in internal energy per unit increase in temperature

$$\therefore \quad C_V = (3/2)R_0 \quad (15.24)$$

Hence, $C_P = C_V + R_0 = (5/2)R_0$ (15.25)

Ratio of these two specific heats is,

$$C_P/C_V = \gamma = 5/3 = 1.67 \quad (15.26)$$

Gas made of rigid diatomic molecules:

$$U = (5/2)R_0T$$

$$C_V = (5/2)R_0 \quad (15.27)$$

$$C_P = (7/2)R_0, \quad (15.28)$$

and $\gamma = 7/5 = 1.40$ (15.29)

Gas made of non-rigid diatomic molecules:

$$U = (7/2)R_0T$$

$$C_V = (7/2)R_0 \quad (15.30)$$

$$C_P = (9/2)R_0, \quad (15.31)$$

and $\gamma = (9/7) = 1.29$ (15.32)

Let us compare these values with the experimental results. From experiments, both on argon and helium, it has been found that

$$C_V = 2.98 \text{ cal/mole-°C}$$

$$C_P = 4.97 \text{ cal/mole-°C},$$

and $\gamma = 1.67$

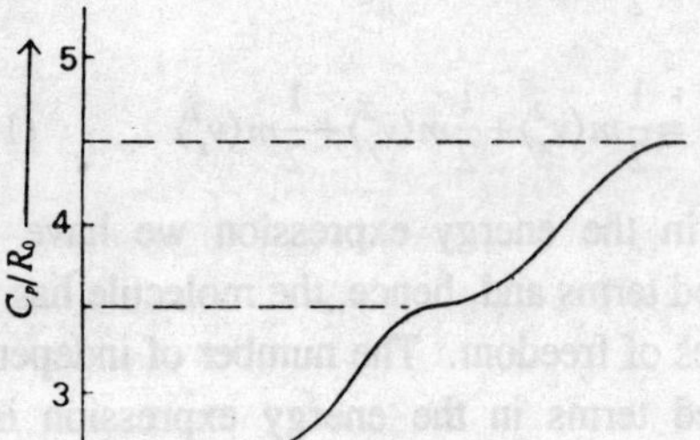

Fig. 15.6 *Variation of specific heat of hydrogen with temperature*

From theory:

$$C_V = 3/2(1.987)$$

$$= 2.98 \text{ cal/mole-°C}$$

$$C_P = C_V + R_0$$

$$= 4.97 \text{ cal/mole-°C}$$

$$\gamma = 1.67$$

Thus, our model adequately explains these specific heats.

But, this model fails in the case of polyatomic molecules. Let us take hydrogen. Experimentally measured specific heat as a function of temperature has been shown in Fig. 15.6. The theory agrees with the experiment only in a limited temperature range (250 − 750 K). At low temperature, C_P approaches $(5/2)R_0$ and at high temperatures it approaches $(9/2)R_0$.

To fit our theory into experiment the only plausible explanation is that at very low temperatures only translational modes absorb energy. As the temperature is raised, the rotational modes also get excited but the molecule remains rigid. As the temperature is further increased, the molecule becomes non-rigid and the energy is accepted by vibrational modes also. But there is one fundamental question. Why so? The classical theory based on Newtonian mechanics has no answer to this question. According to this, all energy should be equally shared by all degrees of freedom. Thus, our theory says that the specific heat should not change with temperature.

Now we have come to a point where our model is failing badly. May be that Newtonian mechanics, as extrapolated to the atomic dimensions, does not hold good in the case of these very small dimensions. Quantum mechanics, to which you will be introduced in Volume Two, explains these experimental results on hydrogen very well.

Example 15.5 We will see in Chapter 31 that the internal energy of an atom is quantized. The first allowed excited state of a hydrogen atom is 10.2 eV above its lowest energy (ground) state. To what temperature should hydrogen gas be raised so that inelastic collisions may excite an appreciable number of atoms to their first excited state?

Let the required temperature be T K. Therefore:

$$(3/2)kT = 10.2 \times 1.6021 \times 10^{-19}\ \text{J},$$

or
$$T = \frac{10.2 \times 1.602 \times 10^{-19}\ \text{J}}{1.381 \times 10^{-23}\ \text{J K}^{-1}} \times \frac{2}{3}$$

$$= 7.88 \times 10^4\ \text{K}$$

We will see ahead that some of the atoms (or molecules) may have energies much larger than the average energy. Therefore, a few atoms may be excited even at somewhat lower temperatures.

15.8 Brownian Motion

We have postulated that matter is atomistic in nature and that these atoms are continuously in thermal motion. The discovery of Robert Brown, an English botanist, in 1827 that pollen granules suspended in water show a continuous random motion when viewed under a high power microscope, is the direct experimental evidence in support of this view. This phenomenon can also be observed if we suspend other small particles in a liquid, such as lamp black (or some dye) in water or a colloidal solution.

It is a fact that Brownian motion can be seen only with smaller particles of size of the order of 10^{-4} cm. Let us take a particle suspended in a fluid. Fluid molecules will strike it from all directions. If the particle is very big, then a very large number of molecules will strike it. Thus all these impulses will balance one another. Even if there is a small unbalanced force, the particle will not move much due to inertia. On the other hand, if the particle is very small, a smaller number of molecules will strike it. Under such circumstances, the unbalanced forces (Fig. 15.7) will be quite dominant and will be able to make this light particle dance to their tune. As the molecular motions are quite random, the motion of the particle will also be random (Fig. 15.7). This Brownian motion goes on for ever. It shows that molecular motions are perpetual motions.

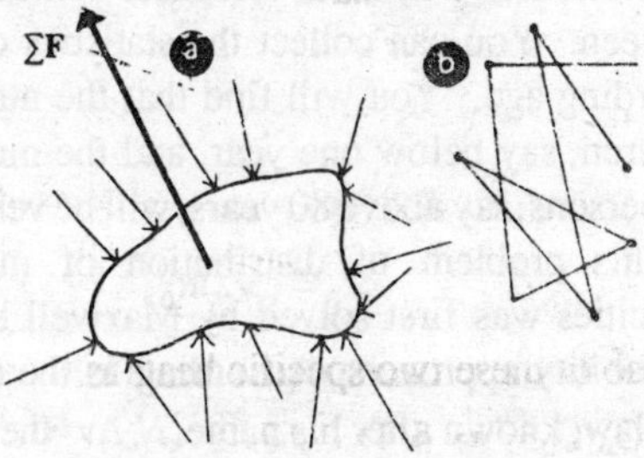

Fig. 15.7 *Brownian motion*

15.9 Distribution of Molecular Velocities

It has been postulated that molecules are in random motion. Further, some theoretical formulae have been derived which agree with the experiment. In Brownian motion, there is a direct evidence of these velocities. Now the question arises whether the molecules are moving with the same speed or their speeds can be anywhere between zero and infinity. One thing we can easily say is that the velocities cannot be the same. The intuitive argument in its favour is the fact that such orderliness is very rarely found in nature. It might be your experience also that one has to put a lot of restrictions to create order. The order of very high degree is found in atoms—the atoms of an element are all alike (isotopes, of course, are there). But it happens under very strict conditions about which you will read in Volume Two. Here, there is no restriction on the velocities of the molecules other than this that they are bound by the walls of the container. Even if we are able, at some moment, to make the molecular velocities equal, this situation will not last long. Due to collisions between the molecules and the walls, the velocities will change. After some time, some molecules will be moving very slowly, while the others will be moving very fast.

Now, the problem before us is to find out the number of molecules between a particular velocity interval. We cannot follow each molecule as their number is very large, of the order of 10^{24}. Therefore, we are bound to follow a statistical approach. One thing we can say quite confidently. The number of molecules with very low velocities and very high velocities will be very small. Most of the molecules will have velocities somewhere in between. You can collect the statistics of a city regarding age. You will find that the number of children, say below one year, and the number of old persons, say above 80 years, will be very small.

This problem of distribution of molecular velocities was first solved by Maxwell by using probability approach. According to the distribution law, known after his name, $N_v \Delta v$ the number of molecules between the speeds v and $v + \Delta v$, is*,

$$N_v \Delta v = \frac{4N}{\sqrt{\pi}}\left(\frac{m}{2kT}\right)^{3/2} v^2 e^{-mv^2/2kT} \Delta v \qquad (15.33)$$

where N is the total number of molecules, and m the mass of one molecule.

One thing should be explained here. It is very essential to take a velocity interval. We may not find any molecule having an exactly specified velocity. If one is told to find out a man whose height is exactly five feet and a half, he may find it difficult to get one. Some will be a little taller while the others will be a little shorter. But if the problem is to find out persons having heights between five feet five inches and five feet six inches, one may find many.

This Maxwellian distribution of molecular speeds is shown in Fig. 15.8. The speed which the maximum number of molecules possess, is known as the most *probable speed* shown as v_m in Fig. 15.8. In this figure are also shown the average, and the root mean square velocities.

Example 15.6 Calculate the kinetic energy of translation of a molecule of nitrogen gas at STP moving with the most probable speed.

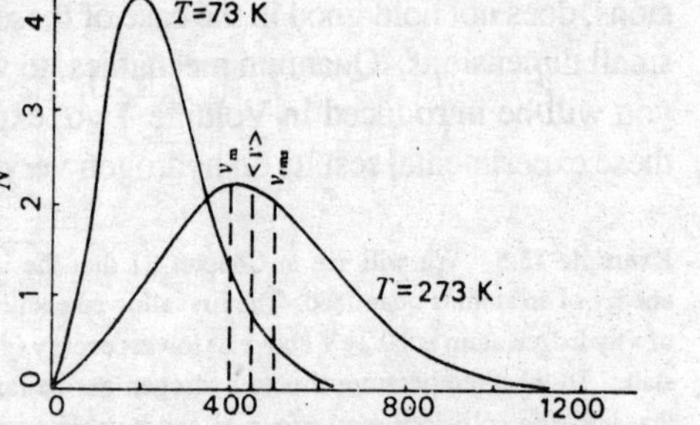

Fig. 15.8 ***Maxwellian distribution of molecular speeds at two temperatures***

*F. Reif, Statistical Physics: ***Berkeley Physics Course***—Vol. 5 (McGraw-Hill Book Co.), 1967.

First we give, without proof, the relations between the root mean square velocity, v_{rms}, the most probable velocity, v_m, and the average velocity $\langle v \rangle$. These are:

$$v_m = \sqrt{2/3}\, v_{rms} = 0.817\, v_{rms}, \qquad (15.34)$$

and

$$\langle v \rangle = \sqrt{8/3\pi}\, v_{rms} = 0.921 v_{rms} \qquad (15.35)$$

Now, from Eq. (15.19), if we have one mole of nitrogen,

$$U = \frac{1}{2}M\langle v^2 \rangle = \frac{3}{2}R_0 T = \frac{3}{2}PV,$$

or

$$\langle v^2 \rangle = \frac{3PV}{M},$$

or

$$v_{rms} = \sqrt{\frac{3PV}{M}}$$

$$= \sqrt{\frac{3 \times 1.013 \times 10^5 \text{ Nm}^{-2} \times 22.4 \times 10^{-3} \text{ m}^3 \text{ mole}^{-1}}{28 \times 10^{-3} \text{ kg mole}^{-1}}}$$

$$= 493 \text{ m s}^{-1}$$

Hence, $v_m = 0.817 \times 493 = 403 \text{ m s}^{-1}$

Mass of one nitrogen molecule,

$$m = \frac{28}{6.02 \times 10^{23}} \text{ g} = 4.65 \times 10^{-26} \text{ kg}$$

Kinetic energy associated with a nitrogen molecule moving with most probable speed is

$$E_m = \frac{1}{2} \times 4.65 \times 10^{-26} \text{ kg} \times (403 \text{ m s}^{-1})^2 = 3.78 \times 10^{-21} \text{ J}$$

$$= \frac{3.78 \times 10^{-21}}{1.60 \times 10^{-19}} \text{ electron volt} = 0.0236 \text{ eV}$$

E_m for a molecule of any gas at STP will be the same, i.e. 0.024 eV. Prove this.

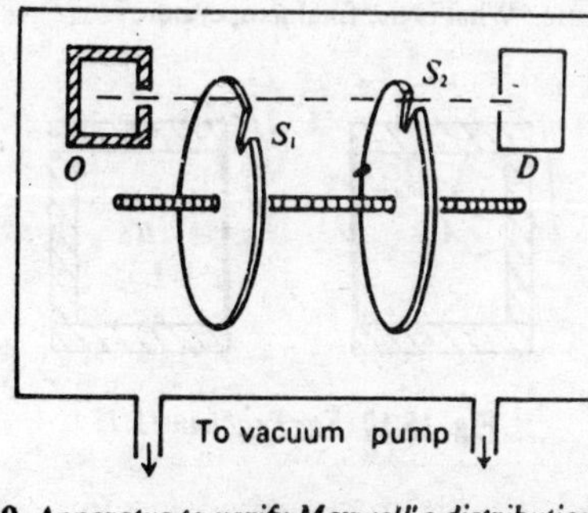

Fig. 15.9 *Apparatus to verify Maxwell's distribution formula*

A direct experimental evidence for Maxwellian distribution can be had by the apparatus shown in Fig. 15.9. This apparatus consists of an oven, a pair of sectored discs having slits S_1 and S_2 mounted on a common shaft, and a detector D. This apparatus is kept in vacuum. Mercury in the oven evaporates and the vapours pass to D through slits S_1 and S_2. When the shaft is stationary, no vapour will reach D because slit S_2 is a few degrees behind slit S_1. When the discs are set in rotation, only those molecules can pass which reach from S_1 to S_2 in a time equal to the time taken by S_2 in coming to the position of S_1 which it had when the molecules left it. Thus, by rotating the shaft with various angular velocities, the molecules of various velocities can be sorted out. The number of these molecules is counted by the detector D. The validity of the Maxwellian distribution has been well confirmed by such experiments.

15.10 Kinetic Temperature

We have seen that if our model of gases is valid, then the gas temperature is proportional to the average, random, translational kinetic energy of the molecules. By extending this concept of temperature from a statistical property associated with the random motion of all the atoms or molecules, of a piece of matter to a description of individual particle energies, we get a new parameter, called the *kinetic temperature*. The kinetic temperature of a particle is that temperature at which the energy of the particles of a perfect gas moving with most probable speed is equal to the energy of that particle.

We have seen in Example (15.6) that the kinetic energy of a particle moving with the most probable speed in a Maxwellian distribution for a collection of particles at 273 K(0°C) is 0.024 eV. Any isolated particle having this much kinetic energy, i.e. 0.024 eV, is said to have a kinetic temperature of 273 K.

Example 15.7 What is the kinetic temperature of an electron having energy 1 keV?

Let this temperature be T K. The kinetic energy of the electron moving with rms speed in an electron gas (assuming it to be perfect) at T K is

$$\frac{1}{2} m_e v_{rms}^2 = \frac{3}{2} kT$$

Therefore, the kinetic energy of an electron moving with most probable speed is

$$E_m = \frac{1}{2} m_e v_m^2$$

$$= \frac{1}{2} m_e \left(\sqrt{\frac{2}{3}} \; v_{rms} \right)^2$$

$$= \frac{2}{3} \times \frac{1}{2} m_e v_{rms}^2 = \frac{2}{3} \times \frac{3}{2} kT,$$

$$\text{or} \quad E_m = kT \qquad (15.36)$$

$$\therefore \quad kT = 1 \text{ keV},$$

$$\text{or} \quad T = \frac{1 \times 10^3 \times 1.60 \times 10^{-19} \text{ J}}{1.38 \times 10^{-23} \text{ J K}^{-1}}$$

$$= 1.16 \times 10^7 \text{ K}$$

$$= 11.6 \text{ million K}$$

Thus, the kinetic temperature of a particle of one electron volt kinetic energy is 1.16×10^4 K.

Discussion and Problems

15.1 Explain why a gas gets cooled (heated) on sudden expansion (compression)?

15.2 If heat is merely a molecular motion, what is the difference between a rapidly moving cold ball and a stationary hot ball?

15.3 Explain why and how friction between moving objects produces heat? Why do we not talk about friction at a molecular level?

15.4 Work and heat are both measures of energy transfer from one system to another. What is the difference between the two?

15.5 What will be the effect if the gas molecules undergo inelastic collisions with the walls of the container? Will they get heated up? Or, will they cool down? What will happen to pressure?

15.6 A rocket is fitted with insulated containers full of a very hot gas. This hot gas, rushing out of a nozzle at the back, drives the rocket forward. Where do the momentum and the energy of the rocket come from? Will the temperature of the ejected gas be greater, equal to, or less than the temperature of the gas inside the rocket? Explain your answer.

15.7 The highest vacuum attained so far in routine is of the order of 10^{-11} mm of mercury. How many molecules are there in one cc of a vessel having such a high vacuum at 0°C?

15.8 To have an idea of how large is the number of atoms or molecules involved in a particular macroscopic process, let us take one mole of a gas. We know that the number of molecules in this amount is 6.022×10^{23}. There are roughly 5 billion people on the earth. How much time will it take for all these people to count these molecules? Assume that the counting is done continuously, one person counting two molecules in one second?

15.9 A gun fires bullets at the rate of 5 rounds per second. The mass of each bullet is 10 g and it moves with a velocity of 450 m/s. These bullets hit a wall where they get stuck up. Calculate the force on the wall.

15.10 Compare the rms speed of a hydrogen molecule with that of a mercury atom at any common temperature. Their mass ratio is : $M_{H_2}/M_{Hg} = 2.016/200.6$

15.11 One mole of oxygen occupies a volume of 24 litres at 20°C and at 1 atm pressure. What is the rms speed of an oxygen molecule? What are the total translational kinetic energy and the total momentum of the molecules?

15.12 Gravitational field at moon's surface is 0.17 times that on the earth. What is the escape velocity? Mean radius of the moon is 1.738×10^3 km. At day time, the temperature at moon's surface goes to about 300°C. What is the rms velocity of an oxygen molecule at this temperature? What is that for xenon? Now, do you expect some atmosphere at the moon? If so, what can be the consituents?

15.13 The specific heat of argon at constant volume is 0.075 kcal/kg-deg. Calculate the mass of an argon atom and the atomic weight of argon.

15.14 15.00 grams of argon are heated from 37°C to 47°C at a constant pressure of 1 atm. How much heat has been transferred to the gas? Out of it, how much amount has been used up in raising the internal energy of argon?

15.15 Let us have two boxes A and B as shown in Fig. 15.10. Box A contains one mole of argon at 90°C, and box B contains one mole of helium at 30°C. The two boxes are joined through diathermic walls, all other walls being adiabatic. What is the final temperature?

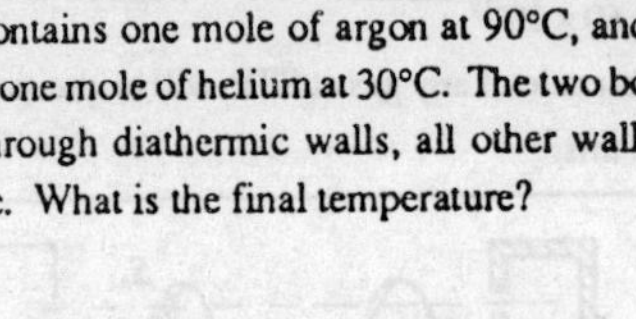

Fig. 15.10 *For Problem 15.15*

If, just after joining, the partition is removed so that the gases get mixed up, what will be the final temperature?

If in box B, we have nitrogen instead of helium, what final temperature do you expect in the above two cases?

15.16 In a certain gas 2/5th of the energy of the molecules is associated with the rotation of the molecules, and the rest 3/5th is associated with the motion of their centres of mass. What is the average translational kinetic energy of one such molecule when the temperature is 27°C? How much energy must be supplied to one mole of this gas at constant volume to raise the temperature by 1°C?

15.17 There are two containers—one having xenon gas and the other having helium gas. Their thermodynamic coordinates are the same. If the two gases are made to communicate with each other, which gas will diffuse faster into the other?

15.18 What is the kinetic temperature of: (*i*) a proton of energy 1 MeV, and (*ii*) a ball of mass 250 g moving with a velocity of 10 m/s?

Chapter 16

Imperfect Gases

16.1 Introduction

The behaviour of perfect gases, i.e. the gases which obey the relation (taking one mole) $PV = R_0T$, has been investigated in the last chapter. If we study the isotherms of such gases the PV-graphs at various temperatures should be equilateral hyperbolas, asymptotic to P- and V-axes (Fig. 16.1). According to the equation $PV = R_0T$ gases should keep on reducing in volume with increase in pressure, and should never liquefy.

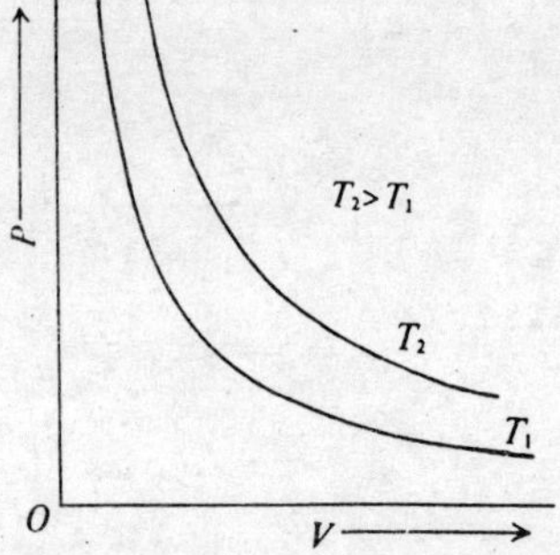

Fig. 16.1 *Isotherms of an ideal gas at two temperatures*

But if we study these isotherms experimentally over wide pressure ranges, as done by Andrews and Amagat separately, we will find that no gas is perfect. One such isotherm for a real gas (CO_2) at 13°C is shown in Fig. 16.2. Starting from the right, we see that as we increase pressure, volume decreases. On reaching point A, volume decreases without change of pressure. This indicates that liquefaction begins at A and continues upto B, where all the gas is liquefied. Beyond B there is negligible decrease in volume even with large increase in pressure, showing thereby that liquids can be compressed very little. As temperature increases, the points A and B approach each other, and at one temperature they coincide. Beyond this temperature, there is no horizontal portion on the curve. This temperature, beyond which a gas cannot be liquefied by merely increasing the pressure, is known as critical temperature. Two isotherms, one at critical temperature (31.4°C) and the other at a higher temperature (48°C), are also shown in Fig. 16.2.

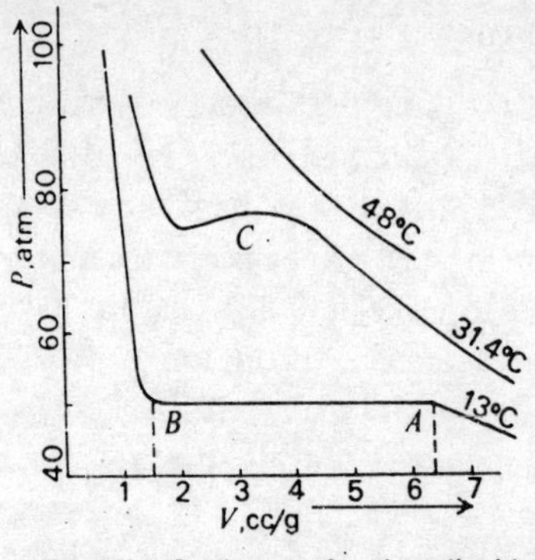

Fig. 16.2 *Isotherms of carbon dioxide*

16.2 van der Waals Equation of State

In the model developed in the last chapter it has been assumed that molecules have negligible size and there are no forces between them. Let us improve upon this model by taking into account the intermolecular forces. We will do it in a simple way as done by van der Waals. The effect of repulsive forces is that the molecules cannot come very near to each other. Thus, the molecules occupy space and we can say that they have finite size. This volume will reduce the volume available for the motion of the molecules. Thus, if b is the effective volume occupied by the molecules, the volume available for free movement will be $(V-b)$ instead of V, the volume we observe. But, van der Waals simplified it too much. He assumed the molecules to be hard spheres. It means that there are no forces of repulsion till the molecules come into contact. On contact the repulsive forces suddenly go to infinity. Actual molecules are much softer than indicated by this.

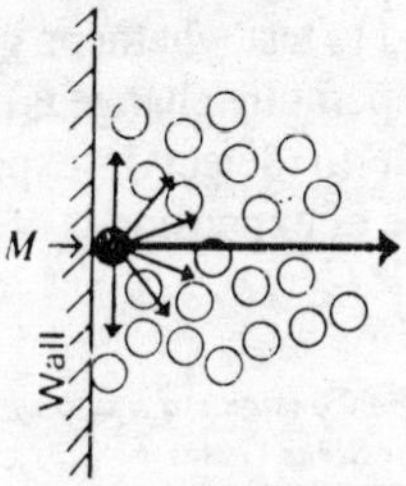

Fig. 16.3 *Pressure exerted by a van der Waals gas*

Let us now investigate the effect of attractive forces. We concentrate on a molecule M which is just striking the wall (Fig. 16.3). If we take the vector sum of the forces exerted on it by the other molecules, the resultant force F will be directed into the gas. It means that the molecule will not be able to strike the wall as hard as it would have done in the absence of the forces of attraction. These forces will affect the molecules only at the boundaries. Inside the gas, these forces will balance one another. Therefore, we can say that the pressure we measure is less than the pressure inside the gas. It is clear that the force on the molecule M is proportional to the number density n of the molecules. Further, the number of molecules striking the wall on unit area around M is also proportional to n. Thus, the net inward force per unit area will be proportional to n^2, or proportional to $1/V^2$ ($n = N/V$, where N is the total number of molecules). Hence, the pressure P in the perfect gas equation should be replaced by $(P + a/V^2)$, where a is a constant. The term a/V^2 is known as the *internal pressure of the gas.*

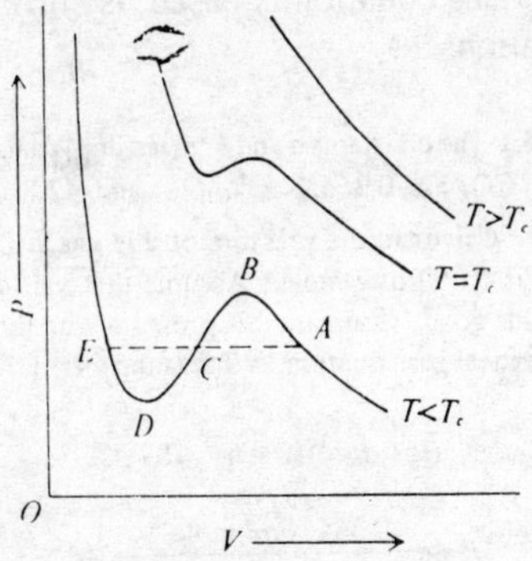

Fig. 16.4 *Isotherms of a van der Waals gas*

Now apply both these corrections to the perfect gas equation. We get

$$(P + a/V^2)(V - b) = R_0T \qquad (16.1)$$

This is the familiar van der Waals equation of state for actual gases. We must admit that the effect of intermolecular forces has been included in a very crude way. We have not at all bothered about the details of the molecular collisions in the presence of these forces. The values of a and b are to be found out by fitting Eq. (16.1) in the experimental isotherms.

Various isotherms as predicted by Eq. (16.1) are drawn in Fig. 16.4. Below critical temperature this theory predicts maxima and minima which are not observed experimentally. Portion AB represents super-saturated vapour and portion ED represents super-heated liquid, the states which are very difficult to get experimentally. BCD represents a self-collapsible state which again cannot be observed. To get the experimental curve from this theoretical curve we draw a line ACE parallel to the V-axis such that the area $ABCA$ equals area $CDEC$. This line ACE will represent the process of liquefaction of the gas.

It should be appreciated that Eq. (16.1) includes qualitatively all features of actual gases. It predicts liquefaction below critical temperature. This equation fits very well in the experiment for small deviations from the perfect gas equation. But over very wide pressure and temperature ranges it fails. The parameters a and b do not remain constant for a particular gas, but they depend on temperature. This is not surprising as the model on which the van der Waals equation is based, is still an over-simplification.

Example 16.1 The constants a and b in van der Waals equation of state for CO_2 are 0.364 N m^4/mole^2 and 42.8 cm^3/mole, respectively. Calculate the pressure of this gas at 0°C, when the volume is 0.560 litre/mole. Assume that van der Waals equation holds good. Compare this pressure with the pressure given by the ideal gas equation. What is the internal pressure?

$$(P + a/V^2)(V - b) = R_0T,$$

or
$$\left[P + \frac{0.364 \text{ N m}^4 \text{ mole}^{-2}}{(0.560 \times 10^{-3} \text{ m}^3 \text{ mole}^{-1})^2}\right] \times$$
$$[0.560 \times 10^{-3} \text{ m}^3 \text{ mole}^{-1} - 42.8 \times 10^{-6} \text{ m}^3 \text{ mole}^{-1}]$$
$$= 8.314 \text{ J K}^{-1} \text{ mole}^{-1} \times 273.15 \text{ K}$$

or $(P + 1.161 \times 10^6 \text{ N m}^{-2}) \times 517.2 \times 10^{-6} \text{ m}^3 \text{ mole}^{-1}$

$$= 2272 \text{ J mole}^{-1}$$

or $P + 1.161 \times 10^5 \text{ N m}^{-2} = 4.391 \times 10^6 \text{ N m}^{-2}$

$\therefore$ $P = 3.23 \times 10^6 \text{ N m}^{-2}$

$$= \frac{3.23 \times 10^6}{1.013 \times 10^5} = 31.9 \text{ atm}$$

If the gas would have been perfect

$$PV = R_0T$$

or
$$P = \frac{8.314 \text{ J K}^{-1} \text{ mole}^{-1} \times 273.15 \text{ K}}{0.560 \times 10^{-3} \text{ m}^3 \text{ mole}^{-2}}$$

$$= 4.05 \times 10^6 \text{ N m}^{-1}$$

$$= 40.0 \text{ atm}$$

Internal pressure $= a/V^2$

$$= 1.161 \times 10^6 \text{ N m}^{-2} = 11.5 \text{ atm}$$

Values of a and b for some other gases are given in Table 16.1.

16.3 Joule-Thomson Expansion

It has been seen in the last chapter that if a perfect gas undergoes free expansion, there is no change in its temperature. But if the same experiment is repeated with imperfect gases, one should expect some change in temperature. If the gases are at low temperatures and low pressures, the long range attractive forces will be predominant. After expansion, along with average distance between the molecules, the potential energy will also increase. If the process is adiabatic, this increase in potential energy will be at the expense of the kinetic energy. This decrease in kinetic energy will be accompanied by decrease in temperature. On the other hand, if temperature is high (molecules move rapidly) and molecular density is also high (high pressure), the molecules will collide quite frequently. During the collision the predominant part is played by the repulsive forces. It means, under such circumstances the short range repulsive forces may be more important. Therefore, after expansion, there may be a decrease in the potential energy and hence heating. Thus free expansion is a very good method to test whether a gas is perfect or not. But the temperature change is so small that it is very difficult to detect it experimentally. Moreover, this type of expansion is of no practical utility.

Table 16.1 *Values of the Constants a and b of van der Waals Equation for Some Gases*

Gas	*a* *Nm^4/mole^2*	*b* *cm^3/mole*
Helium	0.00346	23.7
Argon	0.136	32.0
Oxygen	0.139	32.0
Nitrogen	0.138	38.8
Hydrogen	0.0248	26.4
Ammonia	0.423	37.2

There is another process, known as Joule-Thomson expansion, which yields similar results. In this process a gas at a high pressure is made to seep through a porous plug, or a throttle valve, to a low pressure. The schematic diagram of such an arrangement is shown in Fig. 16.5. The plug is surrounded by heat insulators like asbestos or cotton wool. Thus the process is adiabatic. The function of the porous plug is to control the flow of the gas from high to low pressures. Thus it maintains the pressure difference on its two sides. This process is made cyclic with the help of suction-cum-compression pump shown in Fig. 16.5.

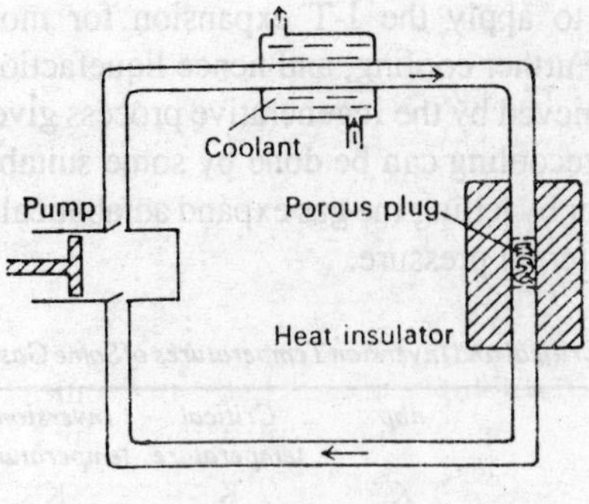

Fig. 16.5 *Porous plug experiment*

Consider a given amount of a gas between the piston and the porous plug [Fig. 16.6(a)]. Let the volume of the gas be V_A and the pressure P_A. Now, let the gas be compressed at this pressure (pressure is kept constant with the help of the piston on the left) to the right through the plug. On the right side the pressure is kept constant at P_B (lower than P_A) with the help of another piston. Let the final volume, after all the gas seeps to the right, be V_B [Fig. 16.6(b)]. Now:

$$\text{Net work done by the gas} = P_B V_B - P_A V_A$$

$$\text{Increase in its internal energy} = U_B - U_A$$

$$\text{Heat given to the gas} = 0$$

because the process is adiabatic. Applying the first law of thermodynamics,

$$\Delta Q = \Delta U + \Delta W$$

or $$0 = U_B - U_A + P_B V_B - P_A V_A$$

or $$P_A V_A + U_A = P_B V_B + U_B \qquad (16.2)$$

Thus during this process the quantity $(U + PV)$ remains constant. You might have realized by now that the quantities which remain unchanged during certain processes are of great interest to the physicists. Therefore, we give it a name *enthalpy*. As we have already pointed out, there may be cooling or heating depending on the initial temperature and pressure. The temperature above which there is always heating is known as *maximum inversion temperature* T_i. This temperature is much higher than the critical temperature.

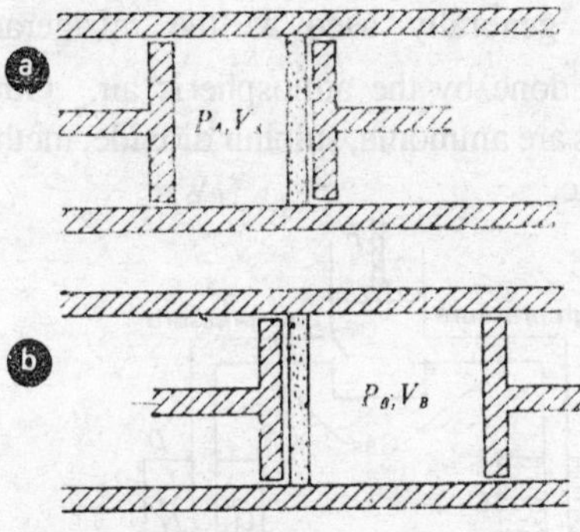

Fig. 16.6 *Transfer of a gas from one constant pressure to another constant pressure*

If the gas is perfect,

$$(PV)_T = \text{constant}, \qquad (16.3)$$

and $$U = \text{constant} \times T \qquad (16.4)$$

Thus, $(U + PV)$ would be a function of temperature only. As $(U + PV)$ remains unchanged during J-T expansion, T also remains unchanged. But in actual gases, both these equations are not true. The J-T expansion experiment can be done with liquids also. Saturated liquids always cool down as a result of this throttling process. Thus Joule-Thomson expansion plays a great role in producing low temperatures, as we will see later.

Question: What is the difference between Joule, Joule-Thomson and adiabatic expansions?

16.4 Refrigerators

The principle of a refrigerator is shown in Fig. 16.7. The gas is compressed to a high pressure with the help of a compressor P. It is cooled down by water circulating through a spiral pipe at C. The gas then gets liquefied. This high pressure liquid passes through a throttle valve T towards a low pressure side. It comes out as a mixture of liquid and vapour and its temperature gets lowered. At D this liquid-vapour mixture takes heat from its surrounding material (thus cooling this stuff), and the remaining liquid also turns into vapour. This vapour is finally sucked by the pump *P*. In home refrigerators, freon (dichloro-difluoro methane, CCl_2F_2) is generally used as the refrigerant. Cooling is done by the atmospheric air. Other refrigerants are ammonia, sulphur dioxide, methyl chloride, etc.

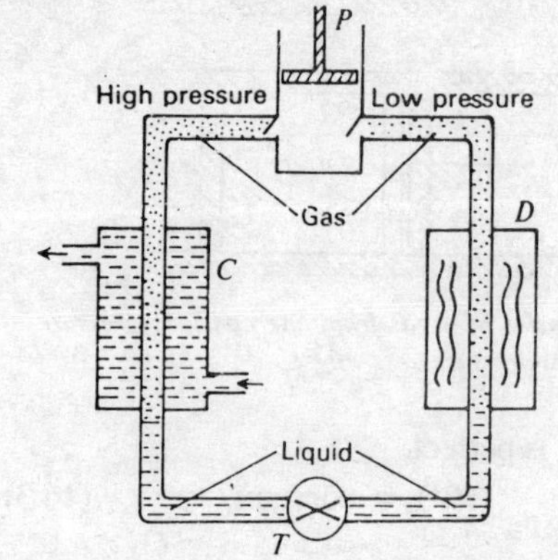

Fig. 16.7 *Principle of a refrigerator*

16.5 Liquefaction of Gases

It has been seen in Section 16.1 that if a gas is to be liquefied by merely applying pressure on it, it has to be cooled below its critical temperature. One can also see from Fig. 16.2 that the lower the temperature of the gas, the lesser the pressure required to liquefy it. So, the simple process would be to cool the gas below its critical temperature by some coolant and then to liquefy it by applying a pressure on it. This method does not work always. In the case of some gases (for example neon and helium), no precooling can bring the temperature below their critical temperatures and, thus, they cannot be liquefied by this method.

Joule-Thomson expansion is a very useful technique to liquefy gases. The temperature below which cooling may occur in this expansion, known as maximum inversion temperature, is much higher than the critical temperature. Critical temperatures and maximum inversion temperatures, along with normal boiling points, have been given in Table 16.2 for some gases. Thus it is easier to precool a gas below this inversion temperature and then to apply the J-T expansion for more cooling. Further cooling, and hence liquefaction, can be achieved by the regenerative process given below. Precooling can be done by some suitable coolants, or by letting the gas expand adiabatically against external pressure.

Table 16.2 *Critical and Inversion Temperatures of Some Gases*

Gas	*nbp* K	*Critical temperature* K	*Inversion temperature* K
Carbon dioxide	195	304	1500
Argon	87	151	723
Nitrogen	77	126	621
Hydrogen	20	33	202
Helium	4	5	40

Flow diagram of the first method of liquefaction is given in Fig. 16.8. Compressor *P* compresses the gas to pressures of the order of 10^2 atmospheres.

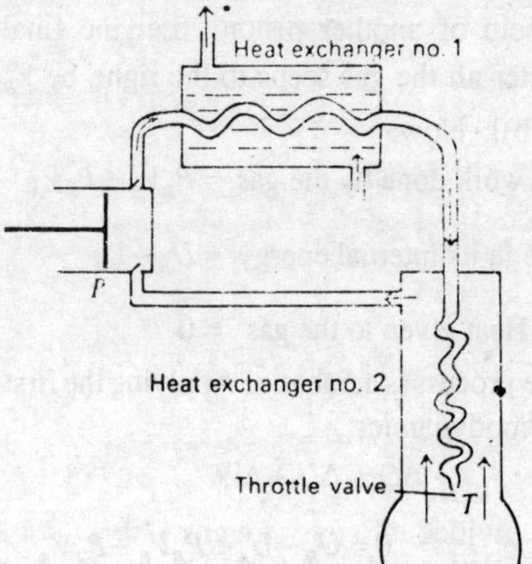

Fig. 16.8 *Principle of regenerative cooling*

After cooling by heat exchanger No. 1, the gas undergoes throttling process at needle valve *T*, where it is further cooled. This cooled gas is allowed to go back to the pump *P*. While going up, it cools the incoming gas in heat exchanger No. 2. This gas gets further cooled after undergoing J-T expansion at *T*. This regenerative process continues till the gas gets liquefied. This process was first used by Linde to liquefy air. In this process the temperature of the coolant in exchanger No. 1 should be below the maximum inversion temperature of the gas which is being liquefied. Thus, to liquefy hydrogen we need liquid air, and to liquefy helium we need both liquid air and liquid hydrogen (see Table 16.2). It means that along with liquid helium plant one has to have the liquid air and liquid hydrogen plants. This is, of course, very expensive. To avoid this, the second method of cooling by expanding the gas adiabatically is used. The principle of this scheme is shown in Fig. 16.9.

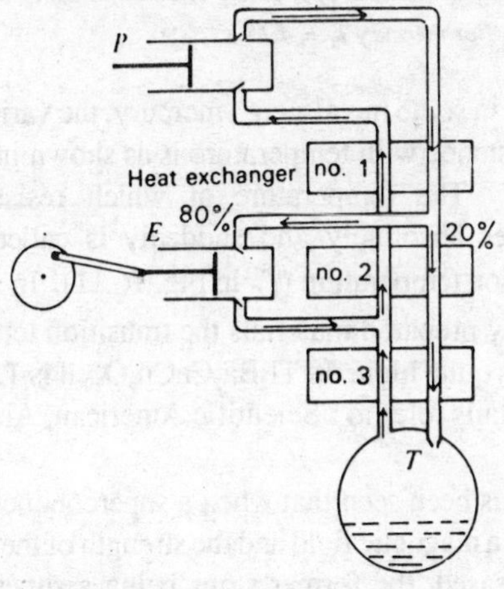

Fig. 16.9 *Liquefaction of helium by combined action of regenerative cooling and adiabatic expansion*

The compressed gas is cooled by an easily available coolant at exchanger No. 1. After this, the gas is divided into two streams. About 80 per cent of the gas goes to an expansion chamber *E*. This is the cylinder of an engine which is actually performing external work. Due to adiabatic expansion the temperature of this gas becomes very low. This 80 per cent of the gas cools the rest of the 20 per cent in exchanger No. 2. This 20 per cent gas then undergoes J-T expansion at *T* and goes up cooling the incoming gas in exchanger No. 3.* After this, it mixes up with the rest of the 80 per cent gas and the whole of the gas goes back to the pump. After some time the gas gets liquefied.

This method was used by Claude to liquefy air. Similar methods were used by Kapitza to liquefy air, hydrogen, and helium. The modern helium liquefier is the one developed by Collins. In this plant helium undergoes adiabatic expansion in a reciprocating engine. The expanded gas then cools the incoming gas. When the temperature, thus, goes below inversion temperature, the gas passes through a needle valve and J-T expansion finally liquefies it.

By boiling helium under reduced pressure, temperatures of the order of 0.1 K have been achieved. By adiabatic demagnetization of paramagnetic salts one can go upto 10^{-3} K. The lowest temperature achieved so far is $\sim 10^{-7}$ K, and that is by adiabatic nuclear demagnetization.

Researches of far reaching consequences have been done at such very low temperatures. This low temperature physics is known as cryogenics.

16.6 Change of Phase

When a system consists of only a single homogeneous substance, it is said to consist of only one phase. On the other hand, if a system is divided into homogeneous and contiguous parts being in physically different states, these parts are called the phases of the system. It has been seen above that below a certain temperature, called the critical temperature, it is possible to liquefy an imperfect gas by the application of suitable pressure. At sufficiently low temperatures when thermal motions further slow down, a liquid can be converted into a solid. The transition of a solid into a liquid is called *fusion,* and of a liquid into a gas is known as *vaporization.* A solid may pass directly to vapour without passing through the liquid state. Such a process is called *sublimation.* The transition temperatures are, however, not fixed, but depend on external pressure. As an example, liquid water can be made to boil at a temperature other than 100°C by varying the pressure suitably.

***In each of these heat exchangers the incoming gas passes through a spiral-shaped pipe. The outgoing gas passes through the jacket containing this spiral. In this way the outgoing gas cools the incoming gas.**

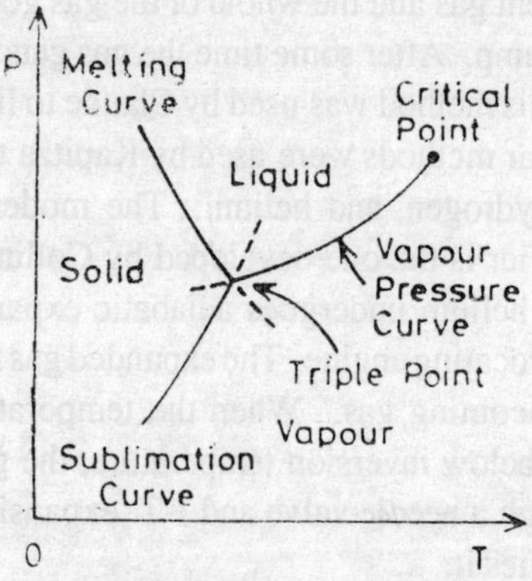

Fig. 16.10 *Phase diagram*

A typical pressure-temperature phase diagram for a system capable of existing in three phases is given in Fig. 16.10. Points on a curve dividing two given phases correspond to phase equilibrium between these two phases. In Fig. 16.10 the point where the three phases coexist is called the *triple point*.

Question: Melting point of ice decreases with increase in pressure. How will the melting curve look like in this case?

16.7 Superconductivity

When the temperature of a substance moves down towards absolute zero, interesting phenomena occur. One of them is the disappearance of resistance of some metals. This phenomenon is called superconductivity, and was discovered by Kamerlingh Onnes in 1911 in the case of mercury.

The resistance of a metal is due to two reasons. The conduction electrons may be obstructed by the atoms vibrating at their sites. These electrons may also be scattered by imperfections in the metal. Generally speaking, imperfections are of two kinds —impurities (chemical imperfections), and physical defects in the metal structure. At very low temperatures the thermal vibrations of the atoms become very small. The only resistance left then, in general, is due to impurity atoms or physical strains in the metal. Therefore, we should expect that if a particular metal is cooled down, its resistance will go down to a limiting value depending upon the purity and physical condition of the metal. The electrical resistance that is left at the lowest temperature is often called the residual resistance.

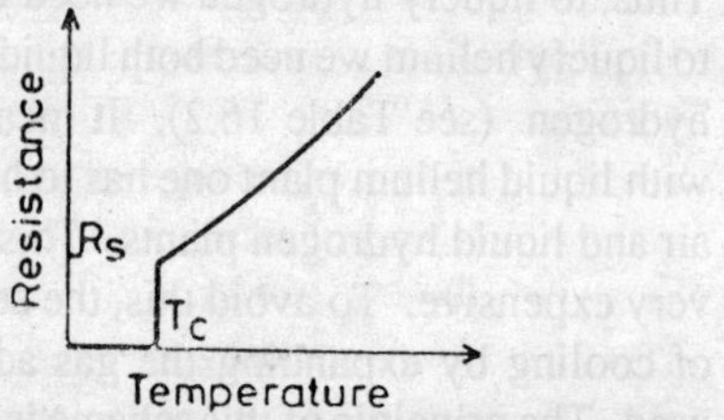

Fig. 16.11 *Variation of resistance of a typical superconducting material as a function of temperature; at very low temperature the resistance does not approach a constant value R_S, but falls to zero. For mercury $T_C = 4.15$ K.*

But in some metals, e.g., mercury, the variation of resistance with temperature is as shown in Fig. 16.11. The temperature at which resistance vanishes absolutely and suddenly is called the transition temperature (T_C in Fig. 16.11). In some recently prepared materials the transition temperature is quite high. In $Tl_2Ba_2CaCu_3O_{10}$ it is 125 K. For details refer to : Scientific American, August, 1990.

It has been seen that when a superconductor is kept in a magnetic field and the strength of the field is increased, the former stops being a superconductor at a certain value of the field. The value of this field depends upon the superconducting material, the temperature (which must, of course, be below the transition temperature of that material), and the size and shape of the superconductor.

The basis of the general quantum theory of superconductivity was laid down by Bardeen, Cooper and Schrieffer in 1957. The central idea of this theory, called the BCS theory, is the attraction between two conduction electrons via the lattice. Below a certain temperature, the disturbance of the lattice by a conduction electron is greater than the thermal motion of the lattice disturbing the electron. One electron interacts with the lattice and deforms it; a second electron adjusts itself to take advantage of the deformation to lower its energy. This amounts to an effective attraction between the two electrons. In some materials, this attraction is greater than the electrostatic repulsive force. If both the electrons are set in motion in the same direction, this will be the lowest energy state available for the electrons. They stay in that state because there is no lower state available for them. Consequently, there will be a permanent net electron current in their direction of motion.

If we imagine heating a superconducting metal at absolute zero, then the superconductive coupling between the electrons via zero-point vibrations* will tend to break down sooner or later by the increasing random thermal vibrations of the atoms. It happens at the transition temperature, and the electrical resistance reappears. As the magnetic field exerts a force on a moving electron, a sufficiently strong magnetic field acting on a superconductor interferes with the electron coupling and breaks it. The resistance of the metal then reappears. The higher the temperature, the lower is the magnetic field required to break the superconductive coupling. This is due to the combination of thermal vibrations with the magnetic field.

As there is no ohmic loss in a superconducting material below transition temperature, power transmission could be highly economical. But the problem is to keep the conductor below the transition temperature. Thus, materials with high transition temperatures would be more useful. Superconducting coils can be used for the fabrication of very strong electromagnets because they pose no problem of heat generation. Superconductors are also useful in very efficient, rapid and miniature switches. If a ring of a superconductor is taken and a current is started in it, this current can go on and on. This property is made use of in fabricating electrical memory circuits. The use of superconducting elements in computers would save a lot of power because there are millions of elements in a computer.

*Due to the uncertainty principle of quantum mechanics, there will be some motion of the atoms even at 0 K. The energy due to this motion is called zero-point energy. But this energy cannot be transferred to any other system because there are no energy states below this.

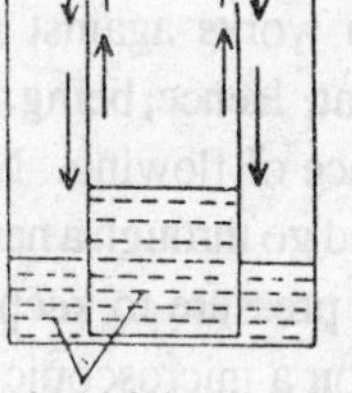

Fig. 16.12 *Surface film creep of liquid helium II from a higher level to a lower level, like flow of an ordinary liquid through a syphon*

16.8 Superfluidity

Another interesting phenomenon that happens at very low temperatures is the vanishing of the viscosity of helium*. Helium liquefies at 4.2 K. If it is cooled below 2.19 K, it starts flowing freely without needing any push to keep it moving. It loses its viscosity and becomes superfluid. If a beaker full of this liquid is kept inside an empty beaker, the liquid will start creeping out of the beaker (Fig. 16.12). This process will continue till the level of the liquid becomes the same in both the beakers. If the outer breaker is full of this helium, then the reverse process will take place.

The properties of liquid helium change dramatically below 2.19 K. Therefore, it is called helium I above this temperature and helium II below it. The changeover point is called the lambda point**. Below this temperature, helium shows enormous heat conductivity—better than many metals.

*This property is exhibited by ordinary helium, called helium-four, which has two protons and two neutrons in its nucleus. The other isotope helium-three, which has two protons and one neutron in its nucleus, exhibits this phenomenon around 1 mK.

**If the rate of flow of helium through a tube is plotted as a function of temperature, the curve shows a sharp turn at 2.19 K, and the graph looks something like an inverted Greek lambda.

Helium remains liquid because it is very light and because of the very weak attractive forces between the atoms*. The effect of zero-point energy, which works against freezing, becomes quite significant. Hence, being a liquid, helium has at least a chance of flowing. Normally, if we try to make a liquid go through a narrow tube, we have to exert some pressure to keep it moving. What we are doing on a microscopic scale is to heat up the liquid a bit. Much of the work we do in pushing the liquid ends up as random energy of the individual atoms. Thus, if a liquid has to encounter viscosity, then a microscopic heat motion has to be generated in that liquid. But we know from quantum mechanics that we can generate the necessary microscopic motion of the atoms only in little energy bundles of a minimum amount. In liquids at ordinary temperatures, where a lot of thermal energy is present, this presents no difficulty because these quanta are so tiny. However, at very low temperatures, where there is very little thermal energy, it may be difficult to generate these particular bundles of energy. Therefore, it might be impossible to produce the right kind of random motion of the helium atoms to absorb the viscous energy. If that is so, there just cannot be any friction.

On the basis of the arguments given above, ^{3}He should show superfluidity in a more pronounced manner because it is lighter than ^{4}He, but it does not. This is partly due to matters pertaining to symmetry. In ^{4}He, we have an even number of particles, while in ^{3}He, we have an odd number of particles.

Discussion and Problems

16.1 Explain why a liquid cools when it evaporates?

16.2 Explain the melting of a solid, its warming to the boiling point, and subsequent conversion into vapour,

16.3 Show that under the limit when volume becomes very large, the van der Waals equation approaches the ideal gas equation.

16.4 Due to intermolecular forces, the internal energy of a gas is partly kinetic and partly potential. How is the temperature related to this internal energy?

16.5 Assuming that gas molecules behave as rigid spheres like billiard balls, show that the effective volume occupied by the molecules (i.e. the co-volume *b*) is four times greater than their actual volume. Hint: Consider binary collisions.

If the molecules are soft, should this volume change with temperature? If so, how?

16.6 Compute the diameter of an argon atom assuming the atoms to be rigid spheres. Use van der Waals constants given in Table 16.1.

16.7 Compute the internal pressure for a given amount of helium at 27°C, when its volume is 1/50 of its value at STP. What will be the internal pressure when the volume is one hundredth of the volume at STP?

Also find the pressures of the gas for these two cases and compare them with their corresponding values if the gas were ideal.

---

*It takes about 30 atm pressure at 1K to solidify helium

Chapter 17

Heat Transfer

17.1 Conduction

Thermal energy can propagate from one place to another by three distinct modes — conduction, convection and radiation. In conduction heat transfer in a medium takes place through particle collisions, while the particles stay at their respective positions. Suppose a solid has non-uniform temperature distribution. Particles in a hotter region will have higher thermal energy than those in the neighbouring colder region. Therefore, during collisions there will, on the average, be transfer of energy from the 'hotter' to the 'colder' particles. This transfer of energy will continue till the temperature becomes uniform.

Let us take a rod of uniform cross-section of area A, and raise the temperature of one end. Let x-axis be taken along the rod with x increasing from the hotter towards the colder side. Take two cross-sections at positions x_1 to x_2. Let, at steady state, temperatures at these cross-sections be T_1 and T_2 $(T_1 > T_2)$. It is found by experiment (after thermally insulating the side walls) that the rate of flow of thermal energy through the rod from the hotter cross-section to the colder cross-section is proportional to both A and the temperature gradient. We can thus write for the heat flow Q per second.

$$Q = KA\frac{T_1 - T_2}{x_2 - x_1} = -KA\frac{\Delta T}{\Delta x} \qquad (17.1)$$

The constant of proportionality K depends on the material of the rod, and is called the *coefficient of thermal conductivity* or simply *thermal conductivity* of the material. Thermal conductivities of some substances are given in Table 17.1.

Table 17.1 *Thermal Conductivities at Room Temperature*

Substance	*K, W/m-K*
Silver	429
Copper	401
Gold	318
Aluminium	237
Brass	110
Steel	50
Glass	0.8
Water	0.61
Wood	0.12 - 0.04
Hydrogen	0.182
Helium	0.152
Oxygen	0.0267
Argon	0.0177

Example 17.1 An annular cylinder has inner radius R_1 and outer radius R_2. With the help of an axial heater its inner surface is maintained at temperature T_1. Thermal conductivity of the material of the cylinder is K, and steady state temperature of the outer surface is T_2. Find the temperature of the cylinder at a distance r from the axis.

Take a coaxial surface of radius r, and length unity. Heat flowing through this surface will be

$$Q = -K \times 2\pi r \times \frac{dT}{dr} \qquad (17.2)$$

$$\therefore \quad \frac{dr}{r} = -\frac{2\pi K}{Q}.dT$$

As at steady state Q is same through all the coaxial surfaces, integration of the above relation gives,

$$\ln(R_2/R_1) = -\frac{2\pi K}{Q}(T_2 - T_1)$$

or

$$Q = \frac{2\pi K(T_1 - T_2)}{\ln(R_2/R_1)} \qquad (17.3)$$

Substituting this value of Q in Eq. (17.2) above, and integrating

$$\int_{R_1}^{r} \frac{dr}{r} = -\frac{2\pi K \ln (R_2/R_1)}{2\pi K (T_1 - T_2)} \int_{T_1}^{T} dT$$

or $$\ln (r/R_1) = -\frac{\ln (R_2/R_1)}{(T_1 - T_2)} (T - T_1),$$

or $$T = T_1 - \frac{\ln (r/R_1)}{\ln (R_2/R_1)} (T_1 - T_2) \qquad (17.4)$$

17.2 Convection

In this process thermal energy is carried from one region of the medium to another by the particles of the medium. Thus convection is possible in fluids but not in solids. Take water (at temperature above 4°C) in a tub. Heat a small area of the bottom of the tub. Water in contact with this part of the tub will get heated up and its density will decrease. As a result of this, pressure at the bottom of the water column will decrease and cold water from its neighbourhood will push in. The warm water will go up and go over to the colder sides. Convection thus set up under gravity is called natural or free convection. Like conduction, convection also tends to level out the temperature in a fluid.

There is no simple mathematical equation for convection. Heat lost or gained by a surface depends on its shape, orientation, and also on its surroundings. Estimation of convection is done partly by experiments and partly by dimensional analysis.

17.3 Radiation

In both conduction and convection the medium plays an active role. But no medium is required in the third mode of heat propagation, the radiation. It is a common experience that even if we place our hand below a hot plate, we feel warm. This transmission of heat cannot be due to conduction, because air is a very good insulator. Also it cannot be due to convection, because convection currents of warm air go up. If we keep a hot body in a vacuum chamber, even then it loses heat. Walls of the enclosure become hot. In such cases the transmission of energy is by radiation.

Properties of this radiation have been thoroughly studied. Radiant energy obeys laws of reflection and refraction. It moves in straight lines, and it can be polarized. Through such observations it is now well established that radiant energy is akin to light. This becomes obvious when we raise the temperature of a filament gradually. After going beyond certain temperature, it starts radiating light. The only difference is that at lower temperatures the wavelength of the radiant energy is very large and is beyond the visible range. Even the sources of visible light lose a lot of energy in this low frequency spectrum. If a thermocouple (see Sec. 22.7) is taken below red colour (infrared) in the spectrum of the sun, galvanometer records deflection there too. Deflection will also be recorded above violet (ultraviolet) colour. The higher the temperature, the more is the energy emitted in short wavelength region of the spectrum.

Matter can be made to emit radiant energy in so many ways. The firefly emits light from its body matter by some chemical process. Electrical discharge through rarefied gases, phosphorescence, fluorescence, etc., are other sources of radiation. But here we will discuss only thermal radiation, i.e. the radiation from heated bodies. The spectral analysis of this radiation can be done by methods similar to those used for light. This has been done by some experimental physicists using fluorspar prisms for dispersion and bolometers* for detection. It is found that the radiant energy is distributed over quite a wide range of wavelengths.

17.4 Some Definitions

It has been seen that various bodies at the same temperature cool down at different rates depending upon the surface conditions. Energy radiated per

---

*Bolometer is a fine resistance element put in the fourth arm of a Wheatstone bridge. When radiation falls on this element, it gets heated up and, thus, its resistance increases. This increase in resistance offsets the balance of the bridge. The amount of this unbalance is a measure of the amount of the radiant energy falling on the resistance element.

second per unit area of a surface is called the *radiant emittance* (or *radiance*), E, of that surface. The unit of E is watt per square metre. The fraction of energy absorbed by a surface falling on it is called its *absorptance* (or *absorptivity*), a. This obviously is a dimensionless quantity. We may also be interested in the energy emitted by a hot body at various radiant frequencies. In that case we define a term monochromatic (or spectral) emittance, E_λ, which is the amount of energy radiated per unit time per unit area in a unit wavelength interval around the wavelength λ. In other words, if $E_\lambda \Delta\lambda$ is the amount of energy radiated per second per unit area in a wavelength interval $\Delta\lambda$ around λ, then E_λ is called the *monochromatic emittance at wavelength* λ. Similarly, if W_λ is the amount of radiant energy falling on the body between the wavelengths λ and $(\lambda + \Delta\lambda)$ and a fraction a_λ of it is absorbed by the body and converted into heat, then a_λ is called the *monochromatic absorptance* of the body for wavelength λ. Both E_λ and a_λ change with wavelength and the nature of this variation depends on the physical conditions of the surface.

17.5 Black-body

A body which absorbs all the incident radiation is called a black-body. No actual surface is perfectly black, there being always some reflection. The closest approach is the lamp black which reflects about one per cent of the incident radiation. But we can have perfectly black bodies indirectly. Get a pitcher and cover it with a cardboard. Make a small hole in the cardboard and look at it. This hole looks black. Try this experiment with other utensils, even tinned from inside. The secret of this blackness is that the light which falls on the hole does not come back. It suffers multiple reflections inside the enclosure till it is absorbed. A similar black-body designed by Fery is shown in Fig. 17.1. The outer wall is an adiabatic one. Inside this wall, there are heater coils to change the temperature of the interior of the enclosure. To avoid normal reflection of the incident light, a projection P is constructed just opposite the opening O. To avoid such chances further and for quicker attainment of equilibrium the inside is coated black.

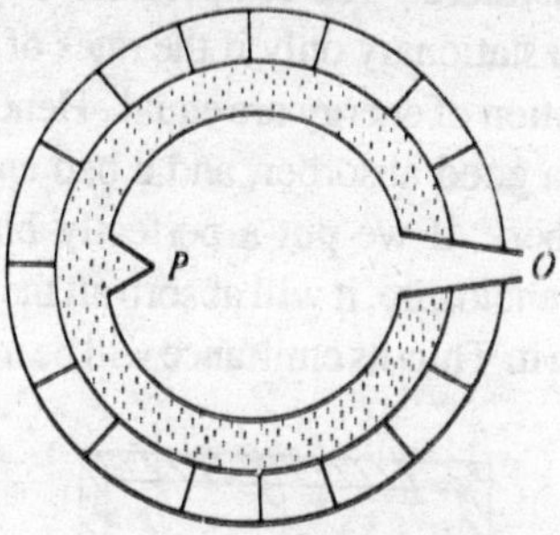

Fig. 17.1 *Fery's black-body*

17.6 Emission and Absorption of Radiation

We have seen that bodies when hot emit radiation and, as a result, their temperature falls. When we switch off light, the filaments of bulbs lose their redness almost immediately. It shows that cooling is quite fast. Now, the question arises that when the radiation loss is so fast, then, why the things do not keep on cooling upto 0 K and radiate out all the internal energy. This should happen if the body under investigation does not receive energy. But it does. The surroundings are also emitting radiation which falls on this body and which it absorbs. When the body is hot it loses more energy by way of radiation than it receives. The net result is cooling. If the body is colder than the surroundings, it will receive more energy and will radiate lesser amounts. This way its temperature will rise. When the temperature of the body is equal to that of the surroundings, the loss and gain balance each other. This is known as Prevost's theory of exchanges.

Let us suspend an object A inside a closed chamber C of adiabatic walls with the help of an insulating thread (Fig. 17.2). The chamber is kept at some constant temperature, say T. The walls of this chamber are continuously emitting and re-absorbing the radiation and, thus, the enclosure is full of radiation. When we put A inside this enclosure, it gets bathed in this radiation. After some time, A will attain the temperature of the enclosure through exchange of radiation with walls of the chamber. If we put another ball having different surface conditions, it will also attain the same temperature. The temperature of an object can remain stationary only if the rates of emission and absorption of energy are equal. Hence, a good emitter is a good absorber, and a bad emitter is a bad absorber. If we put a perfectly black body inside this enclosure, it will absorb all the radiation incident on it. Thus its emittance will be maximum.

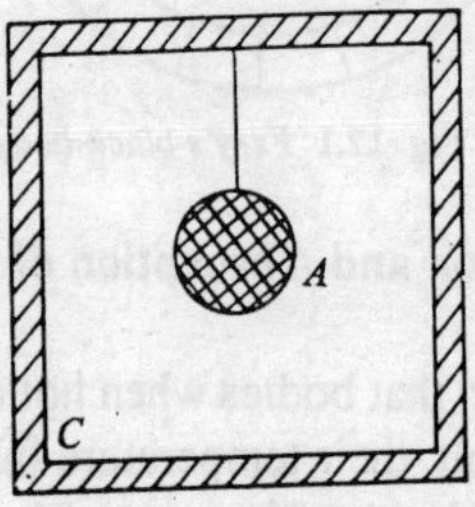

Fig. 17.2 *Any object when put inside an isothermal enclosure attains the temperature of the enclosure*

Let E and a be the emittance and the absorptance, respectively, of body A. Let an amount W of the radiation fall on unit area of A in one second (W is called the irradiance within the enclosure). Out of this a portion aW is absorbed by A, and the rest $(1-a)W$ is reflected or transmitted. Further, A will emit radiation E per second per unit area by virtue of its temperature. Thus, at steady state,

$$E = aW \tag{17.5}$$

But instead of A, if we have a perfectly black body ($a = 1$),

$$E_b = W \tag{17.6}$$

where E_b is the emittance of the black-body. From relations (17.5) and (17.6) we get,

$$E/a = E_b \tag{17.7}$$

Hence, at a given temperature, *the ratio of the emittance to the absorptance of all substances is constant, and is equal to the emittance of a black-body at that temperature*. This is known as Kirchhoff's law.

17.7 Black-body Radiation, or Temperature Radiation

From the above discussion we conclude that the hole of Fery's black-body should be a very good radiator. As it absorbs radiation of all wavelengths, it should also emit radiation of all wavelengths. At first sight it seems that it may not be true. But let us look more closely. If we look into a hole in a hot enclosure, we will receive the radiation from a spot at the inside wall opposite to the hole. In addition, we will get some more radiation. Some of the radiation emitted by other parts of the enclosure which falls on this spot, may come to the hole after reflection. We will interpret this also as if coming from the spot. Thus the hole will look brighter than the rest of the body. This can be observed experimentally. Take an iron pipe (closed at both ends) and make a very small hole near the middle of its wall. Make this pipe red hot and look at the hole. It will look brighter than the rest of the wall. The radiation coming out of this opening is known as *black-body radiation*, or temperature radiation. We can understand it in another way. One can see from Eq. (17.6) that the irradiance within a cavity is equal to the radiant emittance of a black-body at the same temperature. Hence, the radiation within a cavity is a black-body radiation.

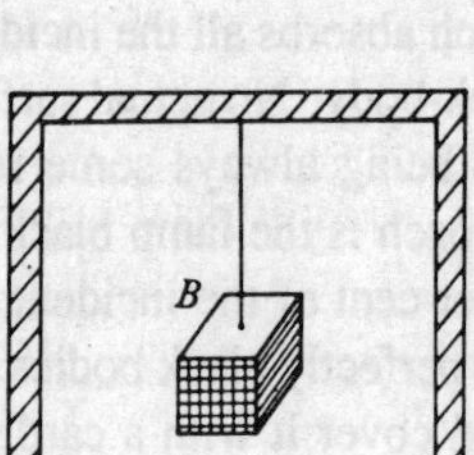

Fig. 17.3 *Temperature of an object inside an isothermal enclosure does not depend upon the nature of its walls or its orientation*

Let us now study the properties of the radiation present inside such an enclosure. Let C be the enclosure having adiabatic walls, and B be a cubic body suspended inside this enclosure with the help of an insulating thread (Fig. 17.3). Let various faces of the cube be of different colours. Thus, different faces of this cube will have different emittances and absorptances. At steady state, the total energy emitted by B equals the total energy absorbed by it. If we now change the orientation of this body, the amount of energy radiated should not change. This is because the temperature of the body is the same as before (equal to that of the enclosure). Hence, the energy absorbed will also remain unchanged. But different faces of the cube have different absorptances. Therefore, the fact that the energy absorbed by B is independent of its position and orientation, shows that the radiation inside C at different places and travelling in different directions is identical in quality as well as in quantity. In brief, the radiation is *isotropic*.

Further, we can make various compartments of different shapes and sizes inside C (which is nothing but having different enclosures at the same temperature), and take B inside each of these turn by turn. Using the same arguments as above, one can say that the radiation inside each of these is identical to the radiation inside any other compartment. Thus, we conclude that the quality and the intensity of the radiation are independent of the shape and size of the enclosure, and the nature of its walls. These are solely the function of temperature and nothing else.

From the above we further conclude that the radiation coming from any object placed inside an enclosure will at steady state be a characteristic of the enclosure. Let us place a mirror inside C. It can reflect only that radiation which falls on it, which is nothing but the radiation inside C.

Now compare the properties of the black-body radiation with the properties of a perfect gas in an enclosure. In gases also, the distribution of velocities and the density of the molecules are identical everywhere inside the enclosure. The distribution of molecular velocities (Maxwellian) is a function of temperature only. It has been observed experimentally that, like gases, radiation also exerts pressure.

17.8 Stefan-Boltzmann Law

Total radiation (summed over all wavelengths) from the surfaces of various bodies as a function of temperature was studied by Tyndall, and Dulong and Petit. On the basis of these observations, it was empirically deduced by Stefan that the rate of emission of radiant energy from a body is proportional to the fourth power of its absolute temperature. Later on, a theoretical derivation of this law was given by Boltzmann. He applied the laws of thermodynamics to the black-body radiation, considering it as an isolated system just like a gas.* Therefore, this law is strictly applicable to radiation emitted by black-bodies, and is called Stefan-Boltzmann law. In a mathematical form this can be expressed as,

$$E_b \propto T^4$$

or
$$E_b = \sigma T^4 \qquad (17.8)$$

Here E_b is the emittance of the black-body summed over all wavelengths ($E_b = \Sigma E_{\lambda b}$), and σ is a constant known as Stefan-Boltzmann constant. The unit of E is watt per square metre and, therefore, σ is expressed in W m^{-2} K^{-4}, and its numerical value is 5.670×10^{-8}.

Let us have a black-body at the same temperature as that of the surroundings, say T_1. The rate of emission of the radiant energy by the body will be the same as the rate of absorption, and each will be equal to σT_1^4 for unit area. If, now, the body is raised to a higher temperature, say T_2, the rate of emission will increase to σT_2^4, while the amount of radiation incident on it from the surroundings will remain the same as before, i.e. σT_1^4. Hence, the net amount of energy lost by black-body per second per unit area will be

$$E_{bnet} = \sigma T_2^4 - \sigma T_1^4$$

or
$$E_{bnet} = \sigma (T_2^4 - T_1^4) \qquad (17.9)$$

*Refer to M.W. Zemansky and R.Dittman, *Heat and Thermodynamics* (McGraw-Hill), 1981.

If the emitter is not black, Eqs. (17.8) and (17.9) will take the following forms

$$E = e\sigma T^4 \qquad (17.10)$$

and

$$E_{net} = e\sigma\,(T_2^4 - T_1^4) \qquad (17.11)$$

Here, e depends on the physical nature of the emitting surface and is known as *relative emittance* (or emissivity) of the surface. It is defined as the ratio of the emittance of a body to the emittance of a black-body. One can conclude from Eq. (17.7) that emissivity is equal to the absorptance of the surface. Numerical values of e for some surfaces are given in Table 17.2.

Table 17.2 Emissivity of Various Substances

Surface	*Temperature range, °C*	*e (or a)*
Copper (polished)	100	0.018
Tungsten (filament)	30-3300	0.032-0.35
Platinum (filament)	30-1200	0.036-0.19
Aluminium (polished)	250-600	0.039-0.06
Iron (polished)	150-1000	0.05-0.37
Rubber (gray)	25	0.86
Lampblack	20-350	0.95
Black-body	–	1

Example 17.2 A metal (tungsten) sphere has a diameter of 4.0 cm and an emissivity of 0.35. Find out the energy lost by radiation in one second when the temperature of the ball is 2927°C.

Energy radiated per second,

$$E = e\sigma A T^4$$

where A is the surface area of the sphere. Therefore:

$$E = 0.35 \times 5.670 \times 10^{-8}\ \text{W m}^{-2}\text{K}^{-4} \times 4\pi(2/100\ \text{m})^2 \times (3200\text{K})^4$$

$$= 0.35 \times 5.670 \times 4\pi \times 4 \times (32)^4 \times 10^{-4}\ \text{W}$$

$$= 10.5\ \text{kW}$$

In a special case when T_2 is not much higher than T_1, Eqs. (17.9) and (17.11) take a simpler form. Let

$$T_2 = T_1 + \Delta T \qquad (17.12)$$

Substituting this value of T_2 in Eq. (17.11),

$$E_{net} = e\sigma[(T_1 + \Delta T)^4 - T_1^4]$$

$$= e\sigma[T_1^4 + 4\Delta T T_1^3 + 6(\Delta T)^2 T_1^2 + 4(\Delta T)^3 T_1 + (\Delta T)^4 - T_1^4]$$

Neglecting terms containing $(\Delta T)^2$ and its higher powers in comparison with other terms (this approximation will hold only if ΔT is small as compared with T_1) we get,

$$E_{net} = (4e\sigma T_1^3)\Delta T \qquad (17.13)$$

Example 17.3 A blackened metal container of negligible heat capacity is filled with water. The container is cubical with each side 5 cm long. It is placed in an evacuated enclosure whose walls are kept at 27°C. In what time will the water temperature change from 30° to 29°C ?

As the container is in vacuum the energy is lost by radiation only. From Eq. (17.13) we see that the energy emitted per second in the beginning is proportional to 3; and at the end it is proportional to 2. Therefore, the average rate at which the energy is being emitted

$$= 4 \times 5.670 \times 10^{-8}\ \text{W m}^{-2}\text{K}^{-4} \times (273 + 27)^3\ \text{K}^3 \times \frac{3+2}{2}\ \text{K} \times \left(\frac{5}{100} \times \frac{5}{100}\right) \times 6\ \text{m}^2$$

$$= 4 \times 5.67 \times 27 \times 2.5 \times 25 \times 6 \times 10^{-6}\ \text{W}$$

$$= 0.230\ \text{J/s}$$

Total energy radiated by water in cooling through 1°C

$$= 125\ \text{g} \times 1\ \text{cal/g-°C} \times 1\text{°C}$$

$$= 125\ \text{cal}$$

$$= 125 \times 4.184\ \text{J}$$

$$= 523\ \text{J}$$

$\therefore$ Time taken for this cooling $= \dfrac{523\ \text{J}}{0.230\ \text{J/s}}$

$$= 2280\ \text{s}$$

$$= 38\ \text{min}$$

Question: If, instead of water, we have a material of specific heat 0.5 cal/g-°C, how much time will it take?

Now, the net radiation emitted per unit area per second is proportional to ΔT, provided the temperature of the surroundings remains constant. Further, the rate of cooling is proportional to E_{net}.

Hence, *the rate of cooling of a body is proportional to the excess of the temperature of the body above the surroundings.* This is Newton's law of cooling. Expressed in mathematical form,

$$-\frac{\Delta T_2}{\Delta t} = \text{constant}\,(T_2 - T_1) \qquad (17.14)$$

The value of the constant depends on the heat capacity of the substance under consideration, and nature of the surface. Eq. (17.14) can be verified experimentally and, further, it can be used to compare the specific heats of two substances*.

Example 17.4 The temperature of a body falls from 30° to 25°C in 5 min. The air temperature is 15°C. Find the temperature after a further 5 min interval.

Let the temperature after the next five minutes be x°C. For the first five minutes

$$\text{Average temperature} = \frac{30+25}{2} = 27.5°\text{C}$$

$$\therefore \quad \text{Average excess temperature} = 27.5 - 15$$

$$= 12.5°\text{C}$$

Average rate of fall of temperature = 5/5 = 1°C/min. Using Eq. (17.14), the average rate of the fall of temperature = $K \times$ average excess temperature, where K is a constant. Therefore:

$$1°\text{C/min} = K \times 12.5°\text{C}$$

or $$K = \frac{1}{12.5} \text{ per min}$$

For the second five minutes

$$\text{Average temperature} = \frac{25+x}{2}°\text{C}$$

$$\text{Average excess temperature} = \frac{25+x}{2} - 15$$

$$= \frac{x-5}{2}°\text{C}$$

Average rate of the fall of temperature

$$= \frac{25-x}{5}°\text{C/min}$$

$$\therefore \quad \frac{25-x}{5} = \frac{1}{12.5} \times \frac{x-5}{2} = \frac{x-5}{25}$$

or $$5(25-x) = x-5$$

or $$x = 21.7$$

$\therefore$ **Final temperature = 21.7°C**

*See B.L. Worsnop and H.T. Flint, *Advanced Practical Physics for Students* (Methuen & Co., London), 1957.

Example 17.5 A copper calorimeter weighing 99.0 g is first filled with water and then with kerosene. Time taken to cool from 45° to 40°C in the two cases is 10 min and 5 min, respectively. The mass of water is 100.0 g and that of kerosene is 80.0 g. Calculate the specific heat of kerosene. The specific heat of copper is 0.094 cal/g-°C, and room temperature is 35°C.

Radiation of energy depends only on surface conditions, e.g. colour, planeness, temperature and area. In the above experiment, all these things are the same. Further, the temperature not being very much different from that of the surroundings, the rate of cooling shall be proportional to the temperature difference. As the temperature differences in the beginning and at the end are also the same in both cases, average rate at which energy is radiated should also be the same.

The average rate at which energy is radiated by water and calorimeter

$$= \frac{(99.0 \times 0.094 + 100.0 \times 1)5}{10} \text{ cal/min}$$

The average rate at which energy is radiated by kerosene and calorimeter

$$= \frac{(99.0 \times 0.094 + 80.0 \times s)5}{5} \text{ cal/min}$$

Here, s is the specific heat of kerosene in cal/g-°C.

$$\therefore \quad \frac{(99.0 \times 0.094 + 100)5}{10} = \frac{(99.0 \times 0.094 + 80.0 \times s)5}{5}$$

or $$\frac{(9.305 + 100)}{2} = (9.305 + 80.0\,s)$$

or $$109.3 = 18.6 + 160\,s$$

or $$s = \frac{90.7}{160} = 0.57$$

$\therefore$ Specific heat of kerosene = 0.57 cal/g-°C

17.9 Spectral Distribution of Black-body Radiation

This problem is one of the most important and exciting problems in physics. The attempts to solve this problem gave rise to a new discipline in physics, which is known as quantum mechanics. To study the black-body radiation experimentally, a small hole is made in an adiabatic enclosure, as is done in Fery's black-body. The distribution of radiation intensity as a function of wavelength is shown in Fig. 17.4 for two temperatures.

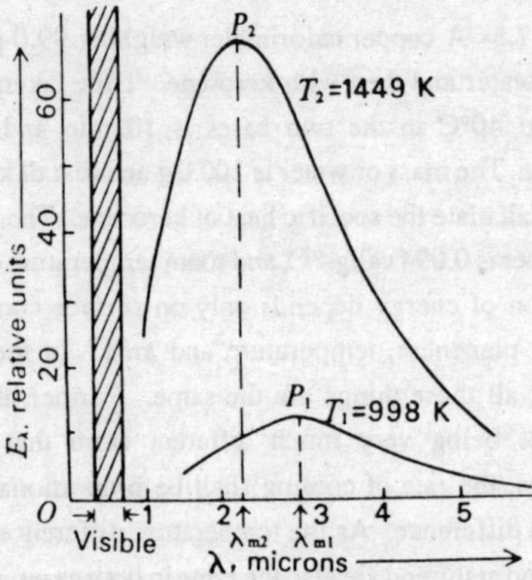

Fig. 17.4 *Distribution of spectral emittance of a black-body as a function of wavelength at two temperatures*

The first attempt to understand black-body radiation was by Wien. Starting from thermodynamics and electromagnetic theory of radiation, he could predict the distribution of energy at any temperature provided it was known at one temperature. To get the curve at temperature T_2 in Fig. 17.4 from the one at temperature T_1, shorten the abscissa in the ratio of T_1/T_2 and increase the corresponding ordinate in the ratio of $(T_2/T_1)^5$.* Thus, the curve will now become taller but will be more closed up. The area under the curve, which represents the total intensity, changes in the ratio $(T_2/T_1)^4$, thus satisfying Stefan-Boltzmann law. The point P_1, corresponding to the maximum emission at temperature T_1, will shift to point P_2 in such a way that

$$\lambda_{m1}T_1 = \lambda_{m2}T_2 = b = \text{constant} \qquad (17.15)$$

where λ_{m1} and λ_{m2} are the wavelengths for which the emission is maximum at temperatures T_1 and T_2, respectively. b is a constant known as Wien's constant, and its value is 2.898×10^{-3} m K. Thus Eq. (17.15) can be used to find out the temperatures of the black bodies.

*See M.N. Saha and B.N. Srivastava, *A Treatise on Heat* (The Indian Press), 1958.

Example 17.6 A furnace is at a temperature of 2000 K. At what wavelength would you expect the maximum intensity of radiation? Is it in the visible region?

From Eq. (17.15)

$$\lambda_m = \frac{b}{T} = \frac{2.898 \times 10^{-3}\ \text{m K}}{2000\ \text{K}}$$

$$= 1.449 \times 10^{-6}\ \text{m}$$

$$= 1.449\ \text{micron}$$

Visible region lies approximately between 4000 Å (0.4 micron) and 8000 Å (0.8 micron). Thus the above wavelength is beyond red colour.

The above law, known as Wien's displacement law, does not solve the basic problem. Why the black-body radiation has this particular wavelength distribution? This problem was first taken up by Rayleigh and then completed by Jeans. Their approach was as follows. Radiation waves get continuously reflected from the walls of the container. During this process of reflection they interact with the atoms of the walls. Thus, at steady state, the radiation will be diffused (going in all possible directions) and will have wavelengths ranging from zero to infinity. Due to reflections, the waves of one particular wavelength will give rise to stationary wave patterns. There will be so many such sets of stationary waves corresponding to each wavelength. This way, they calculated the total number of degrees of freedom and associated with each degree an energy $kT/2$ given by the classical theory. But the final expression so obtained agreed with experiment at higher wavelengths only. Their expression predicts that emittance should go to infinity as wavelength approaches zero, which is against experimental evidence. It is also physically impossible, because the total radiation will, then, be infinite.

There was no flaw in Rayleigh-Jeans treatment. It was exactly in accordance with the classical physics. It occurred to Planck that may be there is some thing basically wrong with our starting point. He abandoned the idea that a system can have any energy between 0 and ∞, or that it can exchange energy by any amount. He postulated that energy can be exchanged only in discrete steps. He assumed that the atoms of the wall behave as electromagnetic oscillators, each having a characteristic frequency. At equilibrium, the energy emitted by these oscillators will be equal to the energy absorbed by them. This is called dynamic equilibrium. Planck postulated that the energy of an oscillator can only be an integral multiple of some minimum energy, say ε, and it can emit radiation in jumps of ε or some integral multiple of it*. Starting with this basic difference, he found that the energy density (energy per unit volume) within the wavelengths λ and $(\lambda+\Delta\lambda)$ of the radiation in a black-body at a temperature T is given by,

$$u_\lambda \Delta\lambda = \frac{8\pi}{\lambda^4}\frac{\varepsilon}{e^{\varepsilon/kT}-1}\Delta\lambda \qquad (17.16)$$

To fit this curve into the experiment, he further assumed that ε must be proportional to the frequency, ν, of the radiation emitted by the oscillator. Or

$$\varepsilon = h\nu \qquad (17.17)$$

where h is a constant now known as Planck's constant. Substituting for ε from Eq. (17.17) in eq. (17.16) we get

$$u_\lambda \Delta\lambda = \frac{8\pi}{\lambda^4}\frac{h\nu}{e^{h\nu/kT}-1}\Delta\lambda$$

But $\nu = c/\lambda$

Hence, $$u_\lambda \Delta\lambda = \frac{8\pi hc}{\lambda^5}\frac{1}{e^{hc/\lambda kT}-1}\Delta\lambda \qquad (17.18)$$

which is Planck's law.

Now, the emittance of a black-body is proportional to the energy density ($E_\lambda = cu_\lambda/4$, where c is the velocity of the radiation). Thus, the dependence of emittance on wavelength will be the same as the dependence of u_λ. It was found that this expression fits into the experiment excellently well. This happened in the year 1900. Five years later Einstein, using the idea of discreteness of energy, could explain another mysterious phenomenon, photoelectric effect. You will read about it in detail in Chapter 33. But the conclusion is that light consists of particles called photons, and the energy of a photon of frequency ν is $h\nu$.*

The value of h is 6.626×10^{-34} J s (unit of angular momentum). One can imagine from this how small the bundle of energy can be.

Question: The wavelength of visible light varies approximately from 8000 to 4000 Å. What is the energy of the photons having these two extreme wavelengths? Express the result in joules, ergs and electron-volts.

As we see now, the black-body enclosure is filled with photons of various wavelengths, and these photons have various energies. Their motion inside the cavity is random. This picture is analogous to the picture of a perfect gas in an enclosure. The molecules of a perfect gas have various velocities and hence various energies. Thus, the problem of finding the number of photons between wavelengths λ and $(\lambda+\Delta\lambda)$, and hence to find u_λ, is analogous to the problem of finding the number of molecules between velocities v and $v+\Delta v$. But there is one big difference. Though the number of atoms is conserved, the number of photons need not be conserved. The problem was tackled in this way be Bose and he got the same result, i.e. Eq. (17.18).

17.10 Pyrometers

We have seen in the preceding sections that the radiation from a black-body depends upon its temperature and nothing else. Thus, the measurement of the radiation emitted by the body can enable us to find its temperature. This can be done in two ways. We can measure the total radiation

*In Planck's postulate this discreteness shown by radiation was only at the time of its emission and absorption. Otherwise the radiation was to behave like waves. It is something like purchasing butter. Though it can be purchased in packets only, but once purchased it can be distributed in any proportions. It was Einstein who, to explain photoelectric effect, postulated that the radiation maintains its bundle-like or photon nature even during propagation.

*It does not mean that it amounts to going back to the corpuscular idea of Newton. The overall behaviour of these photons is wave-like.

and deduce the temperature using Stefan-Boltzmann law. Instruments which work on this principle are called total radiation pyrometers. The measurement of the radiant energy between wavelengths λ and $(\lambda + \Delta\lambda)$, i.e. $E_\lambda \Delta\lambda$, in conjunction with Planck's law [Eq. (17.18)] also enables us to find the temperature. The instruments employing this procedure are called spectral or optical pyrometers.

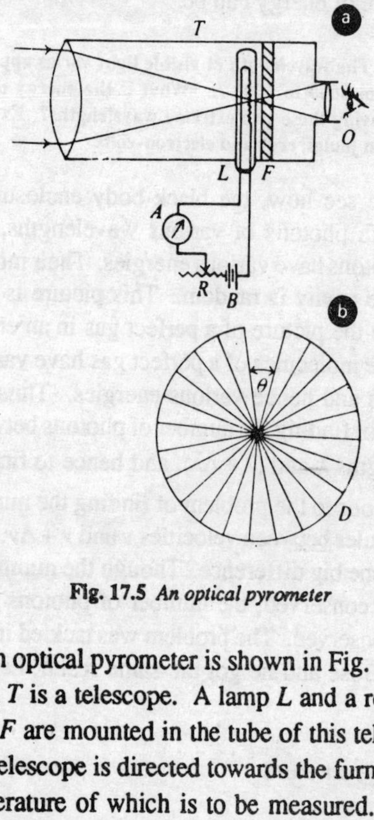

Fig. 17.5 *An optical pyrometer*

An optical pyrometer is shown in Fig. 17.5(a). Here, T is a telescope. A lamp L and a red glass filter F are mounted in the tube of this telescope. The telescope is directed towards the furnace, the temperature of which is to be measured. To an observer O the lamp filament will look dark against the bright background of the furnace. The rheostat R is now adjusted to increase the current in the filament. The current is adjusted to a value such that the brightness of the filament matches with its background. At this stage the filament will not be visible separately. From the previous calibration of the scale of the ammeter, the temperature of the furnace can be immediately read.

To calibrate this instrument, a black-body at the temperature of freezing silver can be taken. The temperature of this body (1234.93 K) is determined with the help of a gas thermometer. While looking at this furnace, the current in the filament is adjusted as already mentioned. Now, another black-body at a higher temperature is sighted through the telescope, and in between the two is kept a sector-shaped disc D shown in Fig. 17.5(b). The current is kept at the same value as before, but the disc is rotated quite fast and the opening in the disc is adjusted in such a way that the filament again disappears. Now, the radiation reaching the telescope will be equal to that which was reaching it from the silver furnace. Let the angular width of this opening be θ. Clearly, the fraction of the radiation passing to the telescope is $\theta/2\pi$. If the filter passes wavelengths between λ and $(\lambda + \Delta\lambda)$, we get,

$$u_\lambda(T_S) = \frac{\theta}{2\pi} u_\lambda(T) \qquad (17.19)$$

where the subscript S stands for silver temperature. Using Eq. (17.18), we get,

$$\frac{1}{e^{hc/\lambda k T_S} - 1} = \frac{\theta}{2\pi} \frac{1}{e^{hc/\lambda k T} - 1}$$

or

$$\frac{e^{hc/\lambda k T} - 1}{e^{hc/\lambda k T_S} - 1} = \frac{\theta}{2\pi} \qquad (17.20)$$

Thus, knowing the wavelength which is transmitted by the filter, the value of T can be found out. Now the disc is removed and the current is reset so that the filament again disappears. At this position of the ammeter needle the temperature determined from Eq. (17.20) is marked. This way, the whole dial of the ammeter is calibrated in terms of temperature. In ITS-90, the optical pyrometers are used above 1234.93 K.

The pyrometers have the advantage that they are not to be raised to the temperature which is being measured. Thus they can measure any temperature where any other thermometer including the thermocouple, cannot work. But pyrometers can measure accurately only the temperatures of black-bodies. If the temperature of a body other than a black-body is measured with the help of these pyrometers, they will give the temperature at which a perfectly black-body would emit the same radiation as the body under investigation. This temperature is called the black-body temperature of the substance, and will be lower than the actual temperature. But corrections can be applied and actual temperature determined.

Discussion and Problems

17.1 Radiant energy with wavelength longer than the visible radiation is often called 'heat radiation'. Do shorter wavelengths not heat the object on which they fall?

17.2 A cube 3.0 cm each side radiates energy at the rate of 20 J/s, when its temperature is 727°C. Determine its emissivity.

17.3 The temperature of a tungsten filament in an incandescent lamp is 2500 K. Find the surface area of the filament of a 60 W lamp. The emissivity of the filament is 0.30.

17.4 Solar radiation received by 1 cm^2 of a black surface held at right-angles to the sun's rays and placed at the mean distance of the earth, provided there were no absorption in the atmosphere or provided the atmosphere were absent, is called the solar constant. It is generally expressed in calories per minute.

Measurements show that the energy received at the earth is 1340 W/m^2. What is the solar constant in calories per cm^2 per minute?

Mean distance of earth from sun is 1.50×10^8 km, and the mean diameter of sun is 1.39×10^6 km. Assuming the sun to be a black-body, calculate its surface temperature.

17.5 A blackened cube of each side 2 cm long is hung in an enclosure kept at 27°C. At what rate would you supply energy to this cube to maintain its temperature at 77 °C. Use both the exact and the approximate formulae and compare the results.

17.6 Two solid spheres of radii a and b are made of the same metal and have similar surfaces. They are heated to the same temperature and are allowed to cool under identical conditions. Compare their initial rates of loss of energy. Also compare their initial rates of fall of temperature.

17.7 A substance takes 3 min in cooling from 50° to 45°C, and takes 5 min in cooling from 45° to 40° C. What is the temperature of the surroundings ? How much time will it take to cool this substance from 40° to 37°C ?

17.8 A copper calorimeter of mass 100.0 g contains 150 cc of turpentine (specific heat 0.60 cal/g-°C, and density 0.87 g/cm^3). It is found to cool at the rate of 1°C/min, when its temperature is 20°C above that of its surroundings. If, now, turpentine is replaced by 150 cc of water what will be the rate of cooling when the temperature is 10°C above that of the surroundings ?

17.9 50 g of water and an equal volume of glycerine (density 1.26 g/cc) are placed, one after another, in the same calorimeter of water equivalent 5 g (i.e. of heat capacity equal to that of 5 g of water). They are found to cool from 45°C to 40°C in 6 and $4\frac{1}{2}$ minuets, respectively.

What is the specific heat of glycerine ?

17.10 Equal volumes of water and paraffin are allowed to cool by turn in the same calorimeter and under the same conditions. Time-temperature readings are given in the following table.

Time (min)	*Water temp* (°C)	*paraffin temp* (°C)
0	60.5	60.1
$\frac{1}{2}$	59.1	58.1
1	57.8	56.2
$1\frac{1}{2}$	56.6	54.4
2	55.5	52.7
$2\frac{1}{2}$	54.5	51.1
3	53.5	49.6
$3\frac{1}{2}$	52.6	48.1
4	51.8	46.9
$4\frac{1}{2}$	51.0	45.7
5	50.3	44.6

Draw temperature versus time graph for both the liquids. Masses of water and paraffin are 100.0 g and 80.0 g respectively. If the thermal capacity of the calorimeter is 13.0 cal/ °C, calculate the mean specific heat of paraffin between 51°and 56°C.

17.11 At what wavelength does the maximum intensity of the radiation from a black-body occur at a temperature of 300 K, 1000 K, 1500 K, 3000 K, 4000 K, 5000 K, 6000 K, and 7000 K ?

With the help of these results, explain the changes in colour with temperature.

17.12 The spectral energy curve of sunlight has a maximum at a wavelength of 4840 Å. Assuming the sun to be a black-body, calculate its surface temperature.

17.13 The wavelength of the light emitted by a sodium lamp is about 6000 Å. What is the frequency ? Calculate the energy of a photon. Express your result in electron volts.

17.14 Assume that a bulb is emitting a monochromatic radiation at 5600 Å. If its wattage is 100 watt, and it converts only 5 per cent of input energy into above radiation, how many photons are emitted by the bulb in one second ?

17.15 A black-body expands isothermally. Will the number of photons in it be conserved ? What happens to gas molecules under similar expansion ?

17.16 A monochromatic radiation is enclosed in a chamber. Will it remain monochromatic as time passes ? If not, what will be the equilibrium state ? Discuss how this equilibrium is attained. How do the molecules attain the equilibrium state if some of them, having uniform velocity, are enclosed in a box ?

17.17 A hollow sphere has inner and outer radii R_1 and R_2, respectively. If at steady state the temperatures of the inner and outer surfaces are T_1 and T_2, respectively, show that the rate of flow of heat is

$$Q = \frac{4\pi K R_1 R_2 (T_1 - T_2)}{R_2 - R_1}$$

and the temperature of the material at a distance r from the çentre is

$$T = T_1 - \frac{(r - R_1)R_2}{(R_2 - R_1)r}(T_1 - T_2)$$

where K is the thermal conductivity of the material.

Appendices

A.1 Vacuum Technique

Vacuum technique is one of the most important techniques for experimental investigations and industrial development. We will give only a brief account of the various devices used to produce and measure high vacua, but an interested reader can refer to the books devoted to this subject.*

A.1.1 Rotary Pump

This pump is used to produce moderately low pressures of the order of 10^{-4} torr (1 torr = 1 mm-Hg). It is based on the principle of displacement of a gas. A portion of the gas inside a vessel is isolated by a rotating disc and is then compressed till it attains a sufficiently high pressure and gets discharged to the atmosphere. As a gas consists of fast moving molecules, more gas comes to occupy that space and, in turn, also gets removed. This process continues and the pressure of the vessel keeps on decreasing.

A typical rotary pump is shown in Fig. A.1.1. The dark-shaded disc rotates clockwise, thus trapping air from the right. S is a stop-cock which connects the pump to the vessel to be evacuated. (When the pump is stopped, this stop-cock is turned to the left to allow the atmospheric air to enter into the pump.) The disc moves immersed in oil. The oil fills the dead space, and thus allows a high

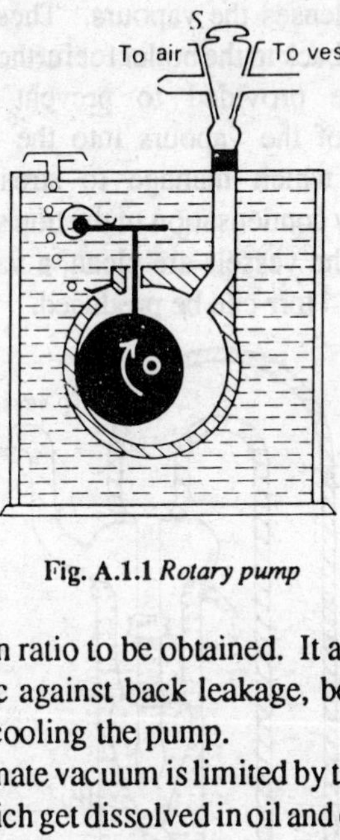

Fig. A.1.1 *Rotary pump*

compression ratio to be obtained. It also seals the moving disc against back leakage, besides lubricating and cooling the pump.

The ultimate vacuum is limited by the gases and vapours which get dissolved in oil and contaminate it. From oil, they go into the vessel being evacuated. Many pumps are accompanied by 'gas ballasts'. A small flow of air, which can be controlled and even stopped when not desired, is allowed into the compression region. This way partial pressure of the vapours being evacuated does not attain high values and, thus, the vapours do not condense in the outlet and get dissolved in oil. In other words, the vapours are carried out by the air.

*See for example: S. Dushman, *Scientific Foundations of Vacuum Technique* (John Wiley), 1962.

A.1.2 Diffusion Pump

To reduce the pressure further, diffusion pumps are used. The vessel is first evacuated by a rotary pump and then the diffusion pump is switched on. It is then continuously backed by the rotary pump. A typical diffusion pump is shown in Fig. A.1.2. Mercury or a low vapour pressure oil (e.g. silicon oil) is heated by heater H. The vapours go up and stream down at the right. This vapour jet sweeps away along with it the molecules entering the pump from the vessel. In Fig. A.1.2, A is the asbestos lining to prevent the cooling of the vapours on their way up. This will give good stream jets without the boiler temperature becoming too high. On the right, the circulation of cold water in the outer jacket condenses the vapours. These condensed vapours go back to the boiler for further circulation. Baffles are provided to prevent direct back streaming of the vapours into the vessel. The molecules which manage to stream back are removed by condensation in the trap shown in Fig. A.1.3. If the vessels are clean, a vacuum of the order of 10^{-8} torr can be produced.

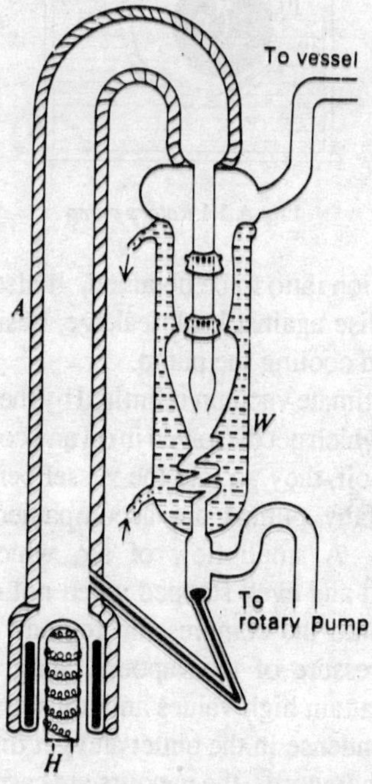

Fig. A.1.2 *Diffusion pump*

A.1.3 McLeod Gauge

This gauge (Fig. A.1.4) is used to measure high vacuum. At the top it is connected to the vessel the pressure of which is to be measured. The mercury in reservoir R is kept below the junction J. When pressure in the vessel acquires equilibrium value (the same will be the pressure in bulb B), the mercury is lifted up. While going up it traps the gas in the bulb B and compresses it into the capillary C. In the left arm A, the mercury level is brought at par with the top of the capillary. Then the mercury level in C reads the pressure written on the scale attached to it. Capillary C' is attached to compensate for the depression in mercury level of C due to surface tension. The mercury in the reservoir R is handled with the help of stop-cock S. When mercury is to be lifted up, the stop-cock is slowly opened to the atmosphere. Mercury is brought down by opening S towards the rotary pump.

Let the initial pressure in B be P (same as that in the vessel), and let V be its volume. Let A be the area of cross-section of the capillary C and let h be its height above mercury level. Thus the gas having a pressure P and volume V is compressed to a pressure h and volume Ah at constant temperature. Therefore,

$$PV = h \times Ah, \quad \text{or} \quad P = h^2A/V$$

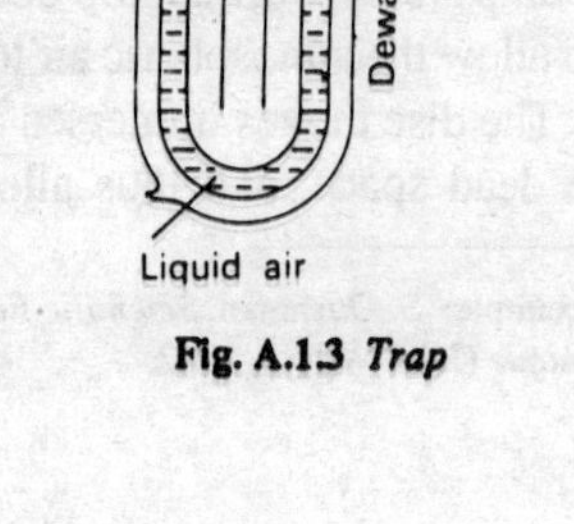

Fig. A.1.3 *Trap*

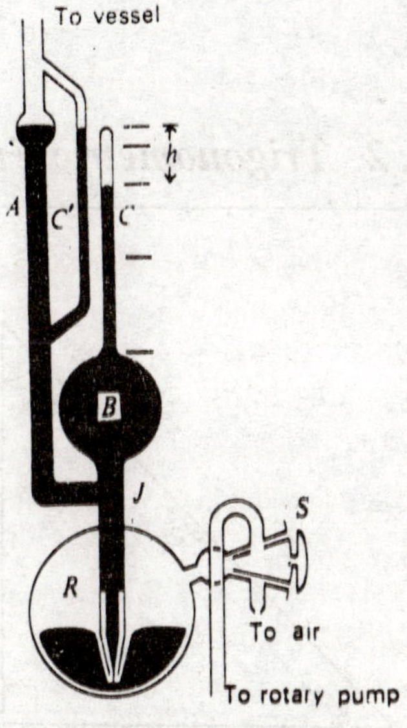

Fig. A.1.4 *McLeod gauge*

One can see from this that higher the volume of the bulb *B*, better is the sensitivity of the gauge.

A.1.4 Pirani Gauge

It consists of a wire (nickel, platinum, or tungsten) suspended in a tube (Fig. A.1.5). This tube is connected to the vacuum system and the wire makes one of the arms of a Wheatstone bridge. When a current is passed through this wire, it gets heated up. But its final temperature will depend

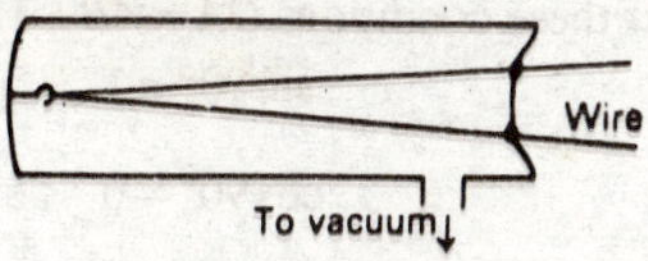

Fig. A.1.5 *Pirani gauge*

upon the pressure of the surrounding gas. If the pressure is more, conduction of heat will be more and, thus, the temperature of the wire will be lower. The reverse will happen if the pressure of the surrounding gas is smaller. Further, the temperature controls the resistance of the wire. Hence, the resistance of the filament is indicative of the pressure of the gas. Thus the changes in pressure either produce a measurable unbalance in the bridge, or are compensated by altering the potential applied to the bridge by a measured amount. This gauge is used to measure low vacua, e.g. those created by rotary pumps. The gauge becomes insensitive when the pressure is so low that heat losses by radiation and to the supports become larger than conduction through the gas.

A.1.5 Ionization Gauge

This gauge is used to measure high vacua, which we get with the help of diffusion pumps. It, like triode, consists of three electrodes (Fig. A.1.6). Electrons from the filament are accelerated by a positive grid until they produce ionization by collision. The ions are produced at a rate proportional to the molecular concentration and, hence, the pressure. These are collected by a negative electrode. Ratio of ion current (recorded by M_1) to electron current (recorded by M_2) is a measure of the gas pressure.

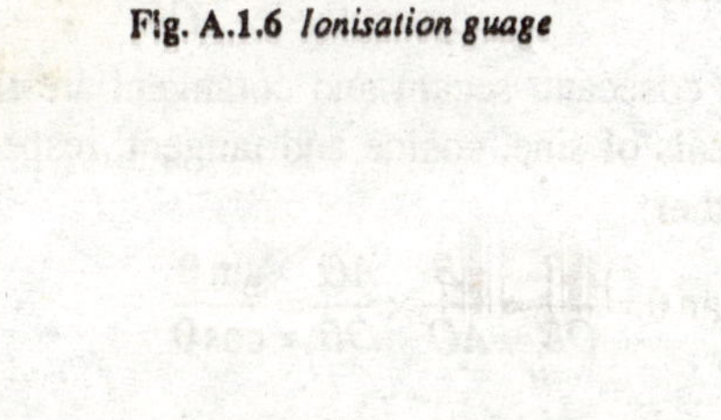

Fig. A.1.6 *Ionisation guage*

A. 2 Trigonometric Functions

Let us take a coordinate system as shown in Fig. A.2.1. OAB is a right-angled triangle with angle B equal to 90°. Let us concentrate on angle AOB and call it θ. Side AB is called the opposite side, side OB is called the adjacent side, and side OA is called the hypotenuse. The trigonometric functions sine, cosine, tangent, cotangent, secant, and cosecant are used to define the characteristics of an angle. By definition:

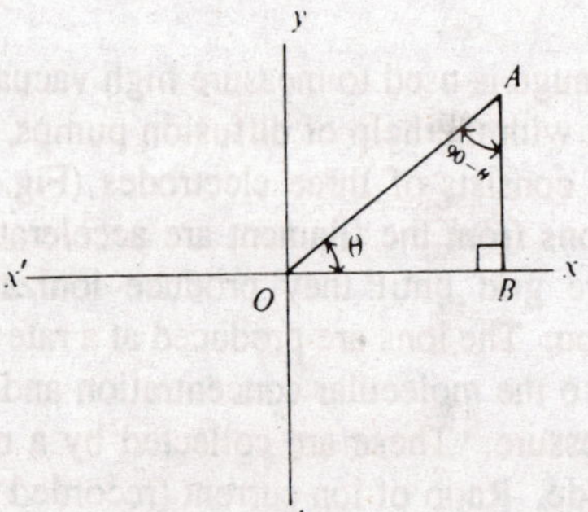

Fig. A.2.1

$$\sin\theta \equiv \frac{\text{opposite side}}{\text{hypotenuse}} = \frac{AB}{OA} \qquad \text{(A.2.1)}$$

$$\cos\theta \equiv \frac{\text{adjacent side}}{\text{hypotenuse}} = \frac{OB}{OA} \qquad \text{(A.2.2)}$$

$$\text{and, } \tan\theta \equiv \frac{\text{opposite side}}{\text{adjacent side}} = \frac{AB}{OB} \qquad \text{(A.2.3)}$$

The cosecant, secant and cotangent are the reciprocals of sine, cosine and tangent, respectively. Further:

$$\tan\theta = \frac{AB}{OB} = \frac{AB}{AO} \times \frac{AO}{OB} = \frac{\sin\theta}{\cos\theta} \qquad \text{(A.2.4)}$$

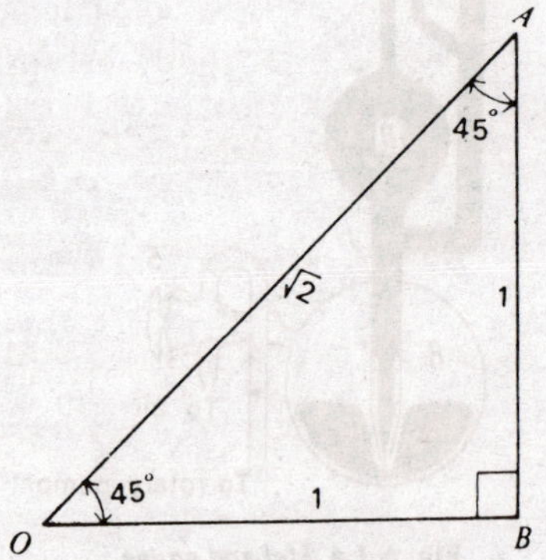

Fig. A.2.2

Therefore, if sine and cosine of an angle are known, all the remaining four functions can be determined.

If, in Fig. A.2.1, A approaches B, then AB and θ both will approach zero, and OA will become equal to OB. Hence:

$$\sin 0° = 0 \qquad \text{(A.2.5)}$$

$$\cos 0° = 1 \qquad \text{(A.2.6)}$$

$$\tan 0° = 0 \qquad \text{(A.2.7)}$$

It is also clear that if A goes on to the y-axis θ will become 90° and base OB will become zero. Under these conditions $OA = AB$. Therefore:

$$\sin 90° = 1 \qquad \text{(A.2.8)}$$

$$\cos 90° = 0 \qquad \text{(A.2.9)}$$

$$\text{and, } \tan 90° = +\infty \qquad \text{(A.2.10)}$$

If in $\Delta OAB, \angle A = \angle O = 45°$, then $AB = OB$. Further, if each of these sides is taken as equal to unity, then hypotenuse OA will be equal to $\sqrt{2}$. This triangle is drawn in Fig. A.2.2 from which we see that

$$\sin 45° = 1/\sqrt{2} \quad \text{(A.2.11)}$$

$$\cos 45° = 1/\sqrt{2} \quad \text{(A.2.12)}$$

and, $$\tan 45° = 1 \quad \text{(A.2.13)}$$

If θ is 30°, then it can be easily shown that the length of the side opposite to it is equal to half the length of the hypotenuse. Therefore, the sides of the triangle will be in the proportion shown in Fig. A.2.3. From this figure it is clear that

$$\sin 30° = 1/2 \quad \text{(A.2.14)}$$

$$\cos 30° = \sqrt{3}/2 \quad \text{(A.2.15)}$$

$$\tan 30° = 1/\sqrt{3} \quad \text{(A.2.16)}$$

$$\sin 60° = \sqrt{3}/2 \quad \text{(A.2.17)}$$

$$\cos 60° = 1/2 \quad \text{(A.2.18)}$$

and, $$\tan 60° = \sqrt{3} \quad \text{(A.2.19)}$$

Let us return to Fig. A.2.1. If line OA is rotated about the origin O, then θ will go through all values between 0° and 360°. We know from convention that OA is always taken as positive; OB is positive only when it is on the right of the axis yOy', but is otherwise negative; and AB is positive when it is above the axis xOx' and is negative when below it. Angle θ is taken as positive when OA rotates in an anti-clockwise direction, and is taken as negative when OA rotates in clockwise direction. The definitions of the trigonometric functions remain the same as in Eqs. (A.2.1)–(A.2.3). From this, one can easily verify that the signs of various trigonometric functions will be as given in Fig. A.2.4.

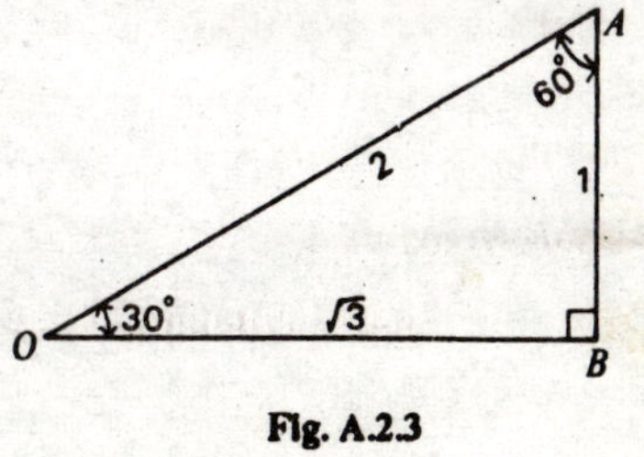

Fig. A.2.3

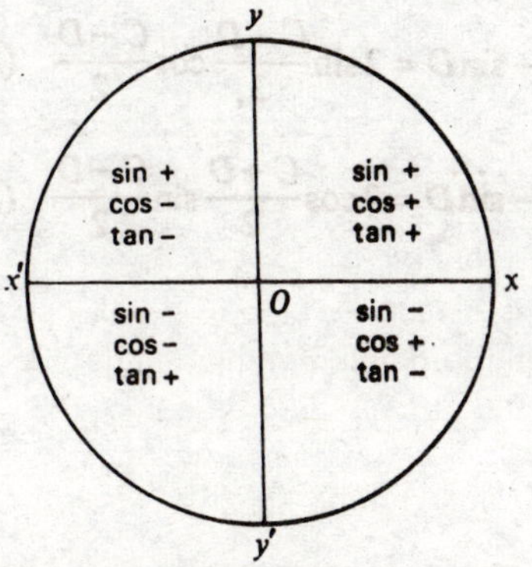

Fig. A.2.4

There are two more useful trigonometric relations which we give here without proof. If there is a triangle ABC as shown in Fig. A.2.5 then:

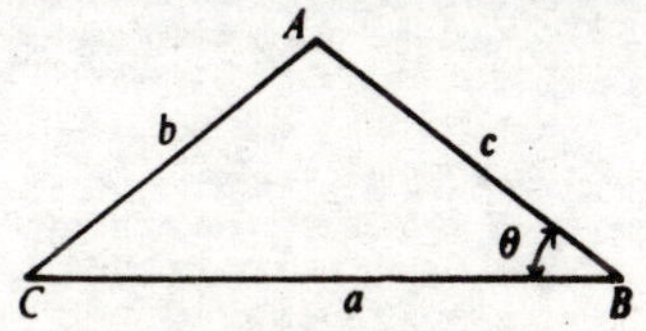

Fig. A.2.5

$$AC^2 = BC^2 + AB^2 - 2AB \times BC \cos\theta \quad \text{(A.2.20)}$$

In a special case when θ = 90°, this relation reduces to

$$AC^2 = BC^2 + AB^2 \quad \text{(A.2.21)}$$

which is the famous Pythagorous theorem.

If the sides opposite to the angles A, B and C are respectively of the length a, b and c, we have:

$$\frac{a}{\sin A} = \frac{b}{\sin B} = \frac{c}{\sin C} \quad \text{(A.2.22)}$$

Relations (A.2.20) and (A.2.22) are known as the law of cosines, and the law of sines, respectively.

Some other identities are:

$$\sin(A+B) = \sin A \cos B + \cos A \sin B \quad \text{(A.2.23)}$$

$$\cos(A+B) = \cos A \cos B - \sin A \sin B \quad \text{(A.2.24)}$$

$$\sin(A-B) = \sin A \cos B - \cos A \sin B \quad \text{(A.2.25)}$$

$$\cos(A-B) = \cos A \cos B + \sin A \sin B \quad \text{(A.2.26)}$$

$$\sin C + \sin D = 2\sin\frac{C+D}{2}\cos\frac{C-D}{2} \qquad (A.2.27)$$

$$\sin C - \sin D = 2\cos\frac{C+D}{2}\sin\frac{C-D}{2} \qquad (A.2.28)$$

$$\cos C + \cos D = 2\cos\frac{C+D}{2}\cos\frac{C-D}{2} \qquad (A.2.29)$$

$$\cos C - \cos D = 2\sin\frac{C+D}{2}\sin\frac{D-C}{2} \qquad (A.2.30)$$

A. 3 Symbols and Notations

A.3.1 Abbreviations of Units*

abA	Ab-ampere (emu of current)
amu	Atomic mass unit
Å	Angstrom unit
AU	Astronomical unit
atm	Atmosphere
BTU	British thermal unit
cal	Calorie
cm	Centimetre
dyn	Dyne
eV	Electron volt
°F	Degree Fahrenheit
ft	Foot
g	Gram
hr	Hour
hp	Horsepower
in	Inch
kcal	Kilocalorie
kg-wt	Kilogram-weight
km	Kilometre
min	Minute
lb	Pound
lb-m̄	Pound mass
mì	Mile
statcoul	Stat-Coulomb (esu of charge)
G	Gauss
yd	Yard
yr	Year

*Also see tables of Chapter 1.

A.3.2 Greek Alphabet

Name	Letter	
	Capital	*Lower*
Alpha	Α	α
Beta	Β	β
Gamma	Γ	γ
Delta	Δ	δ
Epsilon	Ε	ε
Zeta	Ζ	ζ
Eta	Η	η
Theta	Θ	θ
Iota	Ι	ι
Kappa	Κ	κ
Lambda	Λ	λ
Mu	Μ	μ
Nu	Ν	ν
Xi	Ξ	ξ
Omicron	Ο	ο
Pi	Π	π
Rho	Ρ	ρ
Sigma	Σ	σ
Tau	Τ	τ
Upsilon	Υ	υ
Phi	Φ	φ
Chi	Χ	χ
Psi	Ψ	ψ
Omega	Ω	ω

A.3.3 Mathematical Symbols

$=$	is equal to	$<$	is less than
$\cong$	is nearly equal to	$\ll$	is much less than
$\simeq$	is approximately equal to	$\geq$	is greater than or equal to
$\approx$	is roughly equal to	$\leq$	is less than or equal to
$\sim$	is of the order of magnitude of	$\propto$	is proportional to
$\neq$	is not equal to	Σ	the sum of
$\equiv$	is identical to, is defined as	∞	infinity
$\pm$	plus or minus	$\langle x \rangle$	average value of *x*
$>$	is greater than		
$\gg$	is much greater than		

A.3.4 Recommended Unit Prefixes

Multiples		*Prefix*	*Symbol*	*Submultiples*		*Prefix*	*Symbol*
10	$= 10^{1}$	deca	da	0.1	$= 10^{-1}$	deci	d
100	$= 10^{2}$	hecto	h	0.01	$= 10^{-2}$	centi	c
1000	$= 10^{3}$	kilo	k	0.001	$= 10^{-3}$	milli	m
1000000	$= 10^{6}$	mega	M	0.000001	$= 10^{-6}$	micro	μ
1000000000	$= 10^{9}$	giga	G	0.000000001	$= 10^{-9}$	nano	n
1000000000000	$= 10^{12}$	tera	T	0.000000000001	$= 10^{-12}$	pico	p
1000000000000000	$= 10^{15}$	peta	P	0.000000000000001	$= 10^{-15}$	femto	f
1000000000000000000	$= 10^{18}$	exa	E	0.000000000000000001	$= 10^{-18}$	atto	a

A.4 Numerical Constants

A.4.1 Table of Physical Constants

Constant	*Symbol*	*Value*	*Units*	
			SI	*cgs*
Speed of light in vacuum	c	2.998	$\times 10^{8}$ m s^{-1}	$\times 10^{10}$ cm s^{-1}
Elementary charge	e	1.602	10^{-19} C	10^{-20} $cm^{1/2} g^{1/2}$*
		4.803	... ...	10^{-10} $cm^{3/2} g^{1/2} s^{-1}$**
Avogadro's number	N_A	6.022	10^{23} $mole^{-1}$	10^{23} $mole^{-1}$
Faraday constant, $N_A e$	F	9.648	10^{4} C $mole^{-1}$	10^{3} $cm^{1/2} g^{1/2} mole^{-1}$*
Planck's constant	h	6.626	10^{-34} J s	10^{-27} erg s
Electron charge to mass ratio	e/m_e	1.759	10^{11} C kg^{-1}	10^{7} $cm^{1/2} g^{-1/2}$*
		5.273	... ...	10^{17} $cm^{3/2} g^{-1/2} s^{-1}$**
Rydberg constant	R_∞	1.097	10^{7} m^{-1}	10^{5} cm^{-1}
Bohr radius	a_0	5.292	10^{-11} m	10^{-9} cm
Bohr magneton	μ_B	9.274	10^{-24} J T^{-1}	10^{-21} erg $gauss^{-1}$*
Gas constant	R_0	8.314	10^{0} J K^{-1} $mole^{-1}$	10^{7} erg K^{-1} $mole^{-1}$
Boltzmann's constant, R_0/N_A	k	1.381	10^{-23} J K^{-1}	10^{-16} erg K^{-1}
Wien's constant	b	2.898	10^{-3} m K	10^{-1} cm K
Stefan-Boltzmann constant	σ	5.670	10^{-8} W m^{-2} K^{-4}	10^{-5} erg cm^{-2} s^{-1} K^{-4}
Gravitational constant	G	6.672	10^{-11} N m^2 kg^{-2}	10^{-8} dyn cm^2 g^{-2}
Atomic mass unit	amu	1.661	10^{-27} kg	10^{-24} g
Electron rest mass	m_e	9.110	10^{-31} kg	10^{-28} g
Proton rest mass	M_p	1.673	10^{-27} kg	10^{-24} g
Neutron rest mass	M_n	1.675	10^{-27} kg	10^{-24} g
Ratio of mass of proton to mass of electron	M_p/m_e	1836		

*Electromagnetic system **Electrostatic system

Source: International Union of Pure and Applied Physics—**Symbols, Units and Nomenclature in Physics, Physica** A, **39A, Nos.** 1 and 2 (August, 1978). The values given have been rounded off at the fourth significant figure.

A.4.2 Some Useful Data

Molar volume of a perfect gas at STP (Standard temperature and pressure = 0°C, 1 atm)	... .	22.41×10^{-3} m^3 $mole^{-1}$, 22.41 litre $mole^{-1}$, 22.41×10^{3} cm^3 $mole^{-1}$
Density of dry air at STP	... .	1.293 kg m^{-3}, 2.458×10^{-3} slug ft^{-3}
Speed of sound in air at STP	... .	331.4 m s^{-1}, 1089 ft s^{-1}
Density of mercury at 0°C	... .	1.360×10^{4} kg/m^3
Density of mercury at 27°C	... .	1.353×10^{4} kg/m^3
Mass of earth	... .	5.979×10^{24} kg

Mean diameter of earth	1.274×10^4 km
Mean density of earth	5522 kg/m^3
Mass of moon	7.343×10^{22} kg
Mean diameter of moon	3.476×10^3 km
Mass of sun	1.987×10^{30} kg
Mean diameter of sun	1.391×10^6 km
Mean distance of earth from sun, 1 AU	1.496×10^8 km
Mean distance of earth from moon	3.84×10^5 km
Solar constant, S	1340 W m^{-2}, 1.92 cal cm^{-2} min^{-1}
Permeability constant μ_0	1.257×10^{-6} henry m^{-1}
Permittivity constant, ε_0	8.854×10^{-12} farad m^{-1}
Earth's magnetic dipole moment	6.4×10^{21} A m^2
Standard acceleration of free fall	9.807 m s^{-2}, 980.7 cm s^{-2}
Earth's magnetic field at Jodhpur:	
Horizontal component	0.356 gauss
Vertical component	0.286 gauss

A.4.3 Some Conversion Factors

1 micron	$= 10^{-4}$ cm
	$= 10^{-6}$ m
1 angstrom unit	$= 10^{-8}$ cm
	$= 10^{-10}$ m
1 inch	= 2.540 cm
1 mile	= 1.609 km
1 kg	= 2.205 lb-m
	$= 5.610 \times 10^{29}$ MeV
1 slug	= 14.59 kg
1 amu (1/12 of mass of a ${}^{12}_{6}C$ atom)	$= 1.661 \times 10^{-27}$ kg
	= 931.5 MeV
1 newton	= 0.2248 lb-wt
1 horsepower	= 550 foot-pounds/sec
	= 745.7 W
1 calorie	= 4.184 J
1 electron volt (eV)	$= 1.602 \times 10^{-19}$ J
Mechanical equivalent of heat	$= 4.184 \times 10^3$ J/kcal
	= 4.184 J/cal
1 BTU	= 777.9 foot-pounds
	= 1055 J
	= 252.0 cal
Standard atmosphere, 760.0 mm-Hg	$= 1.013 \times 10^5$ N/m^2
	$= 1.013 \times 10^6$ dyn/cm^2
	= 14.70 lb/in^2
1 bar	$= 10^6$ dyn/cm^2
	$= 10^5$ N/m^2
	= 0.1 M Pa
1 gamma	$= 10^{-5}$ gauss

$X°C = (273.15 + X)$ K

$X°C = (\frac{9}{5}X + 32)°F$

Answers to Selected Problems

Chapter 1

1.6 (i) 5 units, 53.1°; $\sqrt{5}$ units, 153.4°
(iii) –2 units; 11**k** (iv) 100.3°

1.10 (i) 3 units; 10 units
(ii) $(3\mathbf{i} + 4\mathbf{j} - \mathbf{k})/\sqrt{26}$; $(4\mathbf{j} + 3\mathbf{k})/5$

Chapter 2

2.6 1.0 km hr^{-1}/s

2.7 0.38 km hr^{-1}/s S-E

2.8 8.2 s; 67 ft; 41 ft/s, 24.5 ft/s

2.9 10 s; 20 s; 150 m

2.10 272 s; 1.1×10^3 km

2.12 15.7 rad s^{-2}; 292 rev

2.15 12.5 cm s^{-2}; 7848 cm; 44′

2.16 (i) 3.2 m s^{-2}, 21° downward
(ii) 116.5 N upward

2.17 106 min

2.21 (i) 10 s (ii) 849 m (iii) 87 m/s (iv) 98 m/s

2.24 2.8 m s^{-2}, 56 N

2.27 70 m/s; double the actual weight

2.30 (i) 1.6 m (ii) 0.90 s (iii) 3.5 m/s

2.31 (i) 1.36 m s^{-2} (ii) 11.3 N

2.32 0.33

Chapter 3

3.5 500 km/hr N-37°W

3.6 15 km/hr; 20.6 km/hr N-14°E

3.12 (i) 4.9 m s^{-2} (ii) 1.5 kg

3.22 $\cong c$, $\cong 4 \times 10^4$ yr

3.25 1.5×10^8 m/s, 520 m

3.28 $x' = -58$ m, $y' = 10$ m, $z' = 1000$ m, $t' = 0.96$ μs

3.31 0.75c

Chapter 4

4.4 4.4 m/s

4.6 10.4 m/s

4.7 6.4×10^{-23} kg m s^{-1}, 76.5° with the direction of electron

4.8 44 km

4.12 6 km/s

4.14 0.30 kg/s

4.16 4.7 m s^{-1} E-45°N; 7.5 m s^{-1} E-26.6° S; 3.7 m s^{-1}
W-26.6° N

4.18 (ii) 2.2 m

Chapter 5

5.4 (i)100 N m^{-1} (ii) 0.5 J
(iii) 50 N m^{-1}, 200 N m^{-1}

5.5 0.11 m

5.6 (i) 0.20 m (ii) 0.71 m s^{-1}

5.8 2.2 J, 0.44 W

5.10 120 m.p.h.

5.14 $(m + M)\sqrt{2gh}/m$

5.19 0.99 c, 18.2

5.20 (i) 116.5 MeV (ii) 0.81 c, 0.20 c

Chapter 6

6.4 (i) 2×10^8 kg m^2 s^{-1}

6.5 123 N

6.7 $2mR^2/3$, spherical shell

6.9 0.25 kg m^2

6.11 2.66×10^{-47} kg m^2

Chapter 7

7.3 $x = 0.014 \sin(40\pi t + \pi/4)$

7.4 $T = 2\pi\sqrt{(M + m)x/mg}$

7.8 (i) 0.25 s (ii) 0.36 s

7.14 $2\pi A(M + m)/Tm$

7.17 256.2 Hz or 255.8 Hz

Chapter 8

8.4 0.010 J

8.6 247 km/hr

8.8 0.87 kg, 10.13 kg

8.10 27 cm-Hg

Chapter 9

9.2 $12\pi\sigma R^2$

9.3 2.0×10^4 dyn/cm^2

9.8 140 N

Chapter 10

10.10 2 π MHz

10.13 3.6/(6n + 1) ms^{-1}, where n = 0, 2, 4 ...

Chapter 11

11.11 40 nanowatt/cm^2, 2.5 nanowatt/cm^2

11.22 2199 Hz, 1834 Hz

11.24 0.76 mach

11.26 28.2°

11.28 40 μW/cm^2

Chapter 12

12.1 150 Hz

12.17 $(1/2\pi)\sqrt{k/m}$, $(1/2\pi)\sqrt{3k/m}$

Chapter 13

13.8 1.27 g

13.9 2.06×10^{-3}g

13.12 0.083 g, 899°C
13.13 8 mm
13.15 (*i*) 2 : 1 (*ii*) 1 atm (*iii*) −123°C

Chapter 14

14.11 8829 J
14.13 0.14°C
14.14 (*ii*) 277.5°C
14.15 (*i*) 74 cal (*ii*) 27 cal, − 87 cal (*iii*) 60 cal (*iv*) 24 cal, 50 cal

Chapter 15

15.7 3.5×10^5
15.11 477 m s^{-1}
15.14 77.9 J, 46.7 J
15.15 60°C, 60°C, < 60°C
15.16 6.2×10^{-21} J, 20.8 J
15.18 (*i*) 1.16×10^{10} K (*ii*) 9.1×10^{23} K

Chapter 16

16.6 2.9 Å
16.7 1.73×10^4 Nm^{-2}; 6.9×10^4 Nm^{-2}; 5.9×10^6 Nm^{-2}, 5.6×10^6 Nm^{-2}; 12.4×10^6 Nm^{-2}, 11.1×10^6 Nm^{-2}

Chapter 17

17.3 9×10^{-5}m^2
17.4 1.922 cal/cm^2-min, 5.76×10^3 K
17.5 0.941 W, 0.735 W
17.7 35°C, 6.4 min
17.12 5.99×10^3 K
17.13 5×10^{14} Hz, 2.1 eV
17.14 1.4×10^{19}

Index

Absolute temperature 192
Absorptance 233
Absorptivity 233
Acceleration 22
 angular 26, 96, 97, 98
 average 22
 centripetal 24, 26
 due to gravity 30
 instantaneous 22
 transverse 26
Accelerated frames 42-44
Acoustic holography 172
Acoustical impedance 163
Acoustics of buildings 173-174
Addition of velocities, Galilean 41
 relativistic 54-55
Adhesive force 141
Adiabatic, demagnetization 227
 expansion 198, 203, 227
  of an ideal gas 209
 nuclear demagnetization 227
 wall 188
Amagat 222
Amorphous substances 136
Ampere 7
Amplitude 105
 of wave 151
 for faintest sound 164
 for loudest painless sound 164
 for superposed waves 153
 of progressive wave 163
Andrew 22
Andromeda 2
Angle of contact 141-142
Angular frequency 105
Angular momentum 92-94, 98
 internal 94
 conservation of 94
 of rigid body 98
 orbital 94
Angular quantities
 acceleration 26, 97, 98
 displacement 24-25
 frequency 105
 impulse 93
 momentum 92-94, 98
 vectors 24-26
 velocity 25, 98, 105
 wave number 152
Anharmonic function 107
Anharmonicity 107, 139
Antinode 178, 179, 183
Aphelion 32
Apogee 32
Archimedes' principle 120-121
Astronomical unit 2
Atomic binding 131
Atomiser 124
Atwood machine 58
Avogadro's law 213
Axial vector 24

Ballistic pendulum 89, 118
Bandwidth 155
Banking of tracks 28
Bardeen 228
Basic interactions 29
 electromagnetic 29
 gravitational 29
 nuclear 29
 weak 29
BCS theory 228
Beat frequency 154
Beats 153-154
Bel 164
Bernoulli's equation 122-123
Big bang theory 2
Binding energy 88-89
 molecular 88
 of earth-satellite system 88
 of nuclear particles 89
 of planets 88
 of water molecule 88
Black body 233
 Fery's 233
Black body (or temperature) radiation 234-235
 spectral distribution of 237-239
Black-hole 57
Bolometer 232
Boltzmann 235
Bow wave 170
Boyle's law 213
Brahe 1, 32
Brownian motion 217
Bulk modulus of elasticity 137, 160, 161, 163

Callender and Griffith's bridge 192
Caloric fluid 196
Calorimetry 199-200
Candela 7
Capillarity 143-144
Capillary (or surface tension) waves 150
Carbon dioxide, isotherms 222
 van der Waals constants of 224
Carburetter 124
Carnot cycle 202-203
Carnot's theorem 204
Cathode ray oscilloscope 115
Celsius scale 192
Centigrade scale 192
Central force 29, 33, 94
Centre of gravity 96
Centre of mass 64-66
 acceleration of 65
 velocity of 65
Centrifugal force 44
Centripetal, acceleration 24, 26
 force 27, 72
Cesium chloride structure 134
Cherenkov radiation 171
Circular membrane, see kettle drum
Claude 227
Clausius 205
Coefficient of, friction 35
 kinetic 35
 static 35
 linear expansion 139
 restitution 82
 viscosity 125
 volume expansion 139
Coherent sources 167
Cohesive force 130, 139
Collinear vectors 13
Collins 227
Collision of two particles 80-83
 elastic 80-82
 head on 80-81
 inelastic 80, 82, 217
 oblique 82-83
Compound (or physical) pendulum 112-113
Compressibility 137
Compression 149, 163
 wave of 159
Concurrent forces 36-37
 coplanar 37
 equillibrium of 36
Conical pendulum 38
Conservation of, angular momentum 94
 energy 199
 linear momentum 62-63
 mechanical energy 77, 80, 85
Conservative force 83-85
Constant volume gas thermometer 190-191
Cooper 228
Coordinate system, centre of mass 66, 67
 laboratory 66, 67
 right-handed 19, 25
Copernicus 1

Cord 185
concordant 185
discordant 185
Cosmic rays 3
Couple 96
Covalent binding 132
Co-volume 230
Critical temperature 222, 226, 227
Cryogenics 227
Crystal, covalent 132
ionic 131
metallic 132
molecular 132-133

Damped SHM 115-116
Decibel 164
Degrees of freedom 176-215
Detergent action 145
Deuteron 89
Dewar flask 190
Diamond structure 135
Diathermic wall 188
Diesel cycle 207
efficiency of 208
Diffraction of sound 165, 168
Diffusion 213
pump 244
Dimensional, analysis 8
equation 8
formula 8
Dimensions 7-9
of some physical quantities 11
Dipole-dipole interaction 130
Dipole-induced dipole interaction 130
Dispersion 179
Dispersion force 130
Dispersive medium 155, 179
Displacement vector 15
Dissipative force 85
Distance measurement by,
parallax method 4-6
triangulation 4-6
Doppler effect 2, 168-170
relativistic for light 170
Dulong 235
Dyne 9, 27

Eddies 128
Eddy currents 126
Efficiency of heat engine 202
Eigenfrequency 178
Eigenfunction 178
Eigenvalue 178
Einstein 48, 50, 239
Einstein's principle of relativity 48
general 56
Elastic collisions 80-82
Elastic constants, table 137
Elastic limit 136
Elasticity ff 136
Electromagnetic interaction 29
Electron affinity 131
Electron volt 88
Electronegativity 132
Elevation head 123
Emissivity 236
of various substances 236
Emittance 233
monochromatic 233
of black body 234
relative 236
Energy,
binding 88-89
molecular 88
of earth-satellite system 88
of nuclear particles 89
of planets 88
of water molecule 88
conservation of mechanical 77, 80, 85
free surface 140
interchange of 80
internal 76, 199
kinetic 75-76
internal 76
rotational 98
of stretched spring 79-80
potential 76-80
rest 87
Enthalpy 225
Entropy 199, 204
Epoch 109
Equation of continuity 122
Equilibrium of rigid body 95-96
Equipartition of energy 215
Erg 72
Errors 10, 12
Escape velocity 69, 79
Ether 45
wind 47
Excess pressure 142

Face centred cubic structure 134
Fahrenheit scale 192
Fermi 4
Fery's black body 233
Fictitious force 43-44
Field,
electromagnetic due to line charge 45
gravitational 30
Filter pump 124
First law of thermodynamics 199
Fluid 119
Fluid motion, streamline 121-122, 126
turbulent 121, 126
Fly-wheel 99, 103
Foot-pound 73
Force, adhesive 141
average 62
central 29, 33, 94
centrifugal 44
centripetal 27
cohesive 130, 139
concept of 26-27
conservative 83-85
fictitious 43-44
frictional 34-36
gravitational 29-30
impulse of 60-62
instantaneous 62
moment of 91
non-conservative 83-85
of buoyancy 121
restoring 107
unit of 27
Force-time curve 61
Forced vibrations 116
Fourier analysis 156-157
Frames of reference ff 39
accelerated 42-44
centre of mass 66, 67
inertial 40
laboratory 66, 67
Free expansion 214
Free surface energy 140
Freon 226
Frequency 105
Friction 34-36
coefficient of 35
kinetic 35
static 35
mechanism of 35-36
rolling 36
sliding 36
Fundamental frequency 178
Fusion 227

Galaxies 2
Galilean, invariance principle 40
transformations 41
and electromagnetism 45
Galileo 1
Galton's whistle 172
Gauge, ionization 245
McLeod 244-245
Pirani 245
pressure 129
General theory of relativity 55-57
Glasses 136
silicate 136
Gravitational, constant 30
field intensity 29, 30

due to a shell 30
force 29-30
head 123
interaction 29
mass 28
potential 78
due to a shell 78-79
red shift 57
Gravity waves 150
Group velocity 155-156

Harmonic, function 105
generation 111
Harmonics 178
Head 123
elevation 123
gravitational 123
pressure 123
velocity 123
Heat 187
capacity 200
conduction 231-232
convection 232
radiation 232
Heisenberg 155
Helium I and II 229
Hexagonal close packed structure 133
Hooke's law 136
Horse power 85
Hydraulic press 120
Hydrogen bond 132

Ideal gas 193-194, 201
temperature scale 191-192
Impact parameter 82
Imperfect gas ff 222
Impulse, angular 93
of force 60-62
Indicator diagram 198
Inelastic collisions 80, 82, 217
Intertial, frame of reference 40
mass 27, 29
Infrasonic waves 171
Intensity of, a wave 152
sound 162-165
Interference of sound waves 166-168
Intermolecular forces 130, 222-223
Internal combustion engines 206-208
Diesel cycle 207
efficiency of 208
Otto cycle 206
efficiency of 207
Internal energy 187, 199
Internal pressure 223
Inversion temperature 225, 226
Ionic binding 131
Ionization, gauge 245

potential 131
Irradiance 234
Irreversible, engine 204
process 205
Irrotational flow 121
Isobaric process 197
Isochoric process 197
Isotherms 189
of CO_2 222
of ideal gas 222
of van der Waals gas 223
Isothermal expansion 198, 202
Isotropic, medium 159
radiation 235
ITS-90 scale 192

Jaeger's method 145
Jeans 238
Joule 72, 196
Joule's expansion, see free expansion
Joule's experiment 209
Joule's law 214
Joule-Thomson expansion 224-225

Kammerlingh Onnes 228
Kapitza 227
Kelvin, 7
scale 192
Kepler 1
Kepler's laws 32, 95
Kettle drum 178
kilocalori 196
Kilogram 7
Kilowatt hour 86
Kinetic energy 75-76
internal 76
rotational 98
Kinetic interpretation of temperature 213-214
Kinetic temperature 219
Kirchhoff's law 234
Kundt's tube 184

Lambda point 229
Laminar flow 125
Laplace's correction 161
Length contraction 50-51
Life time 155
Light, clock 49
year 2
Linde 226
Line of flow 121
Linear momentum 60-62
conservation of 62-63
Liquefaction, of air 226
of gases 226-227
Liquids 135-136

Lissajous figures 113-115, 118
Loaded spring 106-107
Longitudinal waves 148-149
velocity of 159-160
Lorentz transformations 52-53
Loudness 185
Lowest temperature 227

Mach, angle 170
number 170
Magnetostriction 172
Maintained oscillations,
externally 116-117
self 117
Mass, and energy 86-87
gravitational 29
inertial 27, 29
proper 69
reduced 111
relativistic variation of 69
rest 69, 86
unit of 7
Maxwell 218
Maxwell's formula 218
Maxwellian distribution 218
McLeod gauge 244-245
Mechanical energy, conservation of 77, 80, 85
Melde's experiment 181-182
Metallic binding 132
Metre 7
Michelson and Morley experiment 45-48
Milky way 2
Model for a gas 210-211
Modes of vibration 176, 183-184
Modulus of rigidity 137
Molar specific heat 209
Mole 7
Molecular binding 132-133
Moment of, force 91
inertia 96-97
parallel axes theorem 101
perpendicular axes theorem 102
Momentum, angular 92-94, 98
conservation of 94
linear 61
conservation of 62-63
Monochromatic, absorptance 233
emittance 233
Morse potential 107, 110
Motion, circular 23-24, 44
of projectile 32
of rocket 63-64
of satellite 30
Musical interval 185
Musical scales 184-186

diatonic (or true) 185
tempered 186

Nebulae 2
Neutron star 3
Neutrino 69
Newton 9, 27
Newton's, first law of motion 27
law of cooling 237
law of gravitation 30
second law of motion 27, 69
third law of motion 33
Newton's formula for speed of sound in air 161
Laplace's correction 161
Node 179, 183
Non-conservative force 83-85
Non-linear, optics 111
oscillations 111
Note 185
tonic 185
Nuclear interaction 29
Null vector 13

Orbital angular momentum 94
Orthogonal vectors 19
Oscillations, anharmonic 107, 111
externally maintained 116-117
harmonic 105, 107
non-linear 111
of loaded spring 108-109
parametric 117
self maintained 117
Oscillators, crystal controlled 4
microwave 4
Oscilloscope 4, 115
Otto cycle 206
efficiency of 207
Overtones 178

Parsec 2
Parallel vectors 13
Parallax 6
Parametric amplifier 117
Pascal 119
-second 125
Pascal's law 120
Pauli's principle 131
Pendulum, ballistic 89, 118
compound (or physical) 112
conical 38
electrical 4
equivalent simple 113
physical (or compound) 112
seconds 112
simple 111-112
Perfect gas, see ideal gas

Perigee 32
Perihelion 32
Periodic time 105
of compound pendulum 113
of loaded spring 109
of physical pendulum 113
of satellite around earth 31
of simple pendulum 112
Petit 235
Phase 227
change 227-228
of SHM 109
Photoelectric effect 239
Photon 69, 239
Physical pendulum, see compound pendulum
Piezoelectric crystal 172
Pipes, open and closed 182-184
Pirani gauge 245
Pitch 185
Pitot tube 129
Planetary motion, Kepler's laws 32, 95
Planck 238
Planck's, constant 239
law 239
Plastic flow 136
Platinum resistance thermometer 192-193
Poise 125
Poiseulle's formula 126
Poisson ratio 137
Polar vector 24
Polarization 179
circular
anticlockwise 115
clockwise 114
elliptical 115
linear 113
Porous plug experiment 224-225
Position vector 15, 39
of centre of mass 65
Potential energy 76-80
gravitational 77-79
of a shell 78-79
of loaded spring 79-80
zero of 77
Power 85-86
average 85
instantaneous 85
Pressure, drag 126, 127, 128
excess over curved surface 142
gauge 129
head 123
Prevost's theory 233
Principle of equivalence 56
Principle of relativity, Einstein's 48
Newton's 45

Progressive waves 150-151
Projectile motion 32
Proper time interval 50
Ptolemy 1
Pulsar 3
Pump, diffusion 244
filter 124
rotary 243
Pyrometer 239-241
optical (or spectral) 240
total radiation 240

Quality (of sound) 185
Quasar 2
Quasi-static process 196, 202, 203

Radiant emittance (or radiance) 233
Radiation, black body 234-235
isotropic 235
spectral distribution of 237-239
temperature 234-235
Radius of gyration 99
Radioactive clocks 4
Reduced mass 111
Rankine's temperature scale 194
Rarefaction 149
Rayleigh-Jean's law 238
Reference circle, of damped SHM 116
of SHM 108-109
Reflection of sound 165
Refraction of sound 165-166
Refrigerator 225-226
Regenerative cooling 226
Relative emittance 236
Relativistic addition of velocities 54-55
Relativistic variation of mass 67-69
Resonance 116
Rest, energy 87
mass 69, 87
Reverbration period 173
Sabine formula for 174
Reversible, cycle 203
engine 203, 204
process 205
Reynold's number 126, 127
Right-handed screw rule 19, 24
Rigid body 95
angular momentum of 98
equilibrium of 95-96
Rigidity 137
Ripples 150
Rocket 63-64
Root mean square velocity 212, 219
Rotary pump 243
Rotational kinetic energy 98

Sabine formula 174
Satellite 30
geostationary 38
orbital velocity 31
time period 31
weightlessness in 44
Scalar product of vectors 18
Scalar quantities 12
Schreifer 228
Second 7
Second law of thermodynamics 202-205
Seconds pendulum 112
Shear modulus 137
Shock wave 170
SI units 6
Significant figures 10
Silicate glasses 136
Simple harmonic motion 105
amplitude 105
angular 112
angular frequency 105
characteristics 105
(also see salient features)
damped 115-116
epoch 109
frequency 105
of LC circuit 107-108
of loaded spring 106
periodic time 105
phase 109
reference circle for 108-109
salient features 107
superposition of orthogonal 113-115, 118
Simple pendulum 111-112
Simultaneity 48-49
Slug 29
Sodium chloride structure 134
Solar constant 241
Solids, amorphous 136
crystalline 136
glasses 136
Sonar 172
Sonic boom 171
Sonography 172
Sonometer 180
Specific gas constant 194
Specific heat 200
of gases 200-202, 216-217
at constant pressure 201
at constant volume 201
ratio of 216
Speed, molecular
average 219
most probable 218
root mean square 212, 219
Speed of light, constancy of 45
Sprayer 124
Square law detector 154
Standard deviation 12
Stationary wave 178-179
Steam engine 205-206
Rankine cycle 206
Stefan-Boltzmann law 235
Steiner theorem 102
Stoke's law 127
Stream line 121
motion 121-122
Sublimation 227
Superconductivity 228-229
Superfluidity 229-230
Superheated liquid 223
Superposition of orthogonal vibrations 113-115, 118
Superposition of waves 152-153, 166-168
Superposition principle 152-153
Supersaturated vapour 223
Supersonic velocity 170
Surface tension ff 139
Surface waves 150

Tabla 178
Temperature, black body 241
choosing scale 189
concept of 188-189
critical 222, 226
inversion 225, 226
kinetic 219
kinetic interpretation of 213-214
measurement of 189-190
radiation (see black body radiation)
transition 228
Temperature scale, absolute 192
Celcius 192
Fahrenheit 192
ideal gas 191-192
Kelvin 190, 192
Rankine 194
Tensile, strain 136
stress 136
Terminal velocity 128
Theorem of parallel axes 101
perpendicular axes 102
Thermal conductivity, coefficient of 231
table of 231
Thermal, energy 187
equilibrium 188
Thermal expansion 138
Thermodynamic coordinates 187-188
Thermodynamics
first law of 199
second law of 202-205
zeroeth law of 188
Thermometer
gas, constant volume 190-191
platinum resistance 192-193
Thermometric property 189
Time 3
Time dilation 49-50
Time interval
improper 50
proper 50
Time period of satellite 31
Tone 185
Torque 91-92
Torr 243
Torricelli's theorem 123
Torsion modulus 137
Transition temperature 228
Transverse waves 148
velocity of 160-161
Triple point 228
of water 190
Tube of flow 121
Turbulent motion 121-122, 126

Ultrasonics 171-172
Uncertainty principle 155
Unit cell 133
Unit vector 13
Units 6-7
conversion of 9
fundamental 8
SI 6
Universal gas constant 194
Universe, age of 3
edge of 3
mass of 3

van der Waals, equation of state 222-223
force 131, 133
isotherms of gas 223
Vaporization 227
Vector ff 12
angular 24-26
axial 24
collinear 13
components of 14-18
difference of 14
displacement 15
four 17
magnitude of 13
null 13
polar 24
summation of 13-14
transformation under rotation 16-18
unit 13

Vector product 19
Velocity 21
absolute 40
angular 25
average 23
distribution of molecules 218-219
escape 69, 79
group 155-156
head of fluid motion 123
instantaneous 23
of centre of mass 65
of longitudinal wave 159-160
of satellite 31
of sound 161
effect of humidity 162
effect of pressure 161
effect of temperature 162
Laplace's correction 161
Newton's formula 161
of transverse wave 160-161
of wave 155-156
supersonic 170
terminal 128
transverse 26
vector nature of 23
Velocity-time graph 21, 22
Venturimeter 123
Viscosity 121, 125
Viscous force 126, 127
Volume strain 137

Water equivalent 241
Water waves 150
capillary(surface tension) 150
gravity 150
Watt 85
Waves, bow 170
electromagnetic 147
infrasonic 171
intensity of 152
longitudinal 148-149
matter 147
one dimensional 150
parameters 151-152
polarized 179
progressive 150-151
shock 170
stationary 178-179
three dimensional 150
transverse 148
two dimensional 150
ultrasonic 171-172
velocity of 159-160
water 150
wavefront 150
Weak interaction 29
Weight 28
Weightlessness 44-45
Wheatstone bridge 232
Wien's constant 238
Wien's displacement law 238
Work 71-75
unit of 72
Work done, by frictional force 72
in gravitational field 74-75
in circular motion 72
in stretching a spring 73-74

Yield point 136
Young's modulus 136

Zero point, energy 229
vibrations 229
Zero vector, see null vector
Zeroeth law of thermodynamics 188
Zinc blende structure 135